POWER
The Pratt & Whitney Canada Story

FRONT ENDPAPER
P&WC's South Shore facilities include Plant 1, seen in this 1988 aerial view. In the foreground is the company's Research and Development building.

BACK ENDPAPER
Plant 2, also on the South Shore. The large extension in the foreground was built as an armaments plant in World War II. In the lower foreground is the old cafeteria, which is still in use. To the right is the original Walmsley plant where in 1928 James Young leased the first space used by P&WC. (Both, Larry Milberry)

POWER
The Pratt & Whitney Canada Story

Kenneth H. Sullivan & Larry Milberry

CANAV Books

Copyright © CANAV Books, 1989
All rights reserved. No part of this book may be reproduced in any form or by any means without prior written permission of the publisher.

Canadian Cataloguing in Publication Data

Sullivan, Kenneth H. 1922-
   Power

Issued also in French under title: Propulsion.
Includes index.
ISBN 0-921022-01-8

1. Pratt & Whitney Aircraft of Canada – History.
2. Airplanes – Motors – History. I. Milberry, Larry, 1943-

HD9711.2.C24S8 1988 338.7'62913435'0971
                                  C88-095222-9

EDITING AND DESIGN
Robin Brass Studio

KEYBOARDING
Hilary Chick

ADDITIONAL RESEARCH
Kenneth I. Swartz

TRANSLATION INTO FRENCH
Richard Beaudet with Normand Taillefer

JACKET ILLUSTRATION
Tom Bjarnason

PHOTO RETOUCHING
Stephen Ng, SNG Retouching Studio

ADDITIONAL PROOFREADING
Ralph Clint

Printed and bound in Canada by
T.H. Best Printing Company Limited, Toronto

Published by
CANAV Books
Larry Milberry, Publisher
51 Balsam Avenue
Toronto M4E 3B6
Canada

Contents

FOREWORD 7

PREFACE 9

EARLY DAYS 11

THE WAR YEARS 35

THE END OF THE WAR 59

THE FIFTIES 73

THE MOVE INTO TURBINES 113

FLYING THE PT6 138

INDUSTRIAL AND MARINE 195

OVERSEAS INROADS 215

TWINPAC AND SEA KING 230

FROM JT15D TO PW300 249

APPENDICES 313

INDEX 316

Foreword

Writing the history of a company while that company is still making history is a tall order, especially in the aviation industry. Things change quickly. One never knows when the next quantum leap will take place. Between the Wright brothers at Kitty Hawk in 1903 and Neil Armstrong's first step on the moon in 1969, there really were only 66 years. And even fewer years between our start-up in 1928 and our becoming the world leader in gas turbine technology.

Power is the story of a remarkable string of successful ventures and of the odd disappointment, too. It is the story of a company celebrating its first 60 years, vigorous, thriving and eager to tackle the new challenges the next decades will bring. And most of all, *Power* is the story of the people who contributed to the success of this company throughout the years. All told, more than 32,000 employees have come through the doors of Pratt & Whitney Canada since 1928. It is to these people that this book is dedicated.

David Caplan
PRESIDENT
PRATT & WHITNEY CANADA
NOVEMBER 29, 1988

Three key eras in the history of Pratt & Whitney Canada. The top photo shows the famous P&WA R-1340 Wasp engine which the company serviced and assembled in its early days and which brought P&WC into the engine manufacturing business in the early 1950s. Beside the Wasp is an early PT6 gas generator. The PT6 brought the company into the turbine era. Thor Stephenson, company president in the 1960s and 1970s looks on. The second photo shows one of the latest PT6s. Not much larger than its ancestor, it can put out triple the power. At the right is a head-on view of the company's latest product, the PW300 turbofan. (P&WC)

Preface

It was a dark and stormy night in June 1927 as the Wasp-powered Loening seaplane struggled to land on a northern Quebec lake. The plane crashed, killing the pilot. The RCAF recovered the engine and installed it in a Douglas seaplane. The performance of the engine led to an order for additional powerplants, with the proviso that a Canadian facility be established to service these U.S.-produced Wasp engines.

This tale, oft told by P&WC's first president, James Young, provided a good measure of drama in setting the scene that led to the start of the company.

As I searched through the archives in a dusty warehouse, in boxes filled with yellowing newspaper clippings and handwritten account ledgers, it offered a natural beginning to the history of Pratt & Whitney Canada. Alas, two months of research revealed that Mr. Young had been misinformed about the facts surrounding the ill-fated flight.

It had been a clear evening. The plane had been a Wright-powered Vedette flying boat—the Wasp-powered Douglas seaplane was safely at anchor in Montreal. The RCAF purchased the seaplane from the estate of the deceased pilot. The remaining details of Mr. Young's account were accurate.

This was my introduction to the world of history writing. After 33 years of marketing activities I had elected for a change. Downtown Wichita, Kansas, had lost its appeal after some 300 visits to this "air industry capital of the world." Company president Elvie Smith had suggested that I undertake the compilation of P&WC's history over the past 60 years. It was a fascinating challenge as most of the records of the early days had been destroyed. I could find detailed production records for every engine that had passed through the shop but records of the people involved had been disposed of after the normal retention period.

Fortunately, the links with the past were still there. Jim Ross, who had joined the company in 1928, was an invaluable source of reference. John Drummond, Bill Reynolds, Bill Vertilneck, Arthur Pond and Gene Schweitzer bridged the early period with the 1950s. From that point on there was no shortage of people actively involved in all the phases of the operation. Their stories and many names appear throughout the book, but because space is finite, many events and people had to be left out. As my work progressed from my word processor through first, second and third drafts, it became obvious that this was quite different from the writing of business reports; these tend to be brief and to rely greatly upon the inherent knowledge of the subject by the reader.

Larry Milberry and his research assistant Ken Swartz were brought in to add professionalism to the product. I gratefully acknowledge their invaluable participation.

As I discovered in my research, the interpretation of certain facts is bound to generate areas of disagreement. Hopefully these will not detract from the main purpose of this book—to provide a lasting record of the company's first 60 years to past, present and future employees.

Kenneth H. Sullivan

Looking eastward downriver, this mid-fifties aerial view shows the old Navy building (Plant 4) once used for engine overhaul, Plant 2 to the left and Plant 1, along with the Fairchild complex, at top right. Plant 4 today houses a steel fabricator. The woods behind Plant 1 have been largely preserved though much of the vacant land seen here is now developed. Compare this photo with the endpapers. (P&WC Archives)

Early Days

The late 1920s were dramatic years for aviation. Every week seemed to bring some new record-breaking feat or development in technology. Lindbergh and others had flown the Atlantic. Explorers were charting new territory by air. The air mail was stimulating growth in commercial aviation. New aircraft and aero engines were appearing on every side. It was in this atmosphere that Canadian Pratt & Whitney Aircraft Company, Limited* was incorporated in 1928 to assemble and overhaul piston aero engines made by the parent company, Pratt & Whitney Aircraft, of Hartford, Connecticut. Located in Longueuil, near Montreal, P&WC was at first a modest operation with a handful of employees. Like any pioneering enterprise, it was the product of a particular time and place. It began in sales and service, steadily building up a loyal clientele and moved into manufacturing, then into research and development. The company was strongly led from the beginning, and this was vital in bringing it through the decades to its present status as the largest member of Canada's aerospace community. Today P&WC is

---

* The Corporation was incorporated under the name Canadian Pratt & Whitney Aircraft Company, Limited. On December 11, 1962, the name was changed to United Aircraft of Canada Limited. On May 1, 1975, it was changed to Pratt & Whitney Aircraft of Canada Limited. On October 26, 1982, this was shortened to Pratt & Whitney Canada Inc. Throughout this book the company is referred to as simply P&WC.

J.A.D. McCurdy sits at the controls of the *Silver Dart* at Baddeck on February 23, 1909, the day of Canada's first airplane flight. His engine was a 35-hp Curtiss V-8. (K.M. Molson Collection)

the world's leader in the design, development and production of gas turbine engines for business and commuter aircraft.

Canada and Aviation: The Early Years:

Powered, heavier-than-air flight came to Canada in February 1909 when J.A.D. McCurdy flew a half-mile at Baddeck, Nova Scotia. His aircraft was the *Silver Dart,* powered by a 35-hp Curtiss engine. McCurdy was a member of the Aerial

Gibson's aero engine weighed 210 lb and delivered 60 hp. Gibson built it in Victoria and it first flew in September 1910. (K.M. Molson Collection)

Thousands of Canadians became pilots, observers and mechanics in World War I, many training on the Curtiss JN-4 Canuck at bases in Ontario in 1917–18. In this scene at Camp Borden, north of Toronto, JN-4s are taking advantage of ideal flying weather. The JN-4's engine was the 90-hp Curtiss OX-5. (NAM 3577)

Experiment Association headed by Alexander Graham Bell, which had conducted experiments with Glenn Curtiss in New York state before Baddeck. In the following years other Canadians delved into powered flight, usually designing and building their own aircraft. In 1910, W.W. Gibson ran Canada's first indigenous aero engine, which developed 60 hp. Canada's "early birds" tended to work in isolation, if not secrecy. Gibson toiled in British Columbia, the Underwood brothers in Alberta. Beginning in 1910, Percy Reid built several aircraft of his own design in Montreal. He operated in a garage on Bishop Street, then built the first aircraft hangar in Canada at Cartierville. Little was heard of Reid's experiments after 1914. There was not much public awareness that aviation was progressing in Canada until the airplane found its place in warfare.

Canada was still young in 1914 when the First World War broke out. Thousands of Canadians went overseas, mostly into the dreadful trenches, but more than 20,000 served in Britain's Royal Flying Corps and Royal Naval Air Service. Yet few in Canada appreciated the importance of air power. Ottawa resisted a major role in aviation until it agreed late in the war to train pilots, observers and mechanics for Britain at a number of aerodromes in Ontario. Canada supplied the aircraft, which were produced at Canadian Aeroplanes Limited in Toronto. The type built was the Curtiss JN-4 Canuck, with a 90-hp water-cooled Curtiss engine. More than 1200 Canucks were completed in 1917-18, many for export to the US.

During this time Canada also built its first aero engines. These were Sunbeam Arabs, of British design, and a small number were made for Britain at a plant near Toronto. In 1918 Canada formed its first aviation units, the Canadian Air Force and the Royal Canadian Naval Air Service. The CAF, based in England, disbanded in 1920. The RCNAS, established to take over coastal patrol duties off Nova Scotia (which to that time had been conducted by the US Navy), was still training its first cadre when the war ended, and it immediately folded.

The war had pushed aircraft and engines far beyond the pioneering of the early innovators. Britain, France and Germany had led the way in airframe and engine development, with America close behind. Air-cooled rotary, and water-cooled in-line and "V" engines dominated, but air-cooled radial engines appeared shortly after the war. Aircraft makers were quick to introduce new designs around these engines.

In 1919 Ottawa established the Air Board to direct civil and military aviation. At this time, Canada's attitude towards aviation differed from Britain's. The British viewed the task as setting up an air network among European cities. Distances were short, and there would be good ground facilities. In Canada the picture was different, with vast distances and few large centres. Much of the country was remote, with hardly any aerodromes. The Air Board was equipped with 114

The 900-lb 390-hp Liberty of an Air Board HS-2L is slung away from its mounts on this engine change in Northern Ontario in 1922. A replacement rests on a crib. One oldtimer recalled that a lot of time was spent on such operations groping and diving for tools and parts lost overboard! Within a few years, the Liberty was outclassed by modern radials and quickly faded from the scene. (G.R. Hutt via K.M. Molson)

The Liberty engine as fitted to the National Aviation Museum's HS-2L. In Canada the Liberty had various adaptations such as the radiator flaps seen here. Engine makers/users have always been ready to modify powerplants to local conditions. (K.M. Molson)

aircraft from Britain. Ottawa had requested a variety of flying boats and seaplanes for use on the country's lakes and rivers. Canada also took over 12 Curtiss HS-2L flying boats which the US Navy had used in Nova Scotia during the war.

In 1920, the Air Board completed the first trans-Canada flight. Several aircraft were used to cover 3265 miles from Halifax to Vancouver in 11 days (45 hours aloft). This operation gave many Canadians their first close-up look at an airplane. At the same time, the Air Board was putting its fleet to work on a variety of tasks, including such work as aerial photography, forest fire and fisheries patrols, flights into

The Vedette was the most successful of the Canadian Vickers designs from the 1920s. It started with a Rolls-Royce Falcon, but was produced with the Lynx (for the RCAF) and Wright (for civil versions). Only one was fitted with a Wasp. (NAM 6912)

The least successful Canadian Vickers' design was the Velos, seen here on the St. Lawrence. It first flew in July 1928 but was an aerodynamic failure, even though it had two good P&WA Wasps. Moored behind is Western Canada Airways' Super Universal with a Wasp. (DND RE11710-33)

remote Indian reservations and some of the first medical evacuations by air. In 1923 the newly established Department of National Defence assumed the duties of the Air Board. In 1924 the Royal Canadian Air Force came into existence, but its work was mainly civil.

Although many types were flying in Canada in the early 1920s the HS-2L was the one that laid the groundwork for things to come. It was Canada's first widely used bushplane. Tales arose about this big flying boat—how, for example, it took off, flew and landed at 65 mph. Many lessons were learned by trial and error with these aircraft, but perhaps the most important was about power. Pilots and air engineers realized that if aviation was to prosper, engines more efficient than the Liberty in the HS-2L would have to be developed. Though available cheaply on the war-surplus market, the 360-400 hp Liberty, designed for wartime use, was ill suited to commercial use. For one thing, such ex-military engines had very short times between costly overhauls.

Soon after the war Canadian Vickers of Montreal established Canada's first postwar aircraft factory. Following discussions with the RCAF, forestry companies and commercial operators, it designed Canada's first indigenous production aircraft. This was the Vedette, a small forestry patrol flying boat which made its first flight from the St. Lawrence River on November 4, 1924, powered by a 200-hp Rolls-Royce Falcon. The Falcon was soon replaced by a Viper and the Viper by a Wright J-4 as Canadian Vickers sought the best engine. Canada's policy was to favour British products like the Falcon, Viper and Lynx. While the J-4 held promise, it was an air-cooled radial, and Canadian Vickers had reservations about whether or not such an engine, mounted as a pusher on the Vedette, could be properly cooled. Wright gave Canadian Vickers a money-back guarantee that there would be no trouble. The J-4 proved a success and was used on commercial Vedettes, but the RCAF picked the Lynx for its Vedettes.

The Beginning of Pratt & Whitney

One of the leaders in the US aero engine industry during the war was Frederick B. Rentschler. Following the war, he joined a new company, Wright Aeronautical. He brought Charles Lawrance into Wright and began developing Lawrance's designs. One of these became the Wright Whirlwind, the engine that made Wright famous overnight. A Wright would take Charles Lindbergh across the Atlantic and Wrights would soon be the choice of most Canadian operators. There were, however, differences between Rentschler and the board over his interest in research and development and he resigned as president of Wright in September 1924. At the same time, Rentschler was aware of a US Navy requirement for a new engine. Anxious to bid on this, he approached the Pratt and Whitney Tool Company of Hartford, Connecticut, seeking its support. Founded in 1860 by Francis A. Pratt and Amos Whitney, both previously associated with the Colt firearms company, this company's reputation in the machine tool business was second to none. In 1924, Pratt and Whitney Tool Co. had vacant space, dormant machine tools and some excess capital. It agreed to support Rentschler and on July 23, 1925, the Pratt & Whitney Aircraft Company was incorporated with Rentschler as president. It leased space in a corner of the tool company's building. Rentschler was wise in adopting the prestigious Pratt and Whitney name–his engine company now accepted the challenge of upholding that reputation.

Rentschler assembled a team of key personnel, mainly ex-Wright men. The challenge for the Navy contract was to deliver a 350-400 hp engine. Rentschler's goal was to produce an engine of the approximate weight of the Wright entry but 75 hp more powerful. Starting fresh, they were open to innovation. They moved from the standard cast crankshaft to a forged one and introduced a supercharger for the first time in a production engine. Their prototype first ran on December 29, 1925, producing 380 hp. It soon passed the Navy's tests, and on May 5, 1926, before a large crowd, took to the air. The engine was named the Wasp and was soon ordered by the Navy. The Wasp was quickly followed by the 525-hp Hornet.

Commercial aviation was blossoming in the United States, mainly because of the government-backed transcontinental air mail service, begun in 1918. Two companies had the air mail contracts: Boeing Air Transport, between Chicago and San Francisco, and National Air Transport, between Chicago and New York. Boeing used mail planes of its own design, powered first by Liberties, then by Hornets. The Hornet permitted carrying two passengers, besides mail. Thus, Boeing earned extra revenue and laid the groundwork for airline operations to come. In 1929 Bill Boeing, Fred Rentschler and aircraft designer Chance Vought merged their interests to form the United Aircraft and Transportation Company. This brought together two airframe companies (Boeing and Vought), a propeller company (Hamilton Standard) and an engine company (Pratt & Whitney Aircraft). Within a year, UAT comprised 16 companies showing a profit in the millions.

How the First Wasp Came to Canada

The story of the first Wasp in Canada began in 1926 when J. Dalzell McKee of Philadelphia purchased a two-seat Liberty-powered Douglas O-2B. McKee proposed an expedition to Hudson Bay, operating on floats. He and his companion, Omer Wicks, started from Washington on July 15, 1926, and the following day were moored on the St. Lawrence in the Canadian Vickers air harbour. Next day they flew on to Sudbury. When they tried to take off again, the Douglas would not leave the water in its heavy state. McKee abandoned the trip and returned to Montreal. The floats on the O-2B had been tested in salt water, but on the freshwater lake at Sudbury provided less buoyancy than expected.

McKee decided to return the O-2B to Douglas in California. To this time, no single aircraft had flown across Canada. McKee decided to make the trip, accompanied by S/L A. Earl Godfrey, who went along to gain experience for the RCAF. McKee and Godfrey laid out a route following the lakes and rivers which were Canada's natural aerodromes. Meanwhile, new floats were fitted and McKee's Liberty engine was overhauled. The expedition took off from Montreal on September 11, 1926, heading westward and stopping at places

J. Dalzell McKee who, with S/L Earl Godfrey, made the first trans-Canada flight in a single aircraft. (K.M. Molson Collection)

(Above) McKee's Liberty-powered Douglas at NAS Anticostia just before flying on to Montreal in July 1926. The photo below shows the aircraft, with its Wasp, in RCAF colours. Basically an O-2B, it was redesignated MO-2BS, "M" for modified and "S" for seaplane. (US National Archives 80-G-424220)

such as Sudbury, Sioux Lookout and Lac du Bonnet. On September 19, McKee and Godfrey took off from Edmonton and later that day alighted on English Bay, Vancouver. They had been 35 hours aloft. To honour those who had helped with his expedition, McKee instituted the Trans-Canada Trophy for presentation each year to the Canadian making the most outstanding contribution to aviation. More commonly known as the McKee Trophy, it is still Canada's premier aviation award.

From Vancouver, McKee flew on to California. Over the winter Douglas replaced the Liberty with Wasp engine serial number 215. This was the first Pratt & Whitney engine installed in a Douglas airplane. The Wasp provided an instant improvement in performance. Space and weight were saved, allowing the addition of a third cockpit and greater payload. McKee now made plans for another trip. He would fly from Montreal to Edmonton, north to the Arctic coast, on to Alaska and the Yukon, south to Vancouver, then back to Montreal. Two Vedettes would be used as support planes. Among those assisting would be Godfrey and a P&WA technical adviser, William Wheatley. These plans came to an end with McKee's untimely death. On June 9, 1927, he was landing one of the Vedettes on Lac la Peche, north of Montreal, when he crashed

The crash that took McKee's life on Lac la Peche. (NAM 4543)

and was killed. Encouraged by Godfrey, the RCAF bought McKee's plane with a view to using it in aerial survey work. The Wasp would give it better performance over other aircraft in use. However, the aft position of the cockpit made it difficult for the pilot to fly straight survey lines. In the end, this orphan aircraft was struck off strength by the RCAF, handed over to the firefighters at RCAF Station Rockcliffe (near Ottawa) and set alight for firefighting practice. So ended the short career of Canada's first Pratt & Whitney-powered aircraft.

Fairchild Aircraft

The first utility aircraft truly suitable for Canadian bush operations was the Fairchild FC-2, powered by a 200-hp Whirlwind. The first of these appeared in Canada in July 1927. Bush operators and the RCAF alike found them serviceable and economical. Through the efforts of the RCAF, the FC-2 was soon improved. Years later, W/C E.W. Stedman of the RCAF's Aeronautical Engineering Division commented: "We had kept in touch with this Wasp development from the start. We soon decided that this engine was of value to us. We asked the Fairchild Company to produce for our department a machine similar to the FC-2 but fitted with a Wasp engine." The RCAF ordered six 410-hp Wasps directly from P&WA for installation in FC-2s. The first was installed by Fairchild on Long Island, NY, in the fall of 1927, and the others by Canadian Vickers, which was building FC-2s under licence. It was the RCAF which inspired Fairchild to establish a Canadian production plant. At least 15 FC-2Ws (as the Wasp-powered version was known) were in Canadian use by the end of 1928 and P&WA had a valuable foothold in Canada.

Another important bushplane was the Fokker Universal, built at Fokker's US plant. It did fine work with its Wright engine and served on the gruelling Hudson Strait Expedition in 1926. Before long it became the Super Universal, powered by a Wasp. These early bush planes worked in the remotest regions of Canada, where natural resources were being developed. They swarmed into northwestern Ontario in 1926-27 when gold was discovered at Red Lake. On its first month of work there, Western Canada Airways' Universal hauled five tons of freight and 78 passengers. Wherever a bush plane went to work, it soon proved its value, and the Pratt & Whitney reputation was rapidly spreading among operators, who were quick to recognize the Wasp as a major breakthrough.

Canadian Pratt & Whitney

Canadian Pratt & Whitney Aircraft Company, Limited was incorporated on November 29, 1928. Its only prospect was the order for six Wasps from the Department of National Defence, but the new company realized there was potential in a nation where aviation had made such rapid progress. Soon Pratt & Whitney became *the* name with which Canadian pilots and air engineers would trust their lives. The Canadian company

The first Canadian Fairchild with a P&WA Wasp was G-CYYU of the RCAF, seen being refuelled at Grand-Mère, Quebec. (K.M. Molson Collection)

(Right) The Wright Whirlwind was the first great postwar breakthrough in engine technology. Soon after it appeared, Fred Rentschler left his job as president of Wright and formed Pratt & Whitney Aircraft. Here a Wright is being installed in a Fokker Universal during the Hudson Strait Expedition. (PAC PA134951)

Another early Canadian Wasp was in this Fairchild FC-2W purchased by Canadian Transcontinental Airways in January 1928. It is seen on Lac Doré near Chibougamau while supporting prospectors in the field. (via *Canadian Mining Journal*)

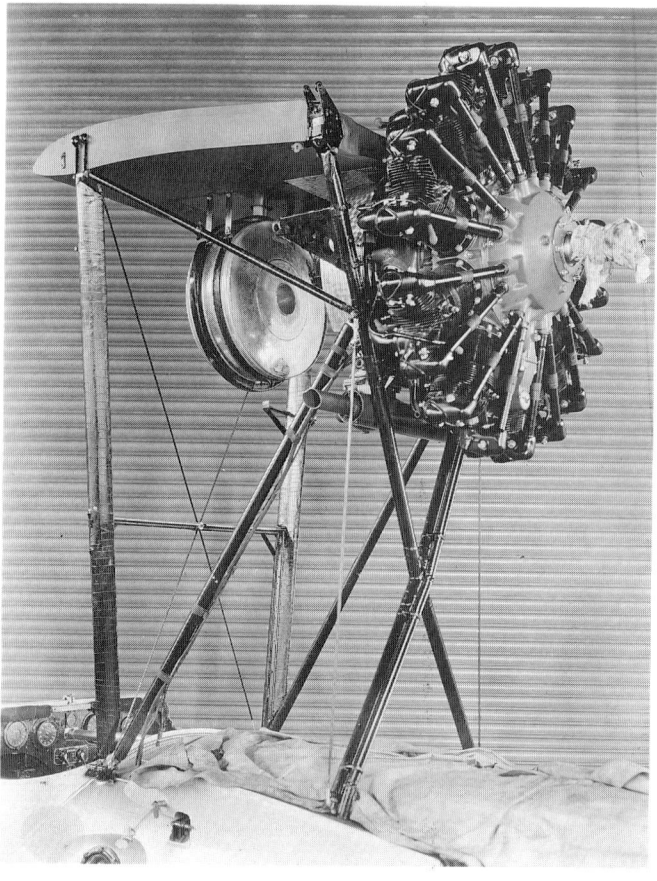

Lovely view of an early Wasp, this one for use in the Vedette. The only Vedette to fly with this engine was CF-OAB of the OPAS, which was converted from a Wright in 1933 and served into 1940. (P&WC Archives)

(Right) CP&WA in the 1930s. At the top is the engine inspection bench. Frank Dando is top centre, discussing matters with James Gallacher. Centre right are George Rose and Henri Prud'homme. The large windows provided ample natural lighting.
(P&WA Archives)

A view in the opposite direction. This is the assembly area and the offices are behind the fence at the rear. Beyond at the right is a corner of Walmsley, which made machinery for paper mills. Several Wasps are seen. In the foreground, starting to build up an engine, is Henri Prud'homme. Next up is Martin Graham. Frank Dando is at the top right with the striped tie. The slight man at top centre is George Rose. Top left is Harold Marr, standing at a drill press. At the bottom left is an engine connected to the electric run-in machine used to slowly turn a newly assembled engine to circulate the oil. Engine packing boxes are to the right. (P&WA Archives)

EARLY DAYS 23

The Pratt & Whitney Co. of Canada offices and factory as they stand today in Dundas. James Young was a senior man with the company in the 1920s, selling small tools and gauges. There was no connection between this company and P&WC, though the two often received each other's mail by mistake! (Robert Finlayson)

James Young, founding president of Canadian Pratt & Whitney, in a photo taken during the Second World War. (P&WA Archives)

would be a sales and service operation, not a manufacturer. From day one, it was P&WC's understanding that it would only prosper by being, as the company put it, "well and favourably known" as a good corporate citizen.

To one man must go the credit for setting P&WC on the road to success. James Young, the company's first president was born in England December 26, 1883, and received his early apprenticeship in the manufacture of precision tools with Armstrong Whitworth in Manchester. In 1905 he moved to the US to join a New York holding company, Niles-Bement-Pond. Representing one of their companies, Pratt and Whitney Tool Co., he travelled throughout Europe and Russia. Niles-Bement-Pond formed a link with the famous tool-maker in Dundas, Ontario, John Bertram and Sons. This brought the Pratt & Whitney name to Canada with construction of a factory on Hatt Street in Dundas. Pratt & Whitney Company of Canada Limited turned out small tools and gauges. In 1913 James Young was posted to Montreal, representing the Dundas companies. When war broke out, he went overseas, serving with the Black Watch in France, where he was twice wounded. After the war Young was general manager of Pond Machine Tool Works in New Jersey. In 1926 he became vice-president of the John Bertram and Pratt & Whitney companies in Dundas. He would become a legend in P&WC. In recent years, Charles Deeds, son of Col. Deeds of P&WA, wrote of Young as "a story book representation of the perfect gentleman, both in appearance and actions. Although dignified at all times, he was warm-hearted and thoughtful, and considerate of others. He behaved in the old school of precision in dress and politeness." Retired RCAF W/C Bill Skelding remarked of Young in 1987, "He was an outstanding person with the highest standards in the aircraft business. He was very influential, and highly regarded by all."

Once P&WA had decided to branch into Canada, Col. Deeds, a director of P&WA and of Niles-Bement-Pond, recommended that Young organize the new company. In August 1928, Young turned all his attention to P&WC. His home at 3 Barat Road in Montreal became temporary company offices. Using his knowledge of Canadian industry, he chose Longueuil on the south shore of the St. Lawrence, opposite Montreal. This was an area where aviation was already entrenched, especially with Canadian Vickers, and the aerodrome at St. Hubert just five miles away. There was a good labour pool and Montreal was the closest large Canadian city to Hartford. Rail service connected the two cities.

Years earlier, Armstrong Whitworth had erected a factory in Longueuil to make cast wheels for railway cars. Later Charles Walmsley and Co. of England took over the plant to manufacture equipment for paper mills. In 1928 Young knew that Walmsley had vacant space. He reached an agreement to lease 4000 square feet from Walmsley, but Dominion Engineering suddenly moved to purchase the plant. Drawn-out negotiations led Young to look for an alternate site in Hamilton. The Dominion Engineering deal finally went ahead and P&WC occupied the eastern corner of the plant. Thomas Shearer Stewart, a blind Montreal lawyer, was most helpful in laying the legal groundwork and he and Young became great friends. Since P&WC would be in sales and service, its first employees were tradesmen, not engineers. Young understood the need for a balanced team and had the ability to pick good people. Four of the originals were former Walmsley employees: Frank Dando, first shop-superintendent; James Gallacher, first chief inspector; Henri Prud'homme, foreman of engine assembly; and James Ross, bookkeeper. As service manager, P&WC hired Martin Graham away from the RCAF. Once Young had his top positions filled, he sent his men to Hartford to learn P&WA's engines and its approach to business.

P&WC had signed a two-year lease with Walmsley, paying a yearly rental of 40¢ per square foot. It opened its doors in early 1929 with 10 employees and paid-up capital of $85,100. Work immediately began outfitting the shop and other facilities. A ceiling of $37,000 was set for alterations and additions. Equipment for assembling/disassembling P&WA Wasp and Hornet engines was procured and plans were made to build two engine test houses. Operations began in February —assembly, overhaul and servicing, with the potential of adding manufacturing should market conditions warrant. Customers liked this arrangement, for they now had a service department in Canada addressing their specific concerns. As for P&WA, the Canadian plant was an answer to the tariffs that favoured British engines. As engine parts were not so heavily taxed, P&WA could ship engine kits to P&WC for assembly, avoiding some engine import tariffs.

In spite of meagre facilities, P&WC was convinced that a

The six original men who came to work for Canadian Pratt & Whitney Aircraft Company, Limited in 1928: James Gallacher, Frank Dando, James Ross, James Young, Henri Prud'homme and Martin Graham. (P&WA Archives)

great future lay ahead. The original board of directors included five Canadians and four Americans. The former, along with James Young, were G. Herrick Duggan, well-known for his engineering expertise and as president of Dominion Bridge and Dominion Engineering; Ross H. McMaster, president of the Steel Company of Canada, a director of the CPR and a leading figure in Canadian business; G. Montegu Black, president of Black & Armstrongs of Winnipeg and prominent in finance and business in Western Canada; and Hubert G. Welsford, general manager of Dominion Engineering Works Ltd., and a wartime aviator. From the US came Fred Rentschler, president of P&WA and of United Aircraft and Transportation Corp; Charles W. Deeds, who had been in aviation since graduating from Harvard, and though not yet 30, was secretary treasurer of P&WA and UAT; Stephen A. McClellan, a wartime aviator; and Joseph F. McCarthy, well known in financial and auditing circles in New York, and associated with P&WA and UAT. The parent company held 70% of the P&WC venture, the rest being spread among the Canadian directors, principally James Young.

The First Year in Business

Having begun with just one Wasp and one Hornet, by year's end in 1929 over 30 Wasp "B" and "C" models has been assembled at Longueuil, and 43 Wasps sold for no less than $7210 each. Considerable overhaul and spare parts business was also concluded, and sales totalled $437,000. Customers were spread from coast to coast. Engine assembly was the major activity. All work was completed following procedures laid down by P&WA. Each engine was built up in the shop, taken across a lane to a small red brick test cell where it was run and checked, then returned to the shop to be partially disassembled and inspected. If all was well, it was reassembled, sent back to the test cell for an acceptance run, crated and shipped. Engines for overhaul were generally taken into the plant after 300-350 hours of use. This was a striking improvement over the 100 or so hours that engines like the Liberty had provided. For overhaul, an engine was torn down and each part inspected. Components showing wear were serviced or replaced, then the parts were reassembled. Testing was similar to that for a new engine.

The earliest photo of P&WC shows a large sign atop the plant bearing the company name. In truth, there was no sign–the photo had been cleverly retouched. A modest sign had been placed near the street, but only after James Ross had cautioned Mr. Young that failure to comply with a municipal by-law requiring a sign could incur a fine. The modest size of P&WC was at times a source of embarrassment, as James Ross recalls: "Once we opened up, there were a lot of visitors coming through. We were a little embarrassed about our small size. A fellow who came in from Colonial Airways one day asked me to show him our stores. I had to inform him that he had just stepped through them!'"

The imposing CP&WA sign-that-never-was atop the Walmsley factory. Two men are wheeling a Wasp from overhaul into the test cell. (P&WA Archives)

Sample deposits to Pratt & Whitney Canada's bank account in 1929 give a good indication of the variety of business done during the company's first year. The initial deposit took place on January 24–$70,000.

February 15	Canadian Vickers	$2485
February 28	Consolidated Mining and Smelting	$5322
April 4	Fairchild	$2323
April 6	Canadian Vickers	$7647
April 12	Starratt Airways	$15025
May 10	Canadian Transcontinental	$2163
May 18	Fairchild	$14004
June 28	Western Canada Airways	$7194
July 8	International Airways	$8296
August 5	James Richardson	$16368
August 13	DND/Vickers	$14308
September 4	James Richardson	$16372
October 16	Vickers/Colonial Airways	$7699
November 1	Western Canada Airways	$11206
December 7	Province of Ontario	$12240

Fairchild in Canada

In May 1929 Longueuil became home to Fairchild Aircraft of Canada Ltd. A mile east of the P&WC plant, it established a small airport and seaplane base on 265 acres. A 140 x 260 foot factory was erected. This was the largest and most modern aircraft plant in Canada and became a favourite place for bush pilots changing their aircraft from floats to skis or vice versa. As a result many aviators came calling at Pratt & Whitney Canada's door.

Although the Fairchild FC-2 had begun with a Wright engine, its fame in Canada came with the Wasp. All 50 of the Fairchilds built at Longueuil were fitted with Wasps and FC-2Ws and Model 71s did outstanding work through the 1930s. Many other bushplanes were P&WA-powered, often converted from other engine types. In its special Canadian issue in May 1930, P&WA's in-house publication, *The Bee-Hive*, listed some of these: the Fokker Super Universal, used on the search in the Arctic for the lost McAlpine Expedition, Boeing flying boats used by Western Canada Airways on the West Coast, Boeing 40B-4s used by WCA on the Prairie air mail, a Junkers W-34 used on aerial photography in BC, and Fairchild 71s freighting in Labrador and Ungava.

Hard Times

Strong sales in 1929 had left P&WC with a healthy bank account, but the stock market crash that year heralded a downturn in company prospects. Even so, a dividend of $6.00 per share was paid to stockholders and, at Mr. Young's suggestion, over $80,000 of profits was invested in tax-free government bonds. These helped keep the company afloat over the coming difficult years. With the Depression, new engine sales dwindled from 43 in 1929 to one in 1932. The company focused on spare parts sales and on engine and overhaul work. Recalls James Ross, who managed the company's books: "It was not until 1934 that the tide turned, and it was only in 1939 that sales returned to the 1929 level."

The loss of business at P&WC was due to cost-cutting imposed by the newly-elected Conservative government of R.B. Bennett. This hit hard at air mail and RCAF budgets. The air mail had been expanding rapidly, with St. Hubert as a hub for mail routes into the US, the Maritimes, and along the Windsor-Quebec City corridor. Flourishing routes served the Prairies between Winnipeg and Calgary, featuring lights installed at 10-mile intervals for night flying. James Richardson's Canadian Airways was hard hit by the air mail cancellations of February 1932. From having spent $1.3 million on contracts in 1931, Ottawa pared back to $225,000, forcing Canadian Airways to reduce staff. By offering excellent rates, P&WC convinced Canadian Airways to send some of the overhaul it usually did in Winnipeg to Longueuil. P&WC expected to make up for its rates in volume of work.

The RCAF suffered a 70% budget reduction for 1931. This forced it to curtail operations and fire 178 officers and men and 110 civilian employees. P&WC could hope for little business from the Air Force and would have to be more ingenious than ever in ferretting out contracts. Mr. Young travelled Canada, visiting the major air services, and this resulted in a new influx of work. At the end of 1930, the worsening economy prompted P&WC to enter the propeller business. Mr. Young concluded an agreement for the rights to sell, service and manufacture Hamilton Standard propellers. Two men were sent to HS in Pittsburgh for training and, in

James Richardson of Winnipeg purchased six Fokker F.14s in 1929. These served the Prairie air mail routes and were the first aircraft in Canada with the P&WA Hornet. Besides carrying mail, the F.14 could accommodate eight passengers. (NAM 6366)

(Right) One of three DH 61 Giant Moths in Canada was this one operated by the OPAS. It arrived with a Bristol Jupiter engine but was converted to a Hornet in 1934. (K.M. Molson Collection)

early 1931, P&WC's new propeller shop had work from several customers. The shop was small but fully equipped. It provided much-needed cash and expertise that would lead to P&WC establishing Canadian Propellers Ltd. during the Second World War.

P&WC remained a spartan place to work. Its offices were in a small area separated from the shop by a fence. This was an exclusively male domain in order to avoid the expense of adding a ladies' washroom. As the Depression continued, the work force was trimmed by finding the stores-keeper a new job at a golf club, leaving the office staff to take over his duties. Martin Graham curtailed much of his travelling to remote regions and spent more time in the shop. Management salaries were cut, starting with 20% from the president's pay. James Ross made regular visits to the safety deposit box to clip the coupons from government bonds, freeing cash to

make up for the slump in engine sales (from which P&WC received a 10% commission).

Montreal was the heart of Canada's aviation industry. Canadian Vickers was working on several projects including licence-manufacturing the Super Universal with a Wasp (it had been planned to use a Bristol Jupiter, but customers demanded the Wasp). Bellanca Aircraft of Canada built a number of CH-300 Pacemakers for the RCAF. These durable utility planes had the Wright Whirlwind. Canadian Wright was part of a Montreal holding company, Aero Engines of Canada, which marketed various products, and there was an interlocking directorship among Canadian Vickers, Bellanca and Canadian Wright.

W.T. Reid, formerly of Canadian Vickers, set up Reid Aircraft Co. in 1928. He built a factory at Cartierville, north of Montreal, to make the two-seat Rambler. Hard times forced

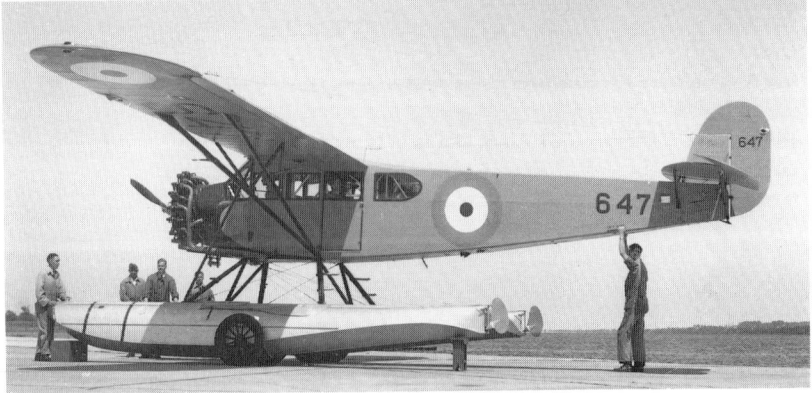

Four Wasp-powered Canadian Fairchild aircraft of the 1930s. Seen at Fort William is an 82 (550-hp Wasp S1H1). Today CF-AXL is on display at the National Aviation Museum in Ottawa. In a northern setting is one of just two 45-80 Sekanis (2 x 420-hp Wasp Junior SB). No. 647 is an RCAF 71C (420-hp Wasp "C"). CF-AUJ, the only commercial Super 71 (550-hp Wasp T1D1), is shown running up at the Fairchild dock in 1935. It later crashed, and the remains are with the Western Canada Aviation Museum in Winnipeg. (CC&F, via *Canadian Mining Journal*, Jack McNulty, PAC PA70856)

The prototype Norseman powered by a Wright R-975. The Wright didn't have the necessary power, and all but four of over 700 Norsemen had P&WA engines. (Jack McNulty)

A standard Wasp-powered Norseman at RCAF Station Borden. (Jack McNulty)

him into a partnership with Curtiss in the US, but the company didn't survive the Depression. Only 43 Ramblers were completed and Reid's hangar was sold to the Noorduyn company, which in 1934 was laying plans for a new bush plane. Fairchild continued its line of Model 71s, adding the all-metal Super 71 with a 525 hp Wasp, of which Canadian Airways took one and the RCAF two. That was as far as the Super 71 went, but the Fairchild 82 was another story. Beginning in 1935, 24 were built for several Canadian companies and export to Mexico, Venezuela and Argentina. Powered by 450- or 600-hp Wasps, the 82s were well liked and some served into the 1950s.

The Norseman

Of all the new Canadian aircraft of the 1930s, the most successful was the Norseman, designed by Bob Noorduyn, a Dutchman with a wealth of experience. He had designed the Universal while with Fokker. He came to Canada with an idea for the ultimate bushplane. He received financing in Montreal from Canadian Wright and promised to deliver an aircraft with all the important features demanded by bush pilots and air engineers. George Parker, who was to design the engine installation for the Norseman, once commented, "A lot of people don't know that the Norseman was originally built to sell Wright engines," but the connection between Wright and Noorduyn was to be short-lived. The prototype made its maiden flight in late 1935 powered by a 420-hp Wright R-975 Whirlwind. Dominion Skyways bought the airplane and three more were built with Wrights, but it soon became clear that the Wright didn't provide the needed power and all further Norsemen carried the 550-600-hp P&WA R-1340 Wasp. Performance was now satisfactory and operators were pleased, though economic conditions limited prewar sales to 23 Norsemen.

George Parker said about the early days of the Norseman, "I remember Mr. Young coming into the plant, and Bob Noorduyn saying, 'Well, we would like to buy your engines, but we just don't have the money to pay for new ones.' Mr. Young agreed to give him credit for two or three of the older R-1340s rated a bit lower than the newest ones. They were reconditioned, and Mr. Young let Noorduyn have them at a good price. On occasion, Mr. Young would come up to the plant to see Bob about payment. There wasn't much money available in those days, and Mr. Young carried Bob for quite some time."

Ferdie Vachon, of the famed Quebec aviation family, knew the Norseman well and writes: "I preferred the Norseman among them all, especially CF-BDF, which I cared for like a jewel. Its trusty P&W engine was an angel, and in three years it never faltered. We used to fly for hours on floats over the Barren Lands, with hardly a lake below, sitting confidently behind that engine. The P&W was a dependable piece of technology for sure. The rough treatment it endured! One day we were flying at -65 degrees F between Fort Smith and Fort Resolution. We lifted off easily, but the oil temperature failed to come up. We pushed on, figuring that we'd get a better reading. We never did, but we made it. Only once did I have trouble starting. That was at Tuk. We had flown in supplies to a ship caught in the ice. The wind blew all night and the battery would not turn the engine next morning. For the first time in three years I brought the battery in to warm it up."

Commercial operators continued to suffer through the early 1930s. Canadian Airways was driven to the brink of collapse before a turnaround saved the day. There was a major upswing in mining in the north and in 1934 Canadian Airways doubled its 1933 freight statistics, carrying more than 5.5 million pounds, a third of all that carried in Canada and more than the total for the US. Canada was leading the world in air freight, and doing this almost exclusively with single-engine aircraft on skis or floats. Within a few years Canada's air freight total would exceed that of all world carriers combined. Meanwhile an important development was taking place. As a "make work" project during the Depression, Ottawa began construction of a nation-wide chain of airfields in anticipation of a trans-Canada air service using modern airliners.

Life in the Bush

While the lore of bush flying is always associated with airplanes, rarely do the engines get the credit they deserve. Gerald L. LeGrave was an air engineer who had joined the Canadian Air Force in 1923. In the Depression he worked for Commercial Airways in Rouyn, Quebec, which operated a Wasp Fairchild: "The Wasp Junior was a beautiful engine. It and the Whirlwind were both reliable. In those days we used to have to pull the plugs every 40 hours, and check the valves and magneto points. When it came to work like this, I always preferred the Wasp. When we were based in Rouyn-Noranda, Martin Graham from Canadian Pratt & Whitney used to come around. He would look through the logbooks to find out what we were doing. If I had let my engine go too long between valve checks, he would want to know why. He was looking after the interests of that engine."

Bill Skelding, a young mechanic in the RCAF, was with a photo detachment in the North: "For the 1930 photo season, I was at Fitzgerald, Alberta, on the Slave River. This base was typical of our many remote operations. We were doing quite a bit of transportation, besides photography, and had two Fairchild floatplanes. One had a Wasp. We operated with a pilot, and a mechanic who was also trained on the big Fairchild aerial camera. "Any aircraft with a Wasp could reach 15,000 feet on photo operations, but the Vedette, one of our aerial photo mainstays, was hard-pressed to reach 5000 and still leave any time for photography. Its rate of climb was very slow. Overall, we mechanics preferred the Wasp to all other engines. It was easy to maintain and had good fuel economy."

Trans-Canada Air Lines

When the Liberals under Mackenzie King won the election of 1935, many breathed a sigh of relief, seeing an end to government restrictions. James Richardson ordered Canada's first modern airliners, Lockheed 10As with P&WA engines, and began training his top Canadian Airways pilots in instrument flying. He had high hopes of launching the first cross-country air service, but a new face appeared on the scene to thwart him. C.D. Howe, an American expatriate who quickly became

TCA's original airliner was the Lockheed 10. These sturdy aircraft had a pair of Wasp Junior SBs. Two original TCA L.10s survive, including CF-TCC, which is in flying condition. In 1986, P&WC assisted in restoring its engines and propellers for Air Canada's 50th anniversary. Here 'TCC visits Plant 5 in April 1986 during its commemorative trans-Canada flight. (Mike Haimes/P&WC)

The larger Lockheed 14, first with Hornets, later with R-1830s, became the backbone of TCA's early fleet. This one (with Wasps) was on a visit to P&WA at Hartford. The TCA personnel are Barney Rawson, Ron George, Herb Hopson, Webb Heenan and Jim Sortie, flanked by two Lockheed reps. (P&WA Archives)

all-powerful in King's cabinet, pushed ahead with a government scheme to link Canada coast-to-coast with an airline. In 1937 Trans-Canada Air Lines was formed with a guaranteed monopoly. Canadian Airways had to abandon its plans, and in August 1937 sold TCA its two Lockheeds and a Stearman mailplane. On September 1, TCA operated its first service on the Vancouver-Seattle route it had purchased from Canadian Airways. Being midway across the country, Winnipeg was made TCA's headquarters and maintenance base. Three more Lockheed 10As were purchased through Lockheed's Canadian agent, Fairchild. These arrived in October to form the basis of TCA's pilot training school.

A Lockheed with Hornets at Malton on June 6, 1939. All TCA Lockheeds had Hamilton Standard propellers supplied through P&WC. (Jack McNulty)

When he formed TCA, C.D. Howe imported a team of airline experts from the US headed by Philip G. Johnson, who had joined Boeing in 1917, was president of Boeing Airplane Company by 1926 and president of United Air Lines in 1931. Johnson and his team returned to the US once TCA was up and running. TCA engineer Jack Dyment recently recorded some of his early memories with the new airline: "The Lockheed 14 had just appeared and the choice was between it and the DC-3. TCA picked the Lockheed with Pratt & Whitney engines as it was faster, had longer range and had better performance at altitude. We had some long-haul routes such as Toronto-Winnipeg with stops en route. We needed the range to reach our alternates, and we wanted the speed because we were flying longer distances than the Americans. TCA also thought that, with more power, the Lockheed 14 would be better suited to the mountain leg. We didn't want to fly *through* the mountains, but clear them by 1000 feet. Most of our early engine mechanics had worked in the North or transferred over from the CNR. They knew plenty about bush operations, but weren't up-to-date on modern engines. So it was natural for TCA to hire Henri Prud'homme from the Canadian Pratt & Whitney Company to be foreman of our engine shop. Henri knew his stuff, was a very good foreman, and was bilingual too. From our first year in business, we always got excellent cooperation from P&WC." Henri Prud'homme stayed with TCA until his retirement.

Martin Graham left P&WC to join TCA at Mr. Young's urging, but later found out that he was not cut out for airline work. Albert Hutt, then a mechanic with Canadian Airways, commented years later: "TCA was just too confining, and too much of a change for a man like Martin. He had had a lot of freedom working for Mr. Young." After a year at TCA, Graham went to P&WA in East Hartford and was then assigned to field services in Europe, where fear of war had prompted Britain and France to order large numbers of P&WA engines. The company needed someone who wasn't American (the US was neutral) and knew the P&WA product. Graham was perfect for the job. In 1940 he was put in charge of P&WA service representatives in Europe, with his office in London.

A scene at Edmonton in March 1938, with a DOT Lockheed 12A (Wasp Junior SBs), a DOT Waco biplane (Jacobs L-4) and a Universal (Wright J-4B) of United Air Transport. UAT was owned by Grant McConachie, who later founded Canadian Pacific Airlines, which brought a great amount of business to Longueuil. Though widely used, Jacobs, Wrights and other types never enjoyed the popularity of the P&WA engines once the Wasp gained its foothold in Canada. CF-CCT is today on show in the National Aviation Museum. It was used on the first dawn-to-dusk trans-Canada flight, July 30, 1937. (CANAV Collection)

P&WC New Engine Sales 1929-39

Year	Wasp	Hornet	Wasp Junior
1929	33	10	
1930	11	6	
1931	5	2	5
1932	1		
1933	1		1
1934	3		2
1935	7		1
1936	12		11
1937	22		9
1938	8	10	1
1939	11		4
Totals	114	28	34

To date, P&WC had focused on sales and service of engines and propellers. No manufacturing had taken place. In 1938, however, the company carried out a study on the feasibility of manufacturing, but there was such a variety of engines and propellers in use in Canada that no single product was in enough demand to support manufacturing. P&WC was now 10 years old. It had grown modestly, from 10 employees to 17 in 1939; and from 4000 square feet of space to 7500. It had done $3 million in business and earned 13% on this; leaving 63% of earnings in the business. Total investment at the end of 1939 was $328,348.

The War Years

In the late 1930s, Pratt & Whitney Canada still depended on commercial bush flying for its business. There was not a lot of work for the RCAF, but TCA was an important new customer. With another war looming, Canada, having no engine manufacturing industry, was vulnerable. While P&WC at first stood to gain little from the government's "buy British" policy or from steps to have British aircraft built in Canada, its products and services were soon to be in demand. From 17 employees and sales of $488,000 in 1939, P&WC would expand to over 440 at work and sales of $8,557,000 in 1943. The war would also give the company its first manufacturing experience.

With unease over the international situation, defence spending in the late 1930s rose and the RCAF began relinquishing its civil duties. In February 1937 the Minister of National Defence stated: "When this government took office in October 1935, there was not a single fighting aeroplane in Canada; there was not a single bomb to be dropped by an aeroplane; there was scarcely any ammunition for the guns." The government still moved cautiously, spending on national interests such as the defence of coastlines, sea lanes and ports. In 1938 Ottawa sent a mission to Washington to purchase $5 million in aircraft and aero engines, when British Prime Minister Chamberlain declared "Peace in our time." The order was cancelled. Between 1936-39, however, the RCAF did receive about 200 new aircraft, including Stranraers for coastal patrol and a few Hurricane fighters.

Industry Gears Up

As the RCAF trained for new roles, Canada's aircraft industry was busy with several projects. After visits to Canada in 1938 by British officials, Canadian Associated Aircraft was formed. It included, in Quebec, Canadian Vickers, Fairchild and Canadian Car and Foundry; and, in Ontario, Fleet Aircraft, National Steel Car and Ottawa Car. These would manufacture components for Hampden bombers, with final assembly in two new plants at Malton (near Toronto) and St. Hubert. The first of 160 Hampden was test flown at St. Hubert in August 1940. Although the Hampden was obsolete, it was a breakthrough for Canadian manufacturers. It was a well-proven design and building and operating it posed no serious problems. It was a safe project, as were others like the Stranraer and Anson. The most modern type was the Hurricane, which Canadian Car began building at Fort William early in the war. Although these were mostly outdated, building them in Canada served a purpose. Valuable technology was transferred to an industry used to wooden and fabric trainers and bushplanes and major new facilities were put in place. An unstated object of the Hampden and Hurricane contracts was to shape public opinion in favour of military support of Britain. While the government

In the late thirties and early forties Canada's aircraft industry was still focused on British products. Here is part of the Hurricane production line at Canadian Car and Foundry at Fort William. The engines are Rolls-Royce Merlins. (CC&F)

was shy about committing Canada to a "British" war, the Air Ministry in London reasoned that the presence of large aircraft factories would subtly educate the public in the need to support Britain.

With the building of British aircraft in Canada, P&WC was left in the background. The types concerned had British engines and propellers. In October 1938 the company minutes noted that "a discussion took place in connection with the situation confronting the corporation due to large British purchases of complete aircraft in Canada and the possibility of changes required to meet the situation." Nonetheless, P&WC was able to do some business. It was, for instance, a sales and service agent for de Havilland propellers.

Canada Finds a Mission

As war clouds gathered, Canada was planning an expanded aircrew training scheme; the RCAF would train 76 new pilots in 1939, some for Britain, with a budget of $6 million, including money for new aircraft and engines. Civilian flying schools would provide elementary training and the RCAF would handle intermediate and advanced stages, using Harvards, the first of which arrived from the US in 1939. When war broke out the RCAF had 14 Harvards and 256 other (mostly obsolete) aircraft.

As soon as the shooting began, the British High Commissioner in Ottawa appealed to Canada, suggesting that a goal for year one of any air training scheme should be 2000 pilots as well as other air crew. On September 17, Mr. King announced an initial agreement, stating, "It will establish Canada as one of the greatest air training centres of the world." Britain soon boosted its requirements to 8000 pilots. By December, the British Commonwealth Air Training Plan came into being, with Britain, Canada, Australia and New Zealand participating. The task of constructing airports, acquiring aircraft and providing accommodations and other needs fell to Canada.

Some 3500 aircraft were called for, including 720 Harvards. Putting over 100,000 aircrew through the BCATP would be one of Canada's major contributions to defeating the enemy. Most of the Harvards were built under licence by Noorduyn at Cartierville. Their Wasp engines were shipped directly from P&WA and Continental (a licensee). They were supported by P&WC, which expanded its overhaul facilities in

Canada's first grand mission during the war was to train aircrew. Dozens of schools were set up across the country, many equipped with Harvards. Repair and overhaul of engines and propellers kept Longueuil working overtime. Field service reps like Gene Schweitzer and Ken Dawson were constantly on the move from base to base, helping the RCAF with daily problems. About 40 Harvards are in view at this station. (via Bill Wheeler)

1940, adding further machinery and tools and building a new test cell. By war's end, Noorduyn had built some 2800 Harvards, 1800 for the RAF, the first 1500 of these paid for by the US. At the same time, it made Norsemen for the US Army Air Force, which from 1943 to 1945 took 767. From 1939 to 1945, Noorduyn saw its work force soar from 110 to over 11,000.

The indispensable Harvard. Some 2800 of these trainers were built by Noorduyn at Cartierville, with engines provided by Hartford and propellers manufactured by P&WC's subsidiary Canadian Propellers Ltd. (S.D. Webb Collection)

Strengthening the Foundation

In the lull between the declaration of war and the government moving the country to a wartime footing, Mr. Young had began readying P&WC for stepped-up production. He knew that success would depend on his care in picking employees. Ronald T. Riley would continue as his assistant. He came from an influential Winnipeg family and had attended Royal

Activity on the tarmac at Noorduyn early in the war. Several Harvards and Norsemen are being worked on. The large biplane is a Stearman 4D Jr. Speed Mail with a Wasp Jr. The small fuselages belong to Fleet Finches in for overhaul. (via Frank Russell)

Military College and the University of Manitoba. After graduating he worked in Ontario for Canadian General Electric. In 1933 he married Margaret Black and it was his father-in-law, G. Montegu Black, who introduced him to Mr. Young. He joined P&WC in 1937, contrary to Mr. Young's usual practice of only hiring people with tool making experience. As James Ross put it, "Mr. Young was very careful with university people. He believed that if a university-trained fellow became supervisor in a mechanical shop, he would have difficulty understanding labour."

Riley went to East Hartford for a year of training. He built an R-1340 on his own and learned the P&WA way of doing things. He earned much respect along the way. Early in the war he enlisted in his old unit from RMC days, and it appeared that P&WC would lose him. But in 1943 he was invalided after a training accident, returned to P&WC and in 1944 became its first full-time vice president. The US was accelerating war production and a P&WA plant was being built in Kansas City, Missouri, to manufacture the "C" version of the R-2800 Double Wasp. Illness created a gap in management at Kansas City, and in August 1944 Riley took leave to go there as vice president and general manager. The plant turned out over 7000 R-2800s, mainly for the P-47 Thunderbolt. Riley returned to Longueuil determined to lead P&WC into manufacturing as soon as an opportunity arose.

Besides Riley, there were other key figures at P&WC when the war began, including Frank Dando, his assistant Harold Marr, and James Ross. New faces soon appeared, among them Frank Santo and Bill Vertilneck. Santo had started his career in a trade school in Calgary, graduating in 1939. Ron Riley offered him a job at 45 cents an hour should Santo come to Montreal: "When I got to Longueuil, I saw all these buildings and said to myself, 'Oh boy, this is a big outfit.' Then I went to the end of the building where P&WC had nothing but a wee corner of a place. I was disappointed, but I took the job. I went to work assembling and overhauling engines. There were 'B' and 'C' Wasps and Wasp Juniors, mainly from Fairchild bushplanes. Frank Dando ran the shop, and explained your job to you. Then you were on your own. There was always someone around who had done it before in case you had any questions,

Ronald T. Riley during a Victory Bond pep rally at P&WC. Riley became the company's first vice-president, then succeeded Mr. Young as president. (P&WC Archives)

(Below) As with all Canadian industry, the emphasis in wartime at P&WC was on victory. This company float is decked out to promote the purchase of Victory Bonds. (P&WC Archives)

THE WAR YEARS 39

John Drummond first served with TCA after graduating from McGill, but joined P&WC in 1940. (via J.W.R. Drummond)

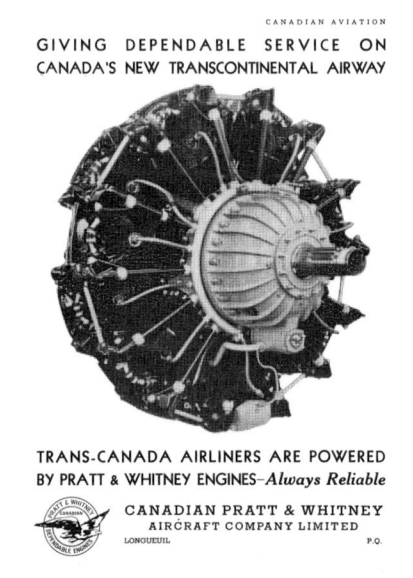

Advertisement in *Canadian Aviation*, January 1939.

but I was familiar with aircraft engines from my training in Calgary. Hartford used to send us up finished engine castings. We would install all the studs, bushings and so on. We assembled many R-1340s for the Harvards and Norsemen. We were overhauling about 100 engines a month, which was heavy for a small shop like ours. It seemed as if we were tearing down walls every month to make it bigger."

Years after retiring from P&WC, Bill Vertilneck confessed that he had got his start "after buying a Model A Ford with money I earned as a 16-year-old pool shark." He tore the car apart and rebuilt it by trial and error. He recalls: "When the war started, an ad appeared in the newspaper saying that if you were single, or married without children, you were obliged to take a job in the war industries. My father-in-law was working at Dominion Engineering in Longueuil, and I went there with him to see about a job. My first impressions were not favourable. The place was dirty and they were only paying about 20 cents an hour. My father-in-law suggested I go next door and ask Pratt & Whitney. I did, and was impressed. I had never seen a place like it. Such clean floors, and everyone was in white overalls. I met Mr. Young and Mr. Ross, who wanted to know if I had any high school education and a lot of other things. I said no to this, and no to that, but they decided to hire me anyway, giving me a job in the engine wash house. The pay was 25 cents. Our shop was very small. There was an area for washing engine parts, two benches for tearing engines down, an inspection department and the assembly line. Washing engines is where I learned about Pratt & Whitney engines. We had to put the parts in tanks to soak, then blow the holes with air pressure. I got to know where those holes were and what each was for. From here I moved to taking engines apart for overhaul, then to inspection and testing. Most of those hired after me had little or no aviation experience, and a lot were straight off the streets."

Early in the war P&WC needed an engineer to oversee expansion. Ron Riley was too busy, so John Drummond was hired from TCA. Drummond had worked for P&WC as a summer student while studying engineering at McGill. He was free to join P&WC after he graduated in 1938, but Mr. Young asked if he would go to TCA instead. Drummond was the first engineer hired by Oliver West for TCA. He recalled recently: "The industry was very small in those days, and Mr.

Young had many friends in it. He undoubtedly knew that TCA was being formed and wanted someone there who knew Pratt & Whitney engines." Drummond installed TCA's first engine test cell in Winnipeg, helped set up its engine shop, then became production engineer. At Mr. Young's request, he returned to P&WC in February 1940. Drummond adds: "Shortly after I got back to Longueuil, I was in charge of several new projects, including construction of new engine test cells and the new propeller repair shop. About this time we hired Ken Dawson and Gene Schweitzer. They worked in the shop, then went out into our field services organization."

Ken Dawson joined P&WC from the Montreal Tramways. He was an intuitive mechanic with a talent for troubleshooting and improvising. For months on end all that was known of him in Longueuil were his field reports and expense accounts! Gene Schweitzer spoke of Dawson in 1988: "Ken was a great fellow, capable and well liked by our customers. When I first met him, he was doing part-time field service work and working in the shop. In early 1941 he turned most of his attention to field service, dealing mainly in the west with Harvards and Ansons. He worked right out to the West Coast until Russ McCormack took over there."

Schweitzer himself became one of P&WC's best-known personalities. He graduated from Curtiss Wright Technical Institute in California in 1940. "I had the option of going to Noorduyn or Pratt & Whitney," he explains. "While it was small, P&WC looked more promising. It had engines, but also propellers and other connections with the parent company. I felt there was greater potential."

While James Ross was treasurer and sales and office manager, he was assisted by J.F. "Jim" Tooley, the general accountant. Tooley had been hired by Ron Riley in 1941 and stayed until 1945. Retired P&WC employee Bob Reive recalled: "Jim was just a young fellow then. He used to take all our money playing craps after work. He was very smart." After the war, Tooley became president of Canadian Aviation Electronics and president of Nordair.

Keeping Pace with Orders

In November 1939, Pratt & Whitney Canada had cashed in the last of the tax-free bonds purchased with the profits of 1929. Dominion of Canada bonds worth $53,000 were sold to net $54,432, and the money was set aside for forthcoming expenditures. The safety deposit box was retired. Mr. Young was always keen to make even a small saving like this. At a board meeting in April 1940, it became clear how the money was to be spent: $22,059 for new equipment, $14,947 for a new engine test house and $11,379 for alterations and an extension to the engine shop.

Early in 1940, the government ordered 100 Harvards and three Norsemen from Noorduyn. On February 20, P&WC signed its first wartime contract with the government, for 100 R-1340 Wasps and 100 Hamilton Standard 12D40 propellers for the Harvards. Six days later P&WC placed its order with parent United Aircraft Corp. As more Harvards and Norsemen were ordered, P&WC sold more Wasps and propellers. War or not, tariffs on aircraft engines had to be paid. The tariff on non-British aircraft was 20% and on complete engines 17%, but there was no tax on parts. There was no tariff on British products.

Another wartime project in Canada was producing PBY Catalina and Canso coastal patrol aircraft. Over 500 were built by Boeing in Vancouver and Canadian Vickers (later Canadair) in Montreal. The PBY had two 1200-hp R-1830 Twin Wasps. In the fall of 1942, P&WC sought a contract to supply 972 US-built R-1830s for the PBYs, but the deal did not go through, although the company did win the contract to supply the propellers and propeller controls. The loss of the sale reflected the new way Ottawa was doing business. Midway through the war it began placing engine and propeller orders directly with the US government rather than with agents like P&WC or US manufacturers.

The Avro Anson

The Anson was the standard multi-engine trainer in the BCATP. British-made Ansons and their Cheetah engines were shipped to Canada for assembly, but Britain soon announced that it could no longer continue. Canada now had to build Ansons domestically. P&WA was at that time unable to supply engines, but C.D. Howe managed to arrange for 2300 Jacobs engines from the US, which were installed on Anson Mk.IIs built at such places as de Havilland Canada. This was a stop-gap airplane, underpowered and, being fabric-covered, very cold in winter.

A hangar-full of BCATP Anson Vs with R-985s. Canadian Car and Foundry and MacDonald Aircraft built 1048 Anson Vs between 1943 and 1945. (Manitoba Archives)

A TCA Lockheed 18 Lodestar (R-1830s), part of a fleet which supplemented the earlier and smaller Lockheed 14s. (Air Canada Archives)

The solution lay with the Anson Mk.V, the prototype of which flew in January 1943. With 450-hp R-985 Wasp Juniors, it had ample power, and its moulded plywood fuselage made it more comfortable in winter. Over a thousand were built. While the Anson Mk.IIs were all scrapped after the war, the Mk.Vs found useful postwar roles in civilian aviation. During the war and after, P&WC supported the Anson Vs with their R-985s and "Ham Standard" propellers.

TCA: Fleet Improvements

When the war started, TCA's fleet was more modern than the RCAF's. It's new transcontinental service was of great importance as countless military, industrial and political personnel were suddenly on the wing, but a problem confronted TCA's engineering department. The Hornet engines in the Lockheed 14 began failing. TCA adopted a shorter time-between-overhaul for the top three cylinders (the ones most likely to fail as they were subject to the most heat) as a temporary solution, and in 1941 contracted with Boeing in Vancouver to convert all 10 Lockheed 14s from the 850-hp Hornets to 1200-hp Twin Wasps. This solved the problem and helped standardize engines, as TCA's Lockheed 18 Lodestars also had the Twin Wasp.

Jack Dyment recently gave his explanation of the Hornet-to-Wasp program: "During this period we cut back the power on the Hornet and the problem cleared up. We had been drawing rated power but that was apparently too much for the engine. We had always been careful not to exceed the recommended rating. On getting favourable results, it was our view that the Hornet, a good little engine, had been over-rated."

The Engine Manufacturing Debate

By 1940 the question of Canada having an aero engine industry was being hotly debated. There were those like the young engineer who, in *Saturday Night*, lamented an order for 9000 Rolls-Royce Merlin engines going to the US. In his view, Canada could build the engines, and thereby add to its technology base. He urged Ottawa to adopt "a more aggressive policy in place of its usual unimaginative mediocrity." To him a Canadian engine industry would ease the foreign exchange situation. He thought it inexcusable for Canada to depend on Britain and the US for engines when it could be self-sufficient and, once the war was over, a competitor in the world market. The engineer with these strong views was Dick

One of the Bolingbrokes converted to Twin Wasp Juniors is seen at the Fairchild airfield in Longueuil. (Jack McNulty)

Guthrie, later to be a vice-president of P&WC.

On the nay side of the argument were people like Ralph Bell, Director-General of Aircraft Production in Ottawa. In his view, an engine industry would overtax the system. There was not enough skilled labour available, and to set up to build 400 examples of just one engine type would take $50 million and 8000 workers. A Canadian industry would still depend on the US and Britain for such components as carburetors. Bell's argument prevailed, and for the duration of the war Canada did no more than assemble and overhaul engines.

One project affected by the shortage of engines from Britain was the Bolingbroke bomber made at Fairchild. The first had been delivered in November 1939 and 100 were on order, but only 17 could be delivered through 1940 as the flow of Mercury engines from Britain had been cut off. This led Fairchild to convert the Bolingbroke to the P&WA Twin Wasp Junior. On the docks in Halifax was a stock of these that had been destined for France (which had by then fallen). These could be diverted to Fairchild.

Ken Dawson and Gene Schweitzer worked on the program, and Schweitzer recalls: "We had to do the normal testing associated with a new installation, such as determining the strength of engine mounts, and what effect the propeller would have on them. We had to watch out for any unusual vibrations.... There were also questions of positioning the propeller in relation to the fuselage; air flow through the cowling and over the cylinders for cooling; and design and efficiency of the exhaust system. It was not a straightforward exchange of one engine for another." The first Bolingbroke with Twin Wasp

Ferry Command at Dorval. This is a typical scene, with aircraft everywhere. As long as the weather permitted, they flowed in and out of Dorval in a steady stream in the thousands. In this view are Lockheed Venturas (R-2800s), two B-24s (R-1830s), three Wright-powered B-25s and a B-17. Service reps from P&WC and later P&WA were permanently assigned to Dorval to tend P&WA engines and Hamilton Standard propellers. (PAC WRF621)

Gene Schweitzer's ID badge when he was a P&WC tech rep with Ferry Command.

Juniors flew in February 1940. The engine did not have adequate power for this installation and, since Mercuries began arriving again, only a few conversions were made. The program had been a good training opportunity for P&WC and had drawn further attention to the dangers of dependence on outside sources of supply.

Ferry Command

By mid-1940, Britain had some 26,000 aircraft on order from US factories. The problem was to get them across the Atlantic. Sea shipment meant time wasted dismantling and reassembling, and they took space that could be used for other valuable cargo. Once at sea, the ships were often torpedoed. The challenge of flying the Atlantic was not a new one. In 1919 the first three trans-Atlantic crossings were made: a US Navy flying boat completed the trip in stages, Alcock and Brown flew non-stop from Newfoundland to Ireland, and the British R-34 airship flew from England to Long Island. By the late 1930s crossings were still newsworthy. Commercial operators like Pan American carried out some proving flights using flying boats and the airship *Hindenburg* made some revenue flights from Germany, but pilots still feared the Atlantic.

In 1940 Britain's Canadian-born Minister of Aircraft Production, Lord Beaverbrook, enlisted staff from Imperial Airways and the help of Sir Edward Beatty, president of the CPR, to launch a service to fly aircraft from North America to Britain. Montreal would be the base for this venture. It had St. Hubert nearby, with its modern airport, and a seaplane harbour downstream from Longueuil at Boucherville. The CPR was to provide maintenance and support. On August 18, 1940, the first contract was signed with Canadian Pacific Air Services (successively replaced by ATFERO, RAF Ferry Command and RAF Transport Command) to fly 50 Lockheed Hudson bombers to Britain. The first seven left Gander on November 10, led by Captain D.C.T. Bennett, and made an uneventful crossing. This proved that Beaverbrook's project could work and soon a steady flow of aircraft was leaving Montreal for the refuelling stop in Gander, before taking off to challenge the Atlantic. Though it would face many dangers, the trans-Atlantic ferry service never looked back and became one of the Allies' great accomplishments.

From its beginning, the ferry service had an association with P&WC. James Ross recalls one of the company's early dealings: "I was in my office one day when a call came from J.W. McConnell. He owned the *Montreal Star* and was one of those recruited to help with ferrying. He said, 'I've got a list of parts here, and wonder if these numbers mean anything to you?' I jotted the numbers down and promised to get back to him. I found that the numbers all referred to parts for P&WA engines and we had them in stock. I called McConnell with the news, and he asked if we could deliver the parts. We used Mr. Young's car and took them over to the *Montreal Star* building."

Gerald L. LeGrave was hired by the CPR to help organize a maintenance and inspection system for aircraft going overseas. When he first went to work at St. Hubert, there were only two others there with air engineer's licences. His helpers were mostly automobile mechanics who knew little about aircraft. Bit by bit they learned, and a smooth-running system developed. LeGrave remembers: "Engine maintenance for Ferry Command was important business, and we gave the engines more attention than the airframe. If an airplane didn't have smooth-running engines, the inspectors wouldn't pass it. There was quite a checklist involved in maintenance. First we would pull all the spark plugs and re-gap them, check the magneto points and so forth. The mechanics would initial their work, then an inspector would check everything. If we had a problem such as a drop in engine rpm, we would depend on the field service representatives. The ones from P&WC knew what they were doing and were really good men to work with. They could always give us the right answers and saved us a lot of precious time. Once cleared, there would be an engine run-up before we turned the airplane over to the flight department. The pilots would test fly each aircraft, then hand over a list of snags to be checked out by the mechanics. The airplane was now ready for delivery. When its number came up on the assignment board and was matched with a crew, it was loaded with radios and survival gear, then sent overseas."

Gene Schweitzer notes that St. Hubert was soon taxed by large numbers of aircraft for overseas: "The first I worked on were Hudsons with Wright Cyclones. I serviced their Hamilton Standard propellers. Later, Hudsons with P&WA R-1830 Twin Wasps began arriving. Most of my early problems at Ferry Command were with propellers—usually leaks or governor problems which I could usually correct, but if not, the propeller had to go to the P&WC shop. We were *the* propeller specialists in Canada. The most interesting project involved modifying a B-24 Liberator for winter flying. When ferrying first began, pilots had to return by ship, or by flying boat via the Azores. This was time-wasting, especially with a pilot shortage. The B-24 with four R-1830s was chosen to equip a return service for the pilots. Nobody had ever operated a scheduled year-round service on the North Atlantic. One day in the spring of 1941 Mr. Young and I went over to Windsor Station to meet 'Punch' Dickins, vice-chairman of CPAS, to discuss how we would handle overhaul and servicing of the R-1830s. On May 1, 1941, the first flight of the return ferry service was operated from Montreal to Blackpool via Gander with Captain Bennett at the controls. The first westbound flight was five days later. The Liberator became the first landplane to pioneer year-round scheduled service on the Atlantic, and paved the way for postwar developments. BOAC took the service over, and soon the Liberators were ferrying people all over the world.

P&WC's test houses during the war. The farthest two handled the R-985 and R-1340. The third was for the R-1830, one of which is seen on the forklift, and the nearest one took the 2000-hp R-2800. (via R. Reive)

"Those of us in Ferry Command on the technical side worked day and night and usually on weekends. Problems had to be quickly and thoroughly handled. We worked on the Liberator through to the spring of 1942. Dave Tennant and Frank Booth had been loaned by TCA to help us with engineering work and came on many test flights. In my view, carburetor heat was the main problem with the R-1830. It was inadequate for de-icing. The designer and builder of the B-24 was Consolidated Vultee, which provided people to work on the program. We also had help from East Hartford. They produced a comprehensive installation manual for the R-1830 and it was great having such a publication at hand.

"Eventually, Ferry Command outgrew St. Hubert, and our operations moved to the new airport at Dorval. TCA took over maintenance of the Liberators. Ferry Command was rapidly expanding and P&WC just didn't have the resources to keep up. Thus, P&WA assigned full-time people to work at Dorval, and we were able to give them some valuable coaching. They were mainly young fellows and very capable."

Growth of the Overhaul Shop

During the Depression, P&WC's overhaul shop covered about 3000 square feet and could handle 10 engines monthly. By 1945, it covered 45,000 square feet and had a monthly capacity of 100 engines. Engines in the shop during the war ranged from the 450-hp Wasp Junior to the 2000-hp R-2800. Several expansions of the shop took place. James Ross recalls: "When war broke out, it took us some time to gear up for increased activity. We had to determine what new machinery and tools were needed and place our orders, with lead-time for delivery taken into consideration. We redeemed the last of our 1929 bonds to get cash to make these purchases. When we needed more labour, we put ads in the papers, and often recruited men who had been discharged from the Army. When men were hard to find, we hired women. That's when we finally spent the money on a women's washroom."

In September 1941 P&WC had signed an agreement with the government covering further expansion of the engine overhaul facilities. Ottawa had bought the Walmsley plant from Dominion Engineering, so P&WC's principal customer was now also its landlord. The main Walmsley plant was turned over to the production of naval guns. P&WC's facilities had expanded into the Walmsley plant and outside to the edges of the property; to gain yet more space the company built a mezzanine above the accessory shop. Here carburetors, Pesco pumps, spark plugs and ignition harnesses were overhauled and tested and engines awaiting parts were stored.

In the shop, white and yellow lines on the wooden block floor marked off aisles and areas where engines for overhaul were dismantled. All parts were cleaned and inspected. Tiny flaws invisible to the eye were detected by an electromagnetic process and defective parts would be scrapped. Engine build then took place in several steps with parts being matched exactly as they had come into the shop, except where worn parts had been replaced. Sub-assemblies were merged at the final assembly line. The completed engine next went into the test cell. Adjustments were made on the spot or back in the shop after the test run. Upon final inspection, the engine was sent for packing and shipping. By 1943 P&WC had four engine test cells in use, two for the R-985 and R-1340, and two for the R-1830 and R-2800.

Jack McPhee provides a few details about overhaul: "It used to take one shift to disassemble an engine, one to wash it, two to inspect it, and three or four on the subassemblies, if you had the replacement parts. Final assembly would take a couple of shifts. If there were no problems, you could rig and run an engine in the test cell in a shift. We tried to get an engine out in 30 days, but if we had problems or ran short of parts it could take longer."

P&WC's shipping department became famous during the war. Frank Kelly, one of TCA's first mechanics, recalls: "Two of the better known people at Canadian Pratt & Whitney were the ones who used to ship us our engines. One was Dando and the other was Pickles. They used to box the engines up and ship them to us, and all the bills were signed 'Dando and Pickles.' We used to get a big laugh out of that."

Many women worked in overhaul. James Ross says: "I discovered that once we hired women to assemble ignition harnesses, we could cut assembly time by about half. Women seemed to have much more dexterity. They were also assigned to mask equipment that had to be plated, and to practically every part of the plant except washing engines that had come in for overhaul." Frank Dando was in charge of all day-to-day matters in overhaul. Jack McPhee states that Dando was a thorough man, but had the discipline of a sergeant major: "He'd come through the shop and if there was anything wrong he'd notice it. He was a fine man, but a tough boss."

The first new R-1340 Wasps assembled from kits at P&WC. From the left are Les Smith, Harold Marr and Frank Santo. (via R. Reive)

R-1340s from East Hartford

Bob Reive started at P&WC as a stenographer when the office staff numbered six stenographers and one typist, all men. With the war, he was placed in charge of production control, responsible for scheduling new engines and overhauls through the shop. He remembers how new R-1340s would arrive from East Hartford: "We had large orders for the R-1340-AN-1 for the Harvard. Hartford manufactured the engines and we assembled and tested them. They would come up in crates. There were many boxes for an engine—one for the crankshaft, others for the power section, blower section, all the gearing, and all sorts of bits and pieces. It was a kit, and we would build it up in 350-400 hours. All the parts were new. Any troubles with fits would have to be repaired by us."

Bill Vertilneck had his share of frustrations working on some of the first R-1340s: "I was working with George Rose, who was badge number one with the company. The blower section on the 1340 was aluminum and the rear section

(Overleaf) The overhaul shop early in the war with R-1340s and R-1830 being assembled or awaiting test. About 40 men can be seen at work. On the left are Frank Santo and Bill Vertilneck with the first R-1830 overhauled by P&WC. Just behind the partition are Aimé Lavallé and Maurice Senez working on one of the rare R-1535 Twin Wasp Juniors. The first two men on the right are Bill Young and Bert Murton. (via R. Reive)

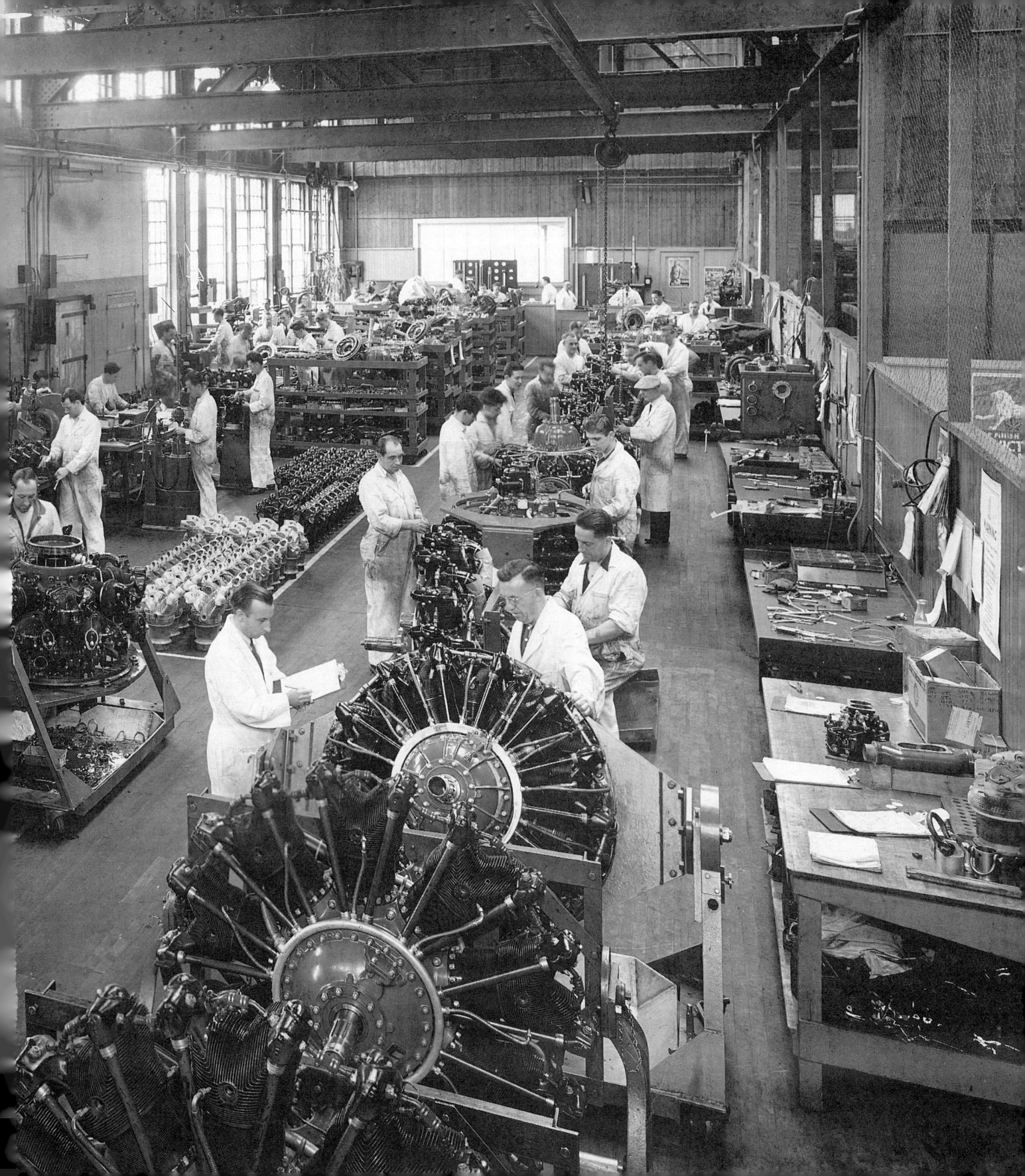

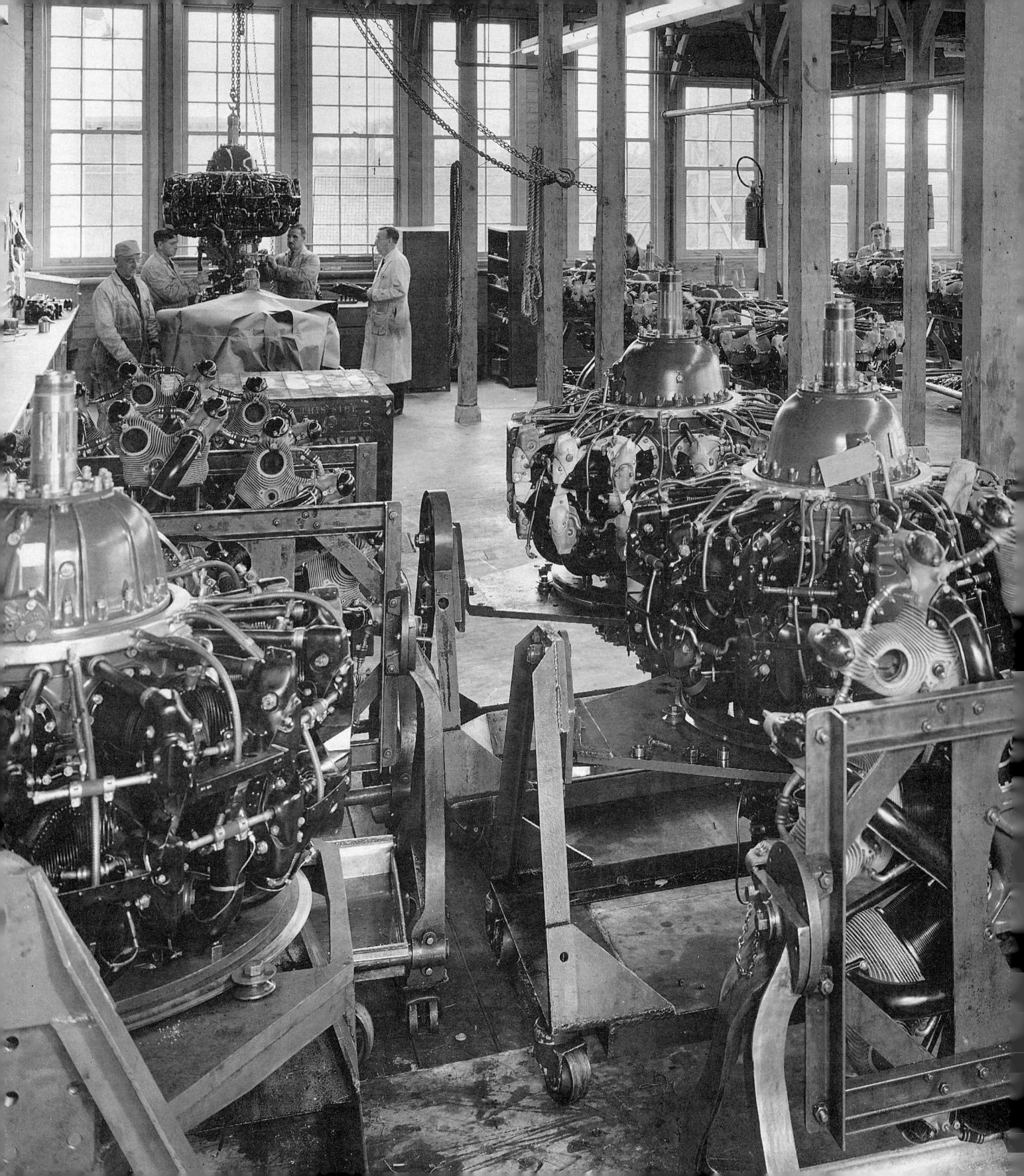

Activity in the box storage extension, October 1943. Engines arrived here for overhaul, and were later crated for shipment. (P&WC Archives)

magnesium. We had to mate the two parts, bore three holes and put in bushings. We had people working on lathes with reamers, but they had never worked on magnesium before. I decided to try it myself. I had never used a reamer in my life. I started making a hole with a standard reamer. I was using a lot of oil, and it was going well. But when I took the reamer out, there was a big groove in the part. George Rose was dismayed! I took another reamer and the same thing happened. George came over and gave me dirt. I said, 'George, show me how it's done.' But he told me to go ahead and try again. I did, and the part had to be scrapped. By now, everyone in the shop was around, and nobody knew what I was doing wrong. I tried another case and I scrapped that too. These parts were worth at least $700 each! I said, 'I'm not doing any more,' and just then Mr. Young came into the shop and asked me, 'What's the trouble?'

"I answered that I was not going to ream any more rear cases and blowers for the 1340. He replied, 'Look Bill, I want you to do one more. You're scared. You're doing it slowly. I want you to take the reamer, put on the oil and push as hard as you can.' I did just that, and the hole turned out as smoothly as could be. That's how we learned." As soon as Bill learned a better way of doing something, he would train a few more men in the shop. They, in turn, would train others. Over 200 new R-1340s were assembled by P&WC in 1941-42, and these were known as "Y" Wasps in honour of James Young.

The Propeller Overhaul Shop

P&WC had to expand its propeller shop. James Perks had worked for Walmsley beginning in 1928, then at another Longueuil factory, Stowell Screw Co. His brothers, John and Harold, also worked at P&WC, as did a dozen other family members. James Perks retired in 1972 as propeller shop foreman: "When I joined the company in 1939 the propeller shop was in a corner of engine overhaul. As business picked up we moved to the garage of the red brick building, but soon outgrew it and moved into a new shop built by the government on Lorne Avenue (now d'Auvergne). John Drummond was in charge of setting up the new facility. Its walls were knocked out several times to meet demand. The propeller shop became a separate organization with two 12-hour shifts. We worked six, even seven days a week.

"We began with a handful of men and ended with about 180. We were hiring people right off the street, and would put each with an experienced man to learn. If a chap didn't work out on one machine we'd try him on another. We worked on Hamilton Standard, de Havilland, Rotol and Fairey-Reed propellers. Some of the DH propellers were damaged by bullets. They were still useful, but you had to radius and polish the holes to prevent cracks from forming. When we tested these propellers, those blades would sure whistle!

"If a blade came into the shop curled, we could repair it by 'cold straightening' if the damage was slight. If it was severe, it had to go into our furnace. Once heated and softened, a blade could be bent back into shape without cracking it. It was first placed under a press, and in a twister to get a rough correction. Next it went back into the furnace, was dunked in water, then left to age. American propellers were aged in the furnace at low temperature. British ones were left on shelves to age for several days. Once a blade was hardened, small adjustments to its angles were made in a hydraulic bending machine. Straightening demanded a lot of judgment. The inspector would check for twists and thickness every six inches to ensure that the blade was within tolerances. Next, blades went into the grinding and finishing section, where nicks and scratches were removed or rounded. A final polish would give the blade a satin or high-shine finish.

Field Service Adventures

P&WC focused its field service on the RCAF, mainly on training and coastal patrol stations. Smaller bush operators had to rely on their own ingenuity to keep their Wasps maintained. Ken Dawson and Russ McCormack spent much time in the West with the RCAF, CPA and Boeing Aircraft of Canada. Dawson was known to Boeing as "a man who could remember engine serial numbers the way most remembered their names." Gene Schweitzer was in eastern Canada and Newfoundland once the P&WA men took over the Ferry Command work at Dorval. He would be gone for six weeks at a time, reporting to Frank Dando and Ron Riley. Schweitzer would visit Eastern Air Command HQ in Halifax to confer with the RCAF's technical people. They would send him to various stations to listen to engineering officers' problems, give lectures and assist in the maintenance hangars. He was dealing mainly with

Every day there were more bent propellers for P&WC's overhaul shop to straighten. Notice the results of this forced landing on the Harvard's prop! In the shop the day the second photo was taken there was no shortage of business. (S.D. Webb Collection, P&WC Archives)

THE WAR YEARS

While P&WC was supporting the RCAF, its traditional commercial operators had to "make do." They did an excellent job and completed some vital airlifts. In the first photo, a Barkley-Grow (Wasp Junior SB), Bellanca Air Cruiser (Wright Cyclone, later re-engined with a Hornet) and Beech A18 (Whirlwind), all of Quebec Airways/Canadian Pacific, are seen at Sept Îles for service on the North Shore. In the second photo are two Wasp-powered Fairchild 71s and a Wright-powered Bellanca of Austin Airways at Sudbury. Jack Austin recalled having to set up a meeting with C.D. Howe when he needed a replacement Wasp. Howe, no doubt appreciating the needs of commercial operators, quickly agreed. He told Jack, "You're going to get your engine. That's the simplest request I've had all day!" (Tony Leriche, Rusty Blakey)

R-1830s in Lodestars, Cansos and Dakotas, and R-2800s in Venturas. There were few real problems, but Schweitzer does recall, "Carburetor icing was always a concern, though the P&WA engines were generally satisfactory when it came to this. Usually problems had to do with the crew not following operating instructions. I would hold briefings with the pilots and maintenance people to clarify such matters.

Messrs Young and Ross

According to Bill Vertilneck, Young and Ross used to watch things very closely around the plant. Vertilneck, looking back in 1987, recalled: "Mr. Young was the most honest man I have ever met. He was a great saver, and to him a dollar was a lot of money. One winter he bought a pair of tire chains and asked me to install them tightly, so there would be no links banging on the fenders. I went to the garage we had in the plant for Mr. Young's and Mr. Riley's cars. I sat there and said to myself, 'If he wants them tight, I'll make them tight!' I let the air out of the tires, cut a link out and put the chains on. Then I refilled the tires and reported to Mr. Young. 'That's what I want, Bill,' he said, and he was as happy as can be all winter.

"One day in spring, Harold Marr called to say that Mr. Young was unhappy about something. I hurried to his office and he said, 'Bill, do you remember the chains you put on my car? It took the garage mechanic nearly two hours to figure out what you had done. It cost me nearly seven dollars to get them removed!' I had cost Mr. Young some money and he wasn't going to let me forget it!"

Canadian Propellers Limited

In 1941 Ottawa asked P&WC to establish a factory to make two-blade controllable pitch and hydromatic Hamilton Stand-

Construction of the Canadian Propellers factory began in June 1941 and was complete within six months. (P&WC Archives)

ard propellers. The plant would be government-financed. On April 1 Canadian Propellers Limited came into being with James Young as president, R.T. Riley vice-president and George M. Black, Jr. (who was married to Riley's sister) secretary treasurer. One hundred shares were issued at $10 each.

Canadian Propellers started with an order for 200 propellers monthly. The first task for Young and Drummond was to hire and train personnel in propeller manufacturing, choose a site, and build and equip a factory. Mr. Young took on most of the hiring and dealt with Ottawa. John Drummond became chief tool and production engineer. Construction began on June 9 on a site on Hochelaga East in Montreal. Work was finished in November, and the first propeller was completed on February 22, 1942, less than a year after the go-ahead. Some of the history of this venture is outlined by Drummond: "Mr. Young's main contribution was selecting well qualified men. He hired R.J. 'Dick' Moffett, previously with the aircraft division of Canadian Vickers. He got Alan G. Day from Hamilton Standard to handle the transfer of licensee information. Then he handed over his responsibilities to George M. Black, Jr.

"My first job was buying machine tools for making parts. All sorts of tools were needed, and we referred to a process

Some of the wartime gang at P&WC. In the back row are Anita Blanchette, Fernande Lafontaine, Marthe Desormeaux, Lois Barfoot, Pauline Cordeau, Raymonde Desormeaux and Jimmy Wilkinson. In front are Elsie Trott, Isabelle McPhee, Thérèse Carrier, Monique Trudel (Bouthillier) and Thérèse Riendeau. Monique stayed with the company until 1988. (via Monique Bouthillier)

Plant manager George M. Black, Jr., C.D. Howe and Mr. Young during a tour of Canadian Propellers Ltd. (P&WC Archives)

sheet that described the steps to produce a part from a forging. We also had to arrange for raw materials such as tubing, bearings and bars of steel. We had one group looking after tooling and another looking after production. My tool and production section supplied all materials, while the shop supplied labour. Occasionally there would be shortages of tools or raw materials, and I would have to go to Ottawa or Wright Field in Ohio to talk with the people who controlled the allocation of scarce resources.

"Besides the 12D40 two-blade propeller for the Harvard, we were to manufacture the three-blade 23E50 for the Canso. We started turning out the 12D40s weeks ahead of schedule, and the government asked us to concentrate on it. We turned over the equipment for the larger product to a US company. We reached a peak of over 1000 12D40s a month. The efficiency of our plant and the quality of our propellers were soon recognized as second-to-none. We shipped propellers out by the carload to Canadian and US destinations. For special occasions, such as turning out our 10,000th propeller, we had a celebration, which C.D. Howe would sometimes attend. The Americans were impressed by our work and proposed awarding us their coveted 'E for Excellence' award. Mr. Howe declined this, as he didn't want any firm singled out when others were doing just as fine work."

During this time, Arthur Pond was Mr. Black's assistant at Canadian Propellers: "Mr. Black was a chartered accountant and an outstanding businessman. He was only about 30, very young in those days for a top executive. You couldn't possibly dislike him. He was bright and fair and never got excited—the ideal person to run a company. Each morning Mr. Black's secretary would bring the mail into his office and hand him the contents of every envelope. That was one way he had of keeping on top of the business. After the war he went to Toronto to join E.P. Taylor in running Canadian Breweries. His son, Conrad, was named after Ron Riley's father.

"Canadian Propellers was run on a management fee of $45,000 per year. This paid the salaries of the managers and executives. They would bill the government for managing the factory, and submit a bill for operating expenses. This paid the salaries of the men in the shop (we had as many as 700 employees). Each month we also invoiced the government for propellers shipped. I was the last employee at Canadian Propellers and helped close the company. War Assets wanted it liquidated. The final accounting was done over the summer of 1945. The factory was sold to Sam Steinberg to be the head office of his grocery chain. I was offered a job at P&WC but decided to finish a degree. Just after exams in the spring of 1946 Mr. Young called me with a job offer. I accepted, and one of my first assignments was looking after personnel and the company's new pension plan. P&WC was one of the first Canadian firms with a formal plan."

With the winding up of Canadian Propellers, the Faculty of Engineering at McGill University was the beneficiary of $4952 surplus from operations. The company had sold $29 million worth of propellers and spares. Hamilton Standard had made a special concession and waived the usual licence fee for manufacture of its products. This represented a saving to the Canadian taxpayer of nearly $530,000. When he surrendered the Canadian Propellers charter to the president of United Aircraft, on September 26, 1947, Mr. Young thanked the corporation for having waived the licence fee and Hamilton Standard for having trained his personnel.

The President's Report

The 1943 President's Report to the shareholders of P&WC provides an inside look at the company in wartime. There were peaks, but also a few valleys as these excerpts show:

"During the year 1943 our sales amounted to $8,556,956 as compared with $7,652,722 during the previous year. The majority of this was military business done directly with the Department of Munitions and Supply, or with other war contractors. However, an amount of $794,318 was commercial business. It is interesting that these are the second highest commercial sales in the history of the company, the highest being $846,729 during the year 1942. The highest pre-war sales were $534,580 during the year 1939 …. The number of engines and propellers overhauled during the year was respectively 1171 and 3814….

"Information furnished us by the Department indicates that our overhaul business will remain at approximately its present level throughout the balance of this year. This is of course subject to change in the event of substantial changes in the European war situation, but present indications are that we will not be required to make any substantial reduction in the number of our employees. This number is at present 441 as compared with 317 at December 31, 1942.

"The most outstanding single event affecting our future is the letting of a contract to Canadian Vickers Limited for the construction of 50 Douglas Model DC-4 aircraft for operation by Trans-Canada Air Lines. The DC-4 is at present fitted with Pratt & Whitney Model R-2000 Twin Wasp engines and Hamilton Standard Hydromatic propellers. If we are successful in obtaining orders for the engines and propellers to be used in these aircraft, this alone will provide a considerable business for several years after the war and will maintain the previous record of 100% Pratt & Whitney Aircraft-built power plants for all of TCA's equipment. We are accordingly devoting every effort to this end.

"Tribute must be paid to all our employees for their loyalty, cooperation and untiring effort throughout the year 1943. One and all have worked without stint, frequently under difficult conditions resulting from new construction or re-arrangement of shops and offices. Their response to all appeals for funds has also been magnificent, and it should be recorded that during the year 1943 the employees of this company subscribed an average of $302.00 to each Victory Loan Bond."

With war production winding down, sales dropped to $3,735,814 in 1944, though net profit was $54,382. Commer-

Six of the principal men at CPL: Jack Wilcox, Alan Day, James Young, Fred Haydon, George M. Black, Jr., and John W.R. Drummond. (P&WC Archives)

Finishing 12D40 propellers at CPL. (P&WC Archives)

Wartime employees at P&WC. In the back are Jim Muir, Jacques Lamare, Eddie Weston, John Ross, Kay Kerwin-McCarthy, Raymond Senez, Joseph Dubreuil, Bill Vertilneck and Bert Murton. The others are Peter Tracy (guard), Joseph Pelletier, Walter Smallritch, Bill McKeown, Normand (Momo) Senez, Sam Hill, Bert Massé, Stu Weir, Antonio Patenaude, unknown, Jack McPhee, unknown, and John Perks. (via Bert Massé)

cial sales fell by about $350,000. A disappointment was TCA's decision to use the Rolls-Royce Merlin engine in its DC-4s, not the R-2000. However, the 1944 Annual Report included this encouraging note: "During the war, people have been much more air minded. This, together with the great advances in the art of flying made under pressure of war, will result, in the postwar period, in a large increase in commercial flying with a resultant demand for new and improved aircraft."

The End of the War

The Allied victory brought changes at P&WC. Employment dropped to 215 by late 1945. Space shrank from 65,025 square feet to 38,667. There was no new work from the RCAF after October and the company readjusted to a peacetime footing. Mr. Young referred to "the trying period which lies ahead wherein the aircraft industry in Canada will undertake re-conversion to commercial production...." To prepare, P&WC had in 1943 set aside $65,000 for "postwar adjustments, inventory obsolescence and other contingencies," and in October 1945 the board authorized $75,000 to meet postwar conditions. Facilities would be smaller, but far more productive than in 1939.

When business at Canadian Propellers levelled off, John Drummond was called back to P&WC to establish a postwar sales plan. He relates: "I returned from Canadian Propellers in 1943 as engineering manager for technical sales. There was little to the organization other than looking after Ferry Command and the RCAF. It was now clear that the war was winding down, and new activities were sure to follow. TCA and other operators would be looking for new aircraft. With this in mind, P&WC looked at the coming market for war-surplus products. Here we were at an advantage. As the European war subsided, the BCATP schools began closing and aircraft, engines, spares and tools became redundant. We could start bidding on them. The US was locked in the war with Japan, so its fighting machine was still tied up.

"The reason we went into surplus was simple. Only the Wasp and Wasp Junior were widely used in commercial aviation in the North. Those had been produced since the war started mainly by licensees. We also knew that thousands of wartime aircraft would soon be on the commercial market. Their operators would need a reliable source of engines and spare parts. The big question for us was whether or not these engines and parts would be legitimate to use on commercial aircraft. Hartford took the view that it didn't want anything to do with that market. It might affect their reputation if they took responsibility for a licensee's product. They also felt the availability of such products would interfere with the sale of new engines and parts.

"I took a different stand—that we had to face the fact that surplus products were a reality, they were going to be used. They would need support, and if we didn't provide it, some other company would. The government was establishing the War Assets Disposal Corporation to handle surplus. I got to know all about it, while the company was laying the groundwork for assuring that the Department of Transport would certify surplus for commercial use. As soon as War Assets came up with bidding procedures, we were ready with our offers to buy.

"I was not privy to the discussions about surplus equipment

A war-surplus Anson V with Wasp Juniors (above). There was a brisk market for these aircraft after the war, and they served many useful peacetime roles, often in the bush. This one was used into the late fifties by the British Columbia government. In the second photo, an Austin Airways Anson V has had one of the inevitable prangs in the North. The bent propellers and damaged engines meant work for P&WC's overhaul department.
(Wilf White, E.A. Kenyon)

that Mr. Young had in Hartford, but he did win P&WA's approval for our plan. As soon as I heard this, I was on the phone to Ottawa to find out the location of new Anson Vs. They were being stored in southern Ontario and, once our bid on them was accepted, I sent Ken Dawson to have the engines, propellers and accessories removed. In Longueuil, the engines were overhauled and checked to see that all parts conformed to our specifications. We sold the first of these engines to Austin Airways in northern Ontario.

"Periodically, I would go to Ottawa to look over the list of engines coming available. At first we got new ones, but later these became harder to find. Parts came from different sources and were in various conditions. We had no other way of pricing our surplus other than by using the wartime price list. That is how we would bill our customers. Then we would determine what our actual costs were and issue quarterly rebate cheques, calculating the cost of parts to us, plus overhead, plus a 15% profit."

One day James Ross received a letter from George Ponsford, head of the Ontario Provincial Air Service, saying: "My faith in mankind has been renewed. I bought all sorts of parts from Canadian Pratt and Whitney and, a few months later, received a credit notice. I think I am going to frame this cheque." The next month, Mr. Ross was balancing his cash book and found that the cheque had not been framed at all but was in the Ontario Treasurer's bank account!

While John Drummond was chasing the Canadian surplus market, James Ross was hunting for US surplus bargains: "I was in Ottawa on business when I met a dealer from New York who told me he was here to sell the government engine parts. I informed Longueuil of this, and we determined that the US government was releasing all sorts of surplus engine parts. We looked into how to corner some of these. We knew that aircraft and engines were going to be a dime a dozen. Our customers would be buying these, but would object if they had to pay regular prices for spares. I hurried down to the US and before long made a deal for a large consignment of parts. Mr. Riley was pleased with this, but Mr. Young was sceptical and said that it sounded too good to be true. We told him it was all in black and white, with every part itemized. To satisfy him, we brought our dealer to Longueuil. We met in the board room, and the first thing our visitor did was to open his brief case and take out several beautiful boxes, each holding a gold pen. He passed these out to us. Now Mr. Young was really sceptical!

"Our deal was soon confirmed, and we realized that the US government had been releasing surplus engine parts, but the dealers were not obliged to pay for them until they were sold. The dealers, however, were slow to pay, and the government was threatening to repossess the parts. Our dealer was in danger of losing his consignments unless he could secure an order showing a major sale. When other dealers heard about our purchase, they swamped us with offers. In the end, we got a good hold on the surplus market south of the border. We could now sell parts at list to our customers, then issue the sales rebates."

The wisdom of P&WC's surplus policy was clear from the start. Early on the company had negotiated excellent deals with US suppliers. Prices began to rise, but P&WC had its supplies guaranteed at low prices. It shared the savings with its clients all over Canada with rebates in the millions. Customers had different ways of showing their appreciation for the rebates. In Austin Airways' case, it shipped a generous load of frozen Northern Ontario fish to Longueuil.

The Canadair Connection

One of the first postwar projects at Pratt & Whitney Canada was overhauling engines for DC-3s being reconditioned by Canadair. It was the Crown company set up in November 1944 to take over from Canadian Vickers, which had closed its aircraft division. C.D. Howe wanted to keep the Canadian Vickers plant at Cartierville in production and hired Benjamin W. Franklin to run Canadair for the government. Howe's strategy was to build a strong peacetime aircraft industry founded mainly on export business and employing the modern plants and skilled labour force built up during the war. Howe agreed that 40 per cent of the profits from foreign sales could be retained by the company to use in financing peacetime growth.

Franklin negotiated to produce a new transport for the RCAF and TCA and he and Ralph Stopps looked into the US government's plan to auction off several of its wartime plants. When the bidding began, they moved in to buy a C-54 plant in Parkridge, Illinois, and a C-47 plant in Oklahoma City. They bought up everything, including complete airframes for $200 a ton, shipped it back to Cartierville by the train- and truck-load, and made a profit sending trucks back south with Canadian products. Canadair was also buying dozens of C-47s at disposal centres in Europe and the US, and began a program to convert them for the airlines. The C-54 airframes were to be converted for the RCAF and TCA and would be known as North Stars.

P&WC's forklift operator, Roger Leroux, lifts a newly-overhauled R-1830 onto Canadair's truck. (P&WC Archives)

One of the 400 or so Canadair DC-3 conversions. This one was sold to DETA of Mozambique, which was replacing its Ju.52 trimotors. (Canadair Ltd. via Bill Anglin)

Activity at Canadair on March 15, 1946. A newly renovated Canso is getting some last-minute attention to one of its R-1830s. In the background are several war-weary C-47 Dakotas waiting their turn on the Canadair line where they were rebuilt to airline configuration. Their R-1830s were overhauled at Longueuil. (CC&F)

Ralph Stopps reported that the C-47s (some 400 were involved) cost as little as $12,000 and later sold for $100,000, and added, "Most of the aircraft we reconditioned had R-1830s, though some had Wright Cyclones. As soon as an aircraft arrived we started dismantling it. The engines were removed on the first day and sent down to P&WC in Longueuil or to Canadian Wright in the east end of the city. We stripped each aircraft and 'zero-timed' the airframe. We would send two engines to Longueuil and the truck would bring back two newly overhauled ones. We reached a turn-around time of 30 days for a 'new' DC-3. At the peak, we had 10,000 people at work." While the DC-3 program kept Pratt & Whitney Canada busy, there was other engine business. Dozens of reconditioned R-1830s were sold to the airlines,

many for export to customers such as BOAC, Iceland Airways, Aer Lingus and Sabena.

The North Star

When the government decided on Rolls-Royce Merlins instead of R-2800s for the Canadair C-4 North Star, spirits at P&WC sank, though there was some compensation as the North Star used Hamilton Standard 43D50 propellers and accessories. An order came on September 13, 1948, for 22 sets of North Star propeller equipment worth $752,353. The North Star first flew in July 1946 and by 1950 was the backbone of three fleets—TCA, the RCAF and BOAC—and was used by CPA on its new trans-Pacific routes.

When a TCA North Star (which was soon to be reallocated to the RCAF) was wrecked, P&WC knew that a replacement would be built and proposed the R-2800. The government concurred, and thus the only Canadair C-5 came into existence. It was fitted out as a VIP transport and handed over to 412 Squadron in Ottawa. Chief test pilot of Canadair was A.J. "Al" Lilly, who recently provided these memories: "The

The exotic Canadian Car and Foundry Burnelli CBY-3 at Cartierville on July 17, 1945. It is just about to fly for the first time, with V.J. "Shorty" Hatton at the controls. P&WC assisted CC&F in the installation of the CBY's R-1830, seen here roaring away. The CBY had several unusual features, including a lift-producing fuselage, but it found no buyers, even after trials on the Iron Ore Co. of Canada airlift from Sept Îles. It later operated in Latin America, and is today part of the aviation museum at Bradley Field, Connecticut. (CC&F)

(Top) The first Canadair North Star taking off from Cartierville in 1946, and displaying its Rolls-Royce Merlins. P&WC had hoped to win the North Star engine contract. The second photo shows the stately C-5 in flight, powered by R-2800s. (Canadair Ltd., DND PL102424)

F/L Wallnutt arrives in Montreal with the first RCAF S-51. (via T. Wallnutt)

Sikorsky pilot Jimmy Viner at the controls of an S-51 while doing demonstrations in Ottawa in 1947. His passengers are A/M Wilf Curtis, Assistant RCMP Commissioner Gagnon, and Skip Eveleth of Sikorsky. (P&WC Archives)

C-5 was a dream to fly. The power-to-weight ratio was about the best of any aircraft I flew, other than fighters. It was like a DC-6 without the extended fuselage. We had just finished building the C-5 and were doing flight trials–cruise control, takeoff under various conditions, etc. During this, we were asked by Ottawa to fly Prime Minister St. Laurent out to open the Calgary Stampede. We supplied the flight crew, as the RCAF wasn't checked out on the C-5. We made the trip July 7-11, 1950. Ken Dawson of P&WC was along as the tech rep. We all enjoyed the trip immensely, and those Pratt & Whitney engines purred like kittens all the way." The sale of six R-2800s for $238,308 was encouraging but no further interest developed in the C-5, mainly because the licence agreement with Douglas prohibited Canadair from selling the aircraft outside Canada and the UK to protect the market for the Douglas DC-6.

Enter the Helicopter

The link between P&WC and Sikorsky, the great name in helicopters, began in 1929 when both companies became affiliated with United Aircraft. Igor Sikorsky, a Russian emigré aeronautical engineer, was then known as a designer of flying boats, and in 1932, P&WC was agent in the sale of one of these to Canadian Airways. There was little further contact between the companies until the first Sikorsky helicopters were sold in Canada. Sikorsky flew his first successful helicopter in 1939. A/V/M Stedman saw his first helicopter at USAAF HQ, Wright Field, Ohio, in March 1942 and started thinking immediately about helicopters in the search and rescue role. He recommended the purchase for the RCAF of six R-6s powered by the R-985 Wasp Junior but was turned down. During the war several Canadians learned to fly the R-4 and served with helicopter units in the Royal Navy and the US Coast Guard, but the RCAF waited until 1947 to order its first "choppers," these being S-51s. Built by Sikorsky at Bridgeport, Connecticut, the S-51 had an R-985.

P&WC became Sikorsky's Canadian agent. First at P&WC in helicopter marketing was Bob Feldsted, a recent McGill engineering graduate. Feldsted left the company in 1947 and Dave Hanchet, also from McGill, and a wartime pilot, replaced him: "When I took over from Bob, the RCAF already had its first S-51s. I went to Bridgeport to look after the transfer. Those were the days when you could pick up the Sunday paper and read the usual trash about helicopters soon being in everyone's backyard. The truth was, they were still developmental. Flying one was plain hard work. It was my job to determine where the helicopter would fit into our marketing picture. We knew that it could go almost anywhere and operate from the smallest clearing. It was adaptable in the bush, but had limited range and was expensive–about $40,000 for a small Bell, $85,000 for our S-51. That was a lot of money in 1947! A Norseman could carry a much greater payload, fly faster and farther, and was cheaper overall. We realized that selling the helicopters wasn't quite the same as selling a bar of soap, and because of the price of our product, we knew that most sales would be to government. Our first was to the DOT. We dealt with J.C. 'Jack' Charleson, who was well-known in aviation and had taken a USAAF helicopter course in 1945. Through his knowledge and interest we sold the DOT an S-51 for use aboard the icebreaker *C.D. Howe*. When the ship sailed north in 1950, the S-51 was tied down on the after deck. It was to be used in scouting ahead in pack ice for the best channel, and for ship-to-shore transport. One day as it was taking off, it flipped over the side and was lost. The accident was caused by a tie-down clip on the helicopter being left attached to the deck.

"We had hoped to sell S-51s to the geological survey in Ottawa to use in field work in remote areas, but the smaller Bells and Hillers were doing that job more economically. In spite of setbacks, the company viewed helicopters as serious business. It earned a 10% commission on sales, which allowed it to carry a stock of spares and to keep people like Charlie Seager, Jock Graham and myself busy. We were 'sales engineers,' an appropriate title, for we dealt with the facts about what a helicopter could do. We had to determine where it might meet an operator's requirements and keep in touch with the market in hope of something developing."

When the RCAF had its first S-51 accident in 1947, P&WC had to rebuild the wreck. The company had never done such a job. Before his death in 1983, Charlie Seager spoke about this experience: "One day John Drummond came to see me at Canadair, where I was working on the DC-3 program. He wanted me to join Pratt & Whitney: 'We've got a challenge here. We've just gotten into the helicopter business and we

"The inglorious fate of my S-51." So reads the caption by Tommy Wallnutt, whose prang of March 28, 1947, brought the first helicopter overhaul work to P&WC. (DND via Tommy Wallnutt)

Gene Schweitzer, Igor Sikorsky, Ron Riley and Dave Hanchet during a tour of Sikorsky. (Sikorsky Aircraft)

An Austin Airways Norseman V at the dock at Sudbury. This vintage Wasp-powered bushplane is preserved with the Canadian Museum of Flight and Transportation in Surrey, BC. (Larry Milbery)

need some help.' I was no engine man, but I agreed to look over the plant. I'll never forget that first visit to Longueuil. They took me into the old storage building at Plant 2 and there was the S-51–in a million pieces. I said to myself that there was no way I would get tangled up in this mess. Here was a pile of junk, and they wanted to make it fly again!

"The DC-3 program was winding down, so I went to my boss to discuss my future with Canadair. I wanted a project engineering job on the upcoming F-86. The company wouldn't make me any promises. John Drummond's offer now looked better, and I decided to accept it. Bob Raven got the job I had wanted on the F-86. He later joined P&WC and we worked together on the Sea King helicopter for the Navy. Nobody at Pratt & Whitney knew much about airframes, let alone helicopters. I went down to Sikorsky to get some basic knowledge and flew back to Longueuil in a new S-51 with an RCAF pilot, Tom Causey. I sat in the back and had the flight of my life. I then set to work on the repair job. Frank Dando was in charge and George Rose and others helped. We worked in whatever space we could find in the engine overhaul shop. When the machine was ready, the RCAF wouldn't test fly it. We had to bring up Sikorsky pilot Jimmy Viner. He was delayed at the border by Customs, which gave him a rough time about bringing a parachute into the country!"

By trial and error, the RCAF learned how to handle the S-51. It had seven on search and rescue, utility and training duties. Over the years they were often back at Longueuil for overhaul and modifications. The last were retired in 1965.

New Bushplanes

The war had ended the supply of new aircraft for Canada's northern operators. By 1945 the pre-war Fairchilds and other types were on their last legs. The surplus market would temporarily relieve this problem but new designs were needed. Several were in the works soon after the war, all featuring Pratt & Whitney engines. One was from Leigh Brintnell, who during the war had managed Aircraft Repair in Edmonton, an overhaul facility serving the RCAF. He took this over in 1945 as Northwest Industries and went to work on the pre-war Bellanca Skyrocket, incorporating several improvements. But Northwest's marketing push fell flat and only a few Skyrockets were completed.

At Cartierville, Bob Noorduyn introduced the Norseman V. It *followed* the Norseman VI. The "V" was not for five but for "Victory." The Norseman V also proved difficult to market, likely because of the many war-surplus Norseman IVs. Even though sales were modest, production continued to 1959, ending a story which had begun 23 years earlier with a Wasp on consignment from Mr. Young.

Fairchild approached the postwar era with three projects. It entered the prefabricated housing market, began designing a twin-engine navigation trainer for the RCAF (powered by

The first Husky leaves the Fairchild dock for its maiden flight on June 14, 1946, piloted by A.M. McKenzie. With its R-985, the Husky was generally considered underpowered for its size. (Eric Bentley)

The Wasp Junior-powered Beaver became one of history's most successful light utility planes. With its better power-to-weight ratio, it killed chances for the Fairchild Husky. For decades its Wasp Junior provided busy times at P&WC in overhaul and spares. (Jack McNulty)

Twin Wasps), and designed a bushplane, the Husky. Fairchild planned to use an R-1340 in the Husky, but the Ontario Provincial Air Service specified the R-985 as a lighter, more economical engine. Fairchild designed the Husky expressly to the OPAS requirement for an airplane with a bulky cabin that could easily carry canoes inside, as opposed to the traditional way of carrying them on the floats. The Husky flew in June 1946 and was certified in September at a gross weight of 6300 pounds. The first was soon at work with Nickel Belt Airways in Sudbury. Soon word was going around that the Husky was underpowered. A note in P&WC's records explained why: "As usual, operators have overloaded the aircraft (as high as 7400 pounds) then complained of poor performance." Fairchild studied alternate engines for the Husky, including a geared R-985, but before a solution was found, the company went bankrupt. It had lost money on the housing venture, the trainer contract was cancelled, and Husky sales did not materialize. Only 12 were built.

The Beaver–The Ultimate Bushplane

In 1987, the de Havilland Canada DHC-2 Beaver was proclaimed one of Canada's top 10 engineering achievements. Over 1600 were built following its first flight in August 1947. Although it was designed with another engine, it gained

THE END OF THE WAR 67

Ed Caswell (left) and Max Nerriere of Orenda make adjustments to Canada's first jet engine, the Chinook. The setting is in the engine test cell at Malton. The Chinook made its first run on March 17, 1948, rated at 2800 pounds of thrust. The Chinook evolved into the successful Orenda series that powered the CF-100 and the Sabre 5 and 6. (Hawker Siddeley Canada 29664)

worldwide fame behind a P&WA R-985 Wasp Junior. The story goes back to 1928 when de Havilland Aircraft of Canada was formed as a branch of its British parent to assemble, sell and service the famous line of de Havilland aircraft. In the early 1930s it imported three D.H.61 Giant Moths, two of which were converted from the Bristol Jupiter to the P&WA R-1690 Hornet. This appears to be the first connection between de Havilland and P&WC, which was the Hornet agent.

De Havilland Canada thrived under the leadership of P.C. "Phil" Garratt. It was busy during the war building Tiger Moths, Ansons and Mosquitos, and entered the postwar years with a run of Fox Moths. It produced its first all-Canadian design, the Chipmunk trainer, in 1946, then turned to the Beaver. It was designed for a 295-330 hp de Havilland Gipsy Queen but there were problems with the engine. R.D. "Dick" Hiscocks, who was on the Beaver team, explains what led to a switch in engines: "Phil Garratt and James Young of P&WC were close friends and would often get together. Garratt could see that the Gipsy Queen hold-up was going to cause problems. Young let us know about the Wasp Juniors he had from the Anson Vs. We also knew that bush operators generally favoured Pratt & Whitney engines, and that the OPAS especially liked the Wasp Junior. The Husky was already flying with one. Garratt decided to scrap the idea of the Gipsy Queen and go for the Wasp Junior. We had already designed the installation for the Gipsy Queen, but were pleased to find that it could be adapted for the R-985."

John Drummond took the first installation drawings to de Havilland and sold the engine for the prototype. He was surprised that the Wasp Junior could be so easily matched to the Beaver. Installation and test flying involved several P&WC people, including Gene Schweitzer, Arden Boland and Jack Gillies. Schweitzer notes: "I used to go down to de Havilland and look at the Beaver mock-up with the Gipsy Queen. Now they wanted the Wasp Junior and that changed the centre of gravity of the aircraft. That meant moving the engine as far back as possible. Engine mounts, the exhaust system and the cowling arrangement became critical as modifications were made to fit the R-985. When flying began, I often went up with test pilot George Neal to take such readings as cylinder temperatures to determine the best cooling arrangement. This information was critical, for the Beaver was

Orenda's test-bed Lancaster with jet engines fitted outboard of the Merlins. This was Canada's first flying test bed for turbines. (Jack McNulty)

designed for steep takeoff and landing angles, and this influenced the flow of cooling air around the engine."

The OPAS's top man, Frank MacDougall, test flew the Beaver and was convinced that this was the right airplane for Ontario. His first flight seems to have set the Beaver on the road to success, ending all hopes for the Husky. Pratt & Whitney Canada had sold the initial zéro-time R-985s and Hamilton Standard propellers for the Beaver. Later DHC purchased engines from War Assets, finding P&WC's prices too high, although P&WC did overhaul some of these for de Havilland. Other customers supplied their own engines. Hudson Bay Mining and Smelting purchased one surplus, had P&WC overhaul it, then had DHC install it in the new Beaver. When the US Army bought over 900 Beavers, it supplied its own engines, but P&WC later became the source of new spares and the Army became the leading customer.

Canada and the Jet Engine

Research into turbine power began in the early 1930s in various European countries. The war brought its first applications in jet fighters. Canada showed an early interest, and A/V/M Stedman toured British research facilities in 1942, including those of Frank Whittle of Power Jets Limited. Whittle's engine had powered the first British jet aircraft on its maiden flight in May 1941. Stedman also visited US facilities, particularly General Electric. While P&WA, Wright and other US manufacturers were restricted to piston engines during the war, GE had a free hand. It adopted British know-how and built the J31 to power America's original jet fighter, the Bell Airacomet. When Stedman suggested that Canada should enter this field, Ralph Bell sent a fact-finding team to England. Following the visit, a facility was established in Winnipeg to test jet engines under low air temperatures approximating those at high altitude. The first jet engine in Canada (a Whittle) was run there in January 1944.

That July Ottawa formed Turbo Research Limited to design and build Canada's first jet engine, taking over work being done in Ottawa, Toronto and Winnipeg by the NRC. In early 1945 engine design got under way but Ottawa sought to privatize Turbo Research, hoping to sell it to a company (likely British or American) with wider knowledge in the field. P&WC was encouraged and James Young and Ron Riley held discussions with C.D. Howe. But the US government had placed a lid on the secrets of its turbine engine R&D. Knowledge could not be exchanged with Canada, so P&WC could not obtain data from Hartford. Personnel from P&WC were not even allowed into areas at P&WA where turbine work was being done.

Turbo Research was acquired by Avro Canada (a branch of Avro in the UK), which had recently taken over Victory Aircraft at Malton. Avro began design of two jet aircraft—a transport and an all-weather fighter. Turbo Research became Avro's Gas Turbine Division and in March 1948 ran the Chinook, Canada's first jet engine. The Gas Turbine Division later became Orenda Engines, which produced a family of successful engines for the CF-100 and F-86, and was developing a powerful new engine, the Iroquois, when the Avro empire collapsed in 1959.

THE END OF THE WAR 69

Engines mounted on stands in the box storage building before going across for disassembly and overhaul.
(P&WC Archives)

Betty Harber works at masking engine parts before electroplating. The masked areas were those not requiring plating. To the right are master rod bearings, masked and ready for plating on the outside.
(P&WC Archives)

Engine inspection. Here any part needing reconditioning was tagged to show the work to be done. Worn parts were disposed of and replacements requisitioned from stores. Everything was thoroughly re-checked before assembly. Rosaire Chetagne is in the foreground. (P&WC Archives)

Doug Boone blasting hard carbon and baked enamel from an engine cylinder. Various types of abrasives were used, depending on the part being cleaned. (P&WC Archives)

THE END OF THE WAR 71

(Above) Crankcase and connecting rods subassembly. A main crankcase, crankshaft, and master and link rod components are being assembled.

(Above right) After overhaul, each engines was taken into a test cell to be run.

(Right) Just before an overhauled engine was shipped to the customer it was washed down on the outside with solvent. Jean Neveu is seen at work in this view. (All P&WC Archives)

The Fifties

Soon after the Beaver, de Havilland Canada began work on a larger aircraft suitable as a Norseman replacement. Dubbed "King Beaver," it would have the Beaver's STOL performance but carry twice the load. Finding an appropriate engine and propeller was vital. The likeliest engines were the Wright R-1300 and the R-1340-S3H1G geared Wasp. According to DHC's calculations, the King Beaver with the Wright would weigh 8200 pounds all-up and carry 2500 pounds over 500 miles. With the Wasp the figures were 7500 pounds/2000 pounds. DHC chose the Wasp based on the P&WA reputation and P&WC's record in product support. The King Beaver remained on the back burner until the OPAS expressed interest and the RCAF came forward with development funds. Construction of one aircraft was authorized in November 1950. The name of the new design was changed to Otter.

Punch Dickins of DHC proposed a direct-drive Wasp and standard Norseman propeller for the Otter but Dick Hiscocks favoured a geared Wasp turning a bigger propeller more slowly. "It all goes back to basic propulsion theory," he said recently. "It's better to move a large mass of air slowly than a small mass quickly. I wanted to use a larger three-blade propeller to obtain maximum static thrust for takeoff. A larger propeller gives higher thrust for a given horsepower." Some of the larger P&WA engines were produced with reduction gearing to meet specific applications but geared Wasps were not common. A few had been made during the war with the gearing attached to the nose of the engine, but by now most of these had been re-converted to direct-drive engines. The Babb Company in California located enough for a start and Phil Garratt decided to go ahead, saying, "I think I can talk my friend Jim Young into making the components. I know he would like to get into the manufacture of engines."

As with the Beaver, engine cooling would be paramount. The Otter would be slow on the climb, which would limit the cooling airflow around the engine, with the chance of overheating just as the engine was working hardest. An exhaust system was devised that sucked air through the engine compartment to cool the cylinders. Gene Schweitzer was active in getting the program rolling, but he turned the reins over to Ken Sullivan, who had joined P&WC in early 1951. Sullivan had begun flying before the war with the Toronto Flying Club, studied aeronautics at Central Technical School in Toronto, then served as a mechanic in the RCAF. After the war he joined the DOT's meteorological service and was posted to Norman Wells, NWT. He was soon back, studying aeronautics at the University of Toronto. For a summer job he was a mechanic with a photo survey company in the North. Sullivan graduated in 1949, went to work for Orenda, but one day received a job offer from P&WC: "I had applied for a job

The prototype Otter with its geared Wasp at Downsview in 1951. Note the smaller area of the fin compared to all subsequent Otters. (DHC)

Conversion kit for the geared R-1340. (P&WC Archives)

at Longueuil in 1948 and they had kept my application on file for two years! There was an opening in sales engineering and the company was prepared to better my salary at Orenda. I thought that P&WC was a more promising company, as it had access to all of Hartford's know-how."

Ken Sullivan recalls his days on the Otter program: "To perfect its engine/propeller installation required many modifications and adjustments. We had a 600-hp engine and had to determine how to modify the gearbox and what gear ratio and propeller to use. We narrowed the choice to two blade types. These varied in shape, twist, and whether the tips were square or rounded. We had to produce performance curves to show just what the aircraft would do under various flight conditions with these blades. The main goal was to achieve the greatest static thrust for takeoff. DHC wanted good takeoff with maximum payload and the floats half submerged. This would be the only way to make a favourable impression on bush operators. We did many tests with various engine-propeller combinations on a test stand. We finally settled on a fairly wide and heavy Hamilton Standard DC-3 type of propeller, but with blade length reduced to improve efficiency. I was at Downsview on December 12, 1951, when George Neal made the first Otter flight, and later flew on the test flights certifying every aspect of engine installation." The Otter was certified on November 5, 1952, and a few days later Hudson Bay Mining and Smelting of Flin Flon, Manitoba, took the first example (CF-GBX). 466 Otters were delivered by 1967 when production ceased, 155 being for the US military and 69 for the RCAF.

Overhauling and converting direct-drive Wasps to geared engines became big business at P&WC, as did the manufacture of reduction gearing once surplus supplies dried up. With large US Army orders for Otters, a steady flow of engines came into the plant. For example, in August 1957 P&WC received an order to convert 84 Wasps to geared configuration and was asked to quote on a further 29. As the R-1340 manufacturing line was active, the Army engines received a "gold plated" overhaul: all worn parts were replaced by ones newly made in Canada.

Helicopter Developments

While the S-51 found little application in Canada except with the RCAF, a larger machine, the S-52 showed more promise. Sikorsky was optimistic about sales and hoped that large orders could get the unit price down to $40,000. Various Canadian operators showed interest, but no orders were placed before production was taken over by the US military to meet Korean War demands. Next from Sikorsky was the S-55, first flown in 1949 with a 550-hp R-1340. Some of the first civilian examples were sold by P&WC.

In 1947 Carl Agar introduced the helicopter to British Columbia. Operating as Okanagan Air Service, he brought a Bell 47 up from Washington to dust orchards in the Okanagan region. This project was not a money-maker, so Agar turned to mining, surveying and construction projects, becoming the father of mountain helicopter flying. On one job in 1949 one of his Bells lifted 400,000 pounds of material to a construction site 3500 feet above sea level. The useful load was only 400 pounds per trip. A larger helicopter would have been useful. In 1950 Agar was working in northwestern BC, where the Aluminum Company of Canada was laying the groundwork for a hydroelectric bauxite-smelting project. The prime contractor, Morrison Knudsen Ltd., teamed up with Agar in 1952 to bring the first S-55 to Canada. It bankrolled the

CF-GHV, Canada's first S-55, at work with a seismic drill rig in northern BC. The S-55, with its economic payload and reliable R-1340, brought a new dimension to helicopter operations in Canada in the early fifties. (via E. Schweitzer)

Standard doings in the North. An Okanagan S-55 undergoes an engine change. A hefty A-frame, tent and Herman Nelson heater were essential equipment for such rigorous maintenance. (NAM 7061)

(Left) Wasp woes! Ross Lennox looks on as his technical experts troubleshoot an unserviceable engine. Problem? A fuel filter clogged with felt that had entered the system during refuelling. Jock Graham of P&WC, shown above with the HBM Otter, was called in to help with the filter problem, which had caused five engine failures in seven days. (via R. Lennox)

machine, and Agar formed Okanagan Helicopters to operate it.

John Fraser "Jock" Graham was Okanagan's base manager at Kemano, where much of the work was centred. A wartime mechanic, his exploits included being shot down, rescued by an enemy submarine, then escaping from a POW camp. Graham worked after the war for Queen Charlotte Airlines before joining Agar in 1950 as his first employee: "Carl Agar and I picked up the first S-55 at Sikorsky in April 1952. While there we watched our machine being completed and took courses from Jim Saunders, a delightful fellow who had flown a Sikorsky boat all over Africa on safari. Alf Stringer, Carl's partner, was also on course, along with a crowd of US Marine mechanics. Meanwhile, Carl took a flying check-out from Jim Chudors. He was not allowed to fly solo until he had accepted title to the aircraft. That took place, according to Sikorsky tradition, at Teterboro Airport, New Jersey, to avoid paying Connecticut sales tax. For some reason, such transfers were always made late on a Friday afternoon.

"Ours was only the second commercial S-55. The first had been delivered to Los Angeles Airways a few weeks earlier and had crashed soon after. This held up our delivery for two weeks due to the investigation. When the day finally came the marketing group was there to see us 'over the fence.' Pilots Jim Viner and Jim Chudors cautioned Carl to keep close to the highways on the way to Vancouver as they would offer a safe place to set down in an emergency. Carl reminded them that the helicopter wouldn't be seeing a highway for three years once it reached Kemano! Jim Chudors flew with us to Teterboro, where all the documents were signed, and that impressive helicopter was now our responsibility."

Agar was soon operating two S-55s at Kemano in establishing camps and resupplying crews blasting footings for the powerlines. Heavy transmission line support towers were broken down into 1500-1800 pound sections and slung to their locations by the S-55s. The S-55 was eventually credited with moving the project ahead by a year. Kemano became the largest commercial helicopter operation in North America in 1953, and a showcase for Sikorsky. Jock Graham went to P&WC in 1953, first in service, then sales. When the company surrendered its Sikorsky agency in 1972, he moved to Bridgeport, Connecticut, (but remained on the P&WC payroll). He retired in 1987 after 37 years in an industry just 40 years old.

His last job was marketing the PT6-powered S-76. Looking back to the old days of the S-55, Graham comments, "The S-55 was an uncomplicated, reliable workhorse. It would compare favourably for serviceability with current models."

Hudson Bay Mining and Smelting had a progressive outlook towards mineral exploration. A pair of Norsemen based at Flin Flon first made it a customer of Pratt & Whitney. Ross Lennox was one of its pilots years before he became chief pilot for P&WC. Lennox grew up in Winnipeg, where he developed a passion for aviation. In 1941, at the age of 17, he joined the RCAF, becoming an instructor in the BCATP, then going overseas to fly Dakotas: "In the spring of 1947 I went north to Flin Flon, where exploration and the inland fishery were booming. The following February, Hudson Bay Mining and Smelting hired me to fly a Norseman on skis, dropping off exploration parties on a lake in the Barren Lands. We freighted in supplies from the Hudson Bay coast using a Norseman with a 600-hp Wasp 'H'. The Norseman was so rugged and the Wasp so reliable that we had no trouble all season. Hudson Bay was the best private company to fly for in Canada. It always had the best equipment. It bought a Grumman Mallard in 1947, one of the first Beavers, and the first production Otter. The Beaver and Otter brought an end to our Norseman. They were lighter for their size and with their STOL wing could operate on smaller lakes. Even though it carried twice the Beaver's load, the Otter could get in and out of the same spots. When it came to moving parties, you could usually place them closer to where they wanted to be and reduce the number of trips needed to establish a camp.

"In 1952 the company made a study of different helicopter types, concluding that the S-55 was best suited to its needs and the company placed its order. I was sent to Okanagan in Vancouver to take the helicopter course. That took $11^1/_2$ weeks and cost the company $10,000! When I got down to Bridgeport, S-55s were being turned out for the Korean War. The place was a beehive of activity. Our machine was still on the line. I couldn't get over how big it was. There was room in the cabin to stand up. When I saw the mock-up of the S-58 I was even more impressed. I did a two-day ground school, then a four-hour flying course learning the basics of the S-55. When that was finished I asked Jim Viner for a letter to take back to the DOT so that I could get helicopter endorsement on my licence. He chuckled and said, 'So you think you know how to fly it!' I set off to deliver our S-55, accompanied by Pete Cornwall of Okanagan. It took us 50 hours of flying to reach McMillan Pass on the Yukon-Northwest Territories border. I found out that I really needed those 50 hours to get the feel of the big helicopter. The S-55 really opened up exploration for Hudson Bay Mining and Smelting. We no longer had to worry about being restricted by freeze-up or break-up as the seasons changed. Carrying an airborne magnetometer, as we did, eliminated much non-productive effort."

With the building of the Mid Canada Line in the early 1950s, the RCAF made a big commitment to helicopters to transport men, equipment and supplies during construction and to resupply and maintain the sites later on. In 1954, 10 S-55s were ordered through P&WC in a contract worth $1.5 million. All activity was undertaken by 108 Communications Flight commanded by S/L R.T. "Bob" Heaslip. Pilots were trained in mountain and bush flying by Okanagan in BC. Heaslip was kept busy ferrying S-55s up from Bridgeport and H-21s from Philadelphia. He recalls: "The S-55s did the initial survey across Canada and were used to locate the best sites for the stations. Next came the airlift of everything needed for construction. The sites were 30 miles apart, with the supply base sometimes 90 miles away. We were battling time to complete the job. The S-55s did their best work in the west. We could count on them for reliability that far away. Besides, it turned out that the loads were smaller there. In the east, larger H-21s and S-58s were used. The maintenance boys were the ones I admired. Once we had a Sikorsky throw a rod between Great Whale and Knob Lake. A Canso flew an engine to the nearest lake, then it was rafted down a river to the helicopter. The mechanics built an 'A' frame and changed the engine right there in the bush."

Helicopter Field Service

From the beginning, helicopters were used in remote areas. Breakdowns and accidents were bound to happen. Carl Agar's first Bell was wrecked within a year, providing a good winter's work for Alf Stringer. Jock Graham's first winter with Agar was spent on a similar project. Once he joined P&WC, Graham would have plenty of work in field service: "Helicopters added a great deal of colour to life around P&WC. The service reps did 'exciting' things, like going into the bush and living in tents. We always made sure to come back with lots of tall tales. Management used to enjoy reading our field reports! In the summer of 1953 an RCAF S-51 crashed on a peak in the Northwest Territories. I was called in to help CPA with the salvage. Flying in to Norman Wells, I had the pilot circle the crash while I studied the wreck and made up a hasty list of spares. Once we reached the site, we righted the S-51, then stripped off all damaged parts. The RCAF flew in everything I needed, including a tail boom and a set of rotors.

"We worked eight days and had a full program. Besides the

Ross Lennox flying Hudson Bay's new S-55 in rugged northwestern terrain.
(via E. Schweitzer)

airframe, we had to do a lot of work on the Wasp Junior, which had been buried in snow all winter. Finally we had the machine back together, and it started OK. As we were high up, there would only be one chance for a successful takeoff, and that would be early in the day while the air was still cold. The RCAF brought in a pilot, and he found that we had done an excellent job. He'd have flown the S-51 all the way back to Manitoba, except, with most of the windows broken, it was a bit too drafty."

Helicopters for the Navy

The Royal Canadian Navy trained its first helicopter pilot at Rivers, Manitoba, in 1950. This was LT George Marlow. His instructor was S/L Bob Heaslip. Marlow was followed by LT J.D. Lowe and they, along with LCDR Dennis Foley (engineer), formed No. 1 Helicopter Flight, RCN, at HMCS *Shearwater*. Of this cadre, Foley was most experienced, having flown the R-4 with the US Coast Guard during the war.

In April 1952 the RCN took delivery of its first Sikorsky

The Piasecki H-21, with a 1425-hp R-1820, was a key contributor on the Mid Canada Line project. This H-21 is slinging a Wasp and a Twin Wasp at Rockcliffe in 1954. (DND PL102771)

Okanagan brought the first S-58 to Canada in 1956. This type, with its 1525-hp Wright R-1820, was an important leap in Sikorksy technology. The S-55 had begun with a 550-hp Wasp, then went to the 800-hp Wright R-1300. P&WC serviced all three engine types. Here an RCAF S-58 delivers a sling load to a makeshift landing site. (DND PL124796)

A mechanic helps to disassemble a mangled RCAF S-51 "somewhere in the boonies." This kind of work was routine for P&WC's field service men. (via J.W.R. Drummond)

The RCN operated the HO4S (naval S-55) beginning in 1952. Its primary mission was ASW. The Navy standardized on the R-1300, and conversions from the Wasp were done by P&WC. (via J.W.R. Drummond)

HO4S-2, the navalized S-55. Marlow and Lowe made the first RCN helicopter carrier landing aboard HMCS *Magnificent* on May 6. The HO4S-2 was for "plane guard" duties, which meant being airborne whenever aircraft were being launched or recovered, ready to rescue airmen in the event of a ditching. In July 1955 the RCN formed its first anti-submarine warfare (ASW) helicopter squadron, equipped with the HO4S-3. This model was similar to the -2 but had a more powerful Wright R-1300. The -3 had sonar that was dunked from a cable to "listen" for submarines and carried torpedoes.

Jock Graham was at Shearwater when the first groundcrew were trained, then accompanied *Magnificent* for 11 weeks on NATO exercises in the Atlantic: "My job was to mother the Navy crews along. I answered their technical questions and handled spare parts and warranty work. There were few problems, and the trip became more of a holiday."

The Navy decided to standardize its HO4S fleet, and P&WC received the contract to convert the -2s to -3s. This business was important to P&WC and paved the way for Canada's first large-scale helicopter assembly line when the company later received a $45 million contact for Sea Kings for the RCN. In the meantime, the helicopter shop at Longueuil had steady business. In February 1956, the P&WC operating committee reported on winterizing six RCAF S-58s, installing tail rotor servos and hoists on HO4Ss, having an order from Okanagan for an S-58, and being on the brink of a deal with Ottawa involving S-58s for the Army and Navy. In the overhaul shop, there was work on R-1340s for US Navy Kaman helicopters.

In 1956 the RCN began investigating the operation of ASW helicopters from smaller warships. If this proved feasible, the Navy could expand its role of protecting the North Atlantic from the threat of Soviet submarines. Tests began in November, with LCDR John MacNeil operating an HO4S on a 40' x 40' landing pad on the stern of HMCS *Buckingham*. MacNeil later commented: "I had been flying helicopters less than a year when I was assigned to the *Buckingham* trials. My only shipborne experience had been flying Bells and a Piasecki HUP that summer aboard HMCS *Labrador* in the Arctic. Our aim was to determine what kind of flying was possible from a small vessel under different sea and weather conditions. We had to consider rolling and pitching of the ship and controllability of the helicopter. Trials began in a calm sea with *Buckingham* anchored off Halifax. Our technique was spartan. Once I landed, deck crew would attach web straps to a ring on the side of the helicopter and to one on the deck. They cinched the straps tight and that was it. Coordination was the key to success. All the straps had to be free for takeoff. Once we were happy with these straightforward trials, we sailed to Bermuda and found some rough waters. On one flight we bent the landing gear struts. My conclusion was that the HO4S with 800 hp was underpowered for such operations."

In September 1957 phase two of the RCN's landing trials began with HMCS *Ottawa*. The Navy borrowed an S-58 (1525 hp) from the RCAF. It weighed about 13,000 pounds all-up, compared with 7,000 for the HO4S. Jock Graham notes: "We sailed to the north of Ireland to do rough water trials for about six weeks. And it was rough! The deck rolled over 30 degrees, pitched 20 and yawed 20. By carefully observing the waves, the pilot could pick a steady period (only about four seconds long!) and land."

Graham took some leave in England while the ship was in Belfast. Later, he caught up with the S-58 and its pilot, F/L Lloyd Cummings, for the flight back to the ship. Flying north towards Coventry, they became lost in bad weather. Cummings landed beside a highway, and Graham, map in hand, approached one of the motorists who had pulled over for a look. Graham determined that they were on course. He thanked the motorist who, looking at the markings on the helicopter asked, "Are you boys from Canada?" Graham replied modestly, "Yes, and it's been quite a trip."

P&WC's operating committee reported in August 1955: "It is rumoured that the RCN and RCAF may get together on their anti-submarine requirements…. At present, the RCN favours the S-58, and studies concerning the possibility of assembly and manufacture of this helicopter in Canada are being pushed ahead." In 1956 helicopter business at P&WC equalled sales of R-1340s to the government. By mid-1957, P&WC learned that the DND might wish as many as 92 S-58s over a 3–4 year period. The company prepared a proposal outlining Canadian content should the helicopters be made in Canada. However, a new government came into power in 1957 and things changed.

John MacNeil hovers his HO4S over the stern landing pad of HMCS *Buckingham*. (DND via D.A. MacNeil)

P&WC's helicopter overhaul line at Longueuil during busy times in the mid-fifties. 9633 is an RCAF H-34A. Two RCN HO4Ss are next, then an S-51. Such maintenance work was "bread and butter" for the company. (P&WC Archives)

The company noted: "The military helicopter picture appears bleak, with the new government questioning all expenditures closely." Turbine helicopters were on the horizon. Sikorsky flew its single-engine S-62 in 1958 and the bigger S-61 flew less than a year later. The RCN now focused on the navalized S-61, designated HSS-2, and Ottawa ordered 18 Vertol 107s with twin GE turbines to be split between the RCAF for search and rescue and the Army for tactical purposes.

Manufacturing Begins

The postwar era was remarkable for Canada's aircraft industry. A whole range of products appeared, including the Beaver, Jetliner, CF-100 and North Star. Canada's first production jet engines were flying, and light planes such as the Fleet Canuck were testing the market. In the five years after the war, the industry produced over $100 million in new products. Sales at P&WC totalled $15.8 million, more than 70 per cent to commercial buyers. Overhaul and surplus sales were the bulk of business, but P&WC was handling new products as an agent—there were HS propellers for the North Star, R-2800s for the C-5 and accessories for the CF-100. Yet, the company wanted to do more. As Ron Riley saw it, "If the company was to continue its growth and development, it must change the character of its business and commence manufacture." This would mean large expenditures for plant and equipment, something that in 1950 was beyond P&WC's capabilities.

John Drummond recalls P&WC's resolve to do something about manufacturing: "There were companies in the industry, particularly Avro with its Gas Turbine Division, which really appeared to be going somewhere. James Young and Ron Riley discussed our goals in many conversations with C.D. Howe and the board. It looked as though we could miss the boat if we didn't do something about getting into manufacturing." Drummond stated that P&WC should form the nucleus of a manufacturing organization capable of expanding to meet the military needs of Canada and/or the US for particular engines. In the short term it should form a division to fabricate parts for P&WA, extend activities to making parts for the Canadian market, and do so at a profit, without financial assistance from the government.

Hartford had delved into turbine research with a prelimi-

The RCAF and Canadian Army received their GE-powered Vertol 107s in 1963. The choice of these over a Sikorsky product delayed P&WC's hope of becoming a helicopter manufacturer. These new Vertols visited Rockcliffe in June 1965.
(Larry Milberry)

nary design in 1941 but was not allowed to carry on this work during the war. As soon as the war ended, it designed the T34 turboprop, which experimentally powered a USAF Globemaster and a US Navy Super Constellation but did not reach production. Although Hartford did not have its own turbine designs ready for early postwar production, it gained experience producing 130 Westinghouse J30s for the Navy's McDonnell Phantom jet fighter, completing this contract in 1947. The Navy realized that the Rolls-Royce Nene was ideal for its new series of jet fighters, and P&WA obtained the licence to build this engine, running its first one in March 1948. It soon had the engine in production as the J42 "Turbo Wasp," with hundreds delivered to the Navy for the Grumman Panther fighter. Working closely with Rolls-Royce, P&WA developed the J42, increasing power by 30 per cent and adding an afterburner. This became the J48, which saw use in aircraft like the Grumman Cougar and Lockheed Starfire.

With the Korean War, P&WA had large orders for jet engines, along with its line of piston types. To keep up, it began subcontracting work out, but subcontractors themselves were extremely busy and P&WA looked to its Canadian subsidiary. The Canadian Propellers success story was well known in Hartford, as was that of Ron Riley in Kansas City a few years earlier. In the fall of 1950 John Drummond went to Hartford to investigate P&WC getting into J48 subcontracting. He took along Bob McLatchie, who was well-versed in machinery and equipment requirements. Under the 1926 US Air Corps Act, the pair, as aliens, were not allowed to visit the

J48 plant. Bill Gwinn of P&WA wrote the Navy asking that P&WC, as a much-needed subcontractor, have access to the J48. He outlined P&WC's war record and its ability to serve in a new capacity. He also sought a waiver to the Buy American Act to allow his company to make purchases in Canada for use on US military contracts. Soon Gwinn was able to release to P&WC data for a number of J48 parts, and the Canadians were allowed to enter offices dealing with the J48, but the plant remained off limits.

P&WC selected 50 parts numbers from the J48 parts list. This equated to 99 separate pieces. Hartford would need 200 sets monthly to meet its target of 150 engines a month for the Navy. P&WC estimated a cost of $782,744 to set up to manufacture the parts, machine tools being the main expense. Some 26,000 square feet of plant space would be needed and a staff of 139, mostly machine hands. Drummond communicated the results of his study to Ron Riley in November 1950. He next went to Ottawa to seek government support for the 1934 Buy American Act waiver and to negotiate to lease space from Canadian Arsenals in the old Walmsley factory. While dealing with the Department of Defence Production, Drummond was asked, "Look, you're planning to make jet engine parts for your parent company. Why not manufacture a complete engine?"

The details left Drummond wondering. Apparently 1000 Wasp engines were needed monthly for the US and Canadian military. Drummond knew that this figure was wildly optimistic: "We agreed to meet a few weeks later, at which time the DDP had its figures straight. The need was for 250 engines yearly for the US and 50 for Canada. Ottawa wanted us to make a proposal. Soon Young and Riley were in Hartford working out details as to space, the cost of machines and equipment, scheduling and so on. We put a proposal together. The government accepted it and asked us to set aside the J48 project. This we did, for we now had a contract that would make us a major manufacturer."

The Wasp program illustrated the changing nature of Canada-US relations. On October 26, 1950, representatives of each country had signed the Canada-US Pact, which removed many of the restrictions of the Buy American Act and offered free trade in the defence sector. Canadian firms could now participate in US defence contracts.

Gearing up for the Wasp

The J48 study had exemplified Ron Riley's low-risk approach to entering manufacturing. Now the company was asked to take on a much larger project. It relished the challenge but knew that the Wasp was a wartime task the benefits of which would soon fade (on the other hand, the Wasp was a mature product and no engineering changes would be needed). At least with the J48, P&WC would have had capital equipment with a future. Once the Wasp contract was fulfilled, though, the company would have modern facilities for overhaul and sales. It might have more space than it needed and that would be a financial burden. Many such questions filled the air as the company grappled with the future.

The letter of intent from Ottawa arrived at P&WC in February 1951 confirming the engine requirement. One of the first issues was how to share the risk of the new plant. Who would pay for land, construction and machinery? The board met on February 22 to review matters. Bill Gwinn and Jack Horner of P&WA had come up for the occasion. In the end, the board stated that P&WC "is authorized and requested to proceed with the procurement of the materials and equipment necessary for the creation of facilities for the production of R-1340-AN-1 engines at an ultimate rate of 200 per month plus 24 per cent spares."

"We proposed that at our expense we buy land and put up a plant," recalls John Drummond. "We felt, however, that the engines would not be produced for very long. We did not have the $7 million needed for machinery, equipment and tools, and also understood the risk of getting tied down to an investment in obsolete equipment. We therefore asked the government to equip the plant. This led to the signing of a capital assistance agreement on August 21, 1951. We agreed that if R-1340 production was cancelled and the company could not economically employ the plant, the government would take it off our hands. We requested and received a capital cost allowance which allowed us to accelerate depreciation of the plant and write off our investment over the cost of the contract. This was a practice common with the US government concerning emergency facilities financed with private funds, but was an innovation in Canada at the time."

On September 5, 1951, an agreement was signed with Ottawa for five years with an order for 1000 engines plus spares, deliveries to begin six months after installation of the necessary facilities and to reach 50 engines per month within six months. The government agreed to pay preproduction expenses and a profit of $1000 for each of the first 100 engines. Thereafter, a target cost per engine was worked out. There would be a profit of five per cent of this, more if costs were below the target. Profits on spares were set at 7.5 per cent. All jigs, tools, dies, gauges, fixtures, instruments and equipment not of a capital nature would be paid for monthly, in advance, by the government.

Engineering

To head engineering on the Wasp project, John Drummond recommended R.H. "Dick" Guthrie. Guthrie had a boyhood fascination with aviation and at 14 entered a national model building contest sponsored by General Motors of Canada, the most prestigious contest in Canada. The task was to build from scratch a scale model of a Napoleonic coach. "Before I even started," says Guthrie, "there was a lot of planning about materials and a schedule. My calculations showed that 1000 hours would be needed. I would have to put in four hours' work a day, and allow for a month at the end for final detailing." In the 1932 contest, Guthrie placed first in Ontario, but there was little financial reward. He set about to better his effort for the 1933 contest and built a second model. He came first nationwide, winning a scholarship of $5000. Guthrie invested it in two years studying architecture at McGill, then went to MIT for three years of aeronautical engineering, graduating in 1939.

Guthrie had always been interested in aviation. He had seen the R-100 at St. Hubert, was at Boucherville when the *Mercury* landed after its trans-Atlantic flight and saw the *Hindenburg* a few hours before it was destroyed in New Jersey. "I first went to P&WC one summer just before going to MIT. I had been to Canadian Vickers and Fairchild looking for a summer job, but considered P&WC a better place for learning. Mr. Young showed me around the shop and offered me a job."

When he returned from MIT, Guthrie went to the NRC in Ottawa to assist in design of the wind tunnel at Montreal Road. He worked in de-icing research and other projects. In

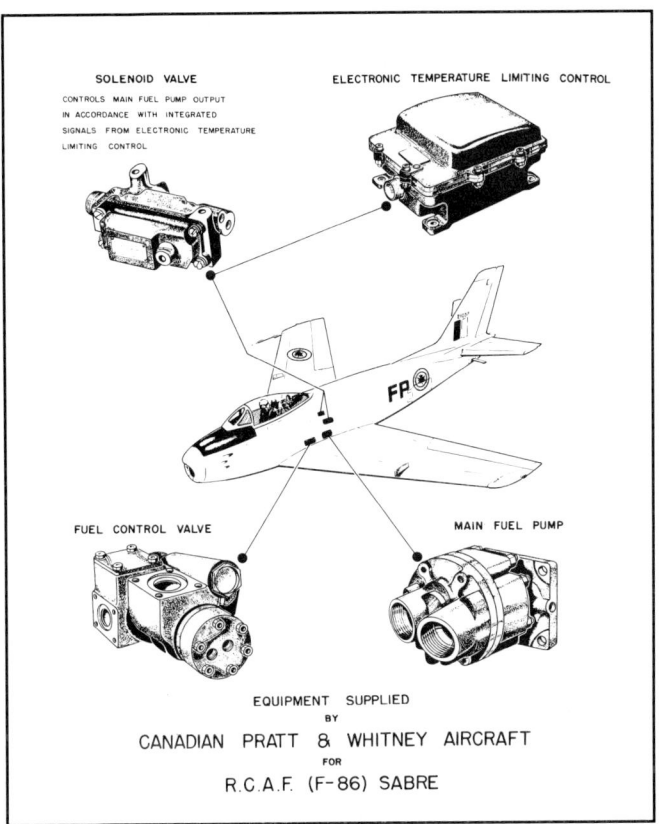

P&WC always sought to offer a diversified product line, and represented such US products as Pesco pumps. Shown here are the "agency" products it supplied for the Canadair Sabre. (P&WC Archives)

The R-1340 Wasp which made P&WC famous as an engine manufacturer. (via J.W.R. Drummond)

1942 he joined the RCAF as an engineering officer and was posted to the repair depot in Calgary. Late in the war he joined Turbo Research, then moved to de Havilland, where he worked on the Chipmunk, Beaver and other projects. Guthrie's experience at DHC taught him that a well-balanced engineering organization was important to a company's future. DHC could well have closed had its engineering people not come up with the Chipmunk and Beaver. When he came to P&WC in 1951 as an engineering manager, he was impressed by the way the company did business: "With Young, Riley and Drummond I got the sense of men in control of the present, and with a careful eye on the future. I wrote to Riley about the importance of hiring draftsmen, designers and project engineers with the right ability and mechanical aptitude, emphasizing the importance of training. Riley knew exactly what he wanted. His attitude was that it was up to each man to make his own future in the company. When he set out to build the R-1340 organization he made it completely separate from overhaul. The skills were different. Besides, overhaul always had its own work, so Riley simply left it to turn a profit as it always had. Each department head built his own team. I would go to Riley with this or that requirement. As long as I could justify it to him the request was approved. I would send my new people to Hartford and depended on them to come back well versed. We learned as we went along."

Guthrie hired his engineering team. There were Hugh Langshur, responsible for liaison with Hartford and for test cells; George Walker, chief inspector, formerly at NRC; George Sanders, quality engineering; Harold Dickie, quality review; and Claude Tatham, materials control lab. One by one, the key slots were filled. The same went in the other departments. Beginning with six men gathered around a table with Ron Riley on May 11, 1951, over 1000 were hired in 18 months to build the first Canadian Wasp.

Night Train to Hartford

Many new P&WC employees spent a period training at the Hartford plant. Thus were they introduced to the *Washingtonian*, the train that carried travellers from Montreal into the US. It departed about 8:30 p.m., with passengers holding tickets with separate stubs for the Canadian National, Central Vermont and two or three other lines. The trip to Hartford could have been made in four hours but stretched to double that. Sleeping car accommodation usually meant an upper berth and one's rest being interrupted by shunting and changing of engines as the train travelled through each of its jurisdictions. Arrival at Hartford at 5:20 a.m. was no cause for celebration. It was too late to take a hotel room and too early to go to work. One usually found a seat in the lobby of the Bond Hotel and dreamily studied the bronze plaque attached about four feet up the lobby wall proclaiming the high-water mark from one of Hartford's memorable floods.

The return to Montreal was about as exciting. The train departed about midnight. Being familiar with Hartford's nightlife, employees enjoyed many a protracted dinner and a lot of movies. One learned to telephone the station about 11:00 p.m. to ask how late the train was going to be. This depended on how much express was being picked up along the way from Washington, but on average was about an hour. The train generally pulled into St. Lambert about 9:00 a.m. Filled with enthusiasm, the travellers from Hartford then hurried by taxi to the office. These were the years when many a father who worked at P&WC was known to his children as a voice over the telephone.

The Program Begins

P&WC acquired 170 acres near the town of Jacques Cartier, a mile east of the Walmsley plant and adjacent to the old Fairchild factory. Drummond drew up a floor plan and T. Pringle and Sons Ltd. was hired as architects. Anglin-Norcross Quebec Ltd. would build the 330,000-square-foot plant. Ground was broken on June 7, 1951, and production began the following May 12. Total cost for land and buildings was $5.2 million (for a company with a net worth of $1.5 million).

A Harvard MBA, Miles Beech, was hired to lay the groundwork for manpower and organization. Five departments were formed: manufacturing (headed by Lew Ord, then Howard Spence) to examine process sheets (summarizing the steps used to make each part) and determine what machine tools and tooling/production gauges would be needed for each operation; purchasing (Miles Beech) to procure raw materials, finished items from suppliers, machine tools, etc.; accounting (Victor W. Tryon) to hold the purse strings and work closely with management and government; personnel (George G.

R.H. "Dick" Guthrie soon after he joined P&WC to head the engineering department in 1952. Shown below is one of his prize-winning Napoleonic coaches. (via R.H. Guthrie, Larry Milberry)

THE FIFTIES 87

Brooks) to coordinate hiring for the various departments; and engineering (R.H. Guthrie) to oversee the transfer of technical data to Canada, ensure that each part drawing complied with specifications, and inspect all raw and finished materials. The basis for the operation was a licence agreement among P&WA, P&WC and Ottawa. Hartford, with wartime licensee experience, re-established its licence office to facilitate the transfer of technical details, based on the specifications for the R-1340 which the three participants had agreed upon. Coordination of manufacturing at Longueuil was in the hands of an operating committee chaired by Ron Riley which first met on May 7, 1951.

As the R-1340 hadn't been manufactured since the war, there were few at Hartford to help with P&WC's many questions. Frank Dando came out of retirement to provide his technical expertise. Workers, hardly any with experience, were trained to perform a particular operation on a particular machine. One task was to decide which parts to make at Longueuil and which to subcontract out. Subcontractors then had to be lined up, such as Alcoa to supply cast cylinder heads, Bethlehem Steel (cylinder barrels), Wyman-Gordon (shaft forgings) and Thompson (valves). Procurement of acceptable raw materials and certain finished parts from reliable suppliers would be a key to success at P&WC. Another challenge was laying down the manufacturing processes. All process sheets, drawings and specifications had to be reviewed and modified for the Canadian operation and approved by the RCAF, which became the certifying authority. P&WC obtained 90 per cent of the process sheets used early in the war by P&WA and 70 per cent of those used by Continental in licence production of the Wasp.

Since P&WC had received little tooling from Hartford, the process sheets helped in determining the tools for each job. There were some 5400 operations required to make 134 manufactured items. With the Korean War, the supply of machine tools was tight. Locating them was a priority for Bob McLatchie, C.M. McGregor and Howard Spence. The Department of Defence Production was also involved, as it was paying for many of the machines. Warehoused tools were examined in government plants all over the US and some hard-to-get ones were borrowed from the US Navy. Each machine had to be proved before entering operation. The floor plan was arranged so that raw materials and purchased items were received through the west side of the plant and progressed eastward towards the shipping bays, from which completed engines and parts left the plant.

The famous symbol of Pratt & Whitney engines was the eagle. It was affixed to every P&WA engine and was synonymous with the company's reputation for dependability. It came as a surprise in Longueuil when P&WA decreed about this time that the eagle was the sole property of Hartford and could not be transferred to P&WC's engines. Ron Riley had a quiet word with Bill Gwinn, general manager of United Aircraft. Gwinn appreciated that P&WC could build a quality engine and had won the right to display the famous Pratt & Whitney emblem.

Inspection

The inspection department ensured that everything coming in and going out of the plant complied with engineering specifications. Jim Bird was part of the R-1340 inspection team: "I was hired by George Walker in 1952 and sent to Hartford to attend P&WA's training school. We had about 75 staff there at any one time. Most later held supervisory positions, training our people in Canada. It was a great experience being in Hartford and learning the American way of life. Our hosts were very intense and dedicated people. When we began the R-1340 program, about 20 per cent of those in the plant belonged to inspection. We had people to plan and schedule inspections, a laboratory to inspect all materials and dimensional inspectors to determine the precise measurement of manufactured parts to make sure that each complied with the blueprint or the inspection method sheet. Every step in assembly and test of an engine was signed off by an inspector. The test cells were run by engineers, but each test was observed by an inspector. Each part we made was like a gem in a jeweller's display window."

The materials control laboratory was run by Claude Tatham, formerly of the Steel Company of Canada. In the lab worked a radiographer, chemist, metallurgist and spectrographer. Harry McGee, a physics graduate, came from NRC: "I was hired in April 1952 to set up the spectrography lab for the Wasp program. The new plant was still under construction, so we were next door in the Leyland Building (formerly

The beginnings of the Jacques Cartier plant (later known as Plant 1), seen first on August 9, 1951, then on October 12 that year. (P&WC Archives)

Floor plan for the Jacques Cartier plant as worked out by John Drummond. (via J.W.R. Drummond)

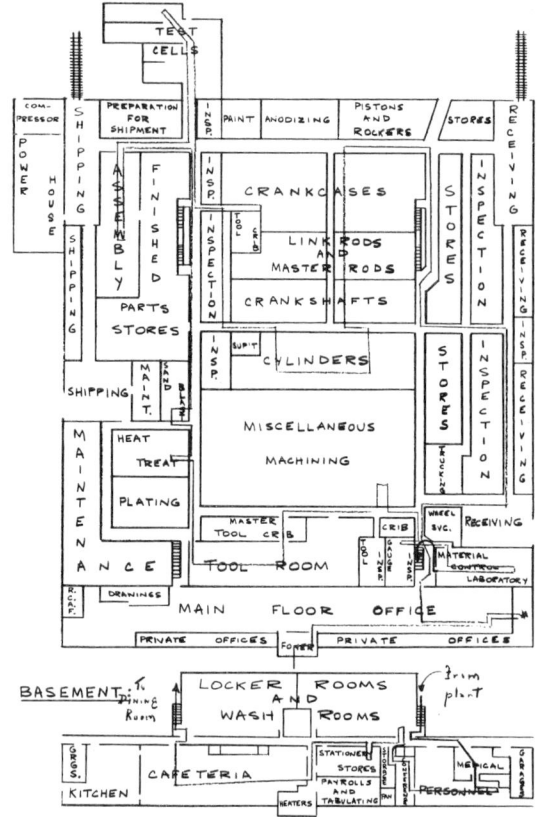

Workmen installing the first machines in the Jacques Cartier plant. (P&WC Archives)

The newly finished Jacques Cartier plant in early 1952. The smaller view shows the plant following addition of a second floor to the front offices, the plant extension to the rear and the Eland test cell in the far corner. (via Keith Stewart, J.W.R. Drummond)

DEPT No
4631
CYLINDER HEAD SEC
1st B.H. BIRD
2nd P.E. CHAMBERLAND
3rd R. BOURQUE
9205
PMC 9205

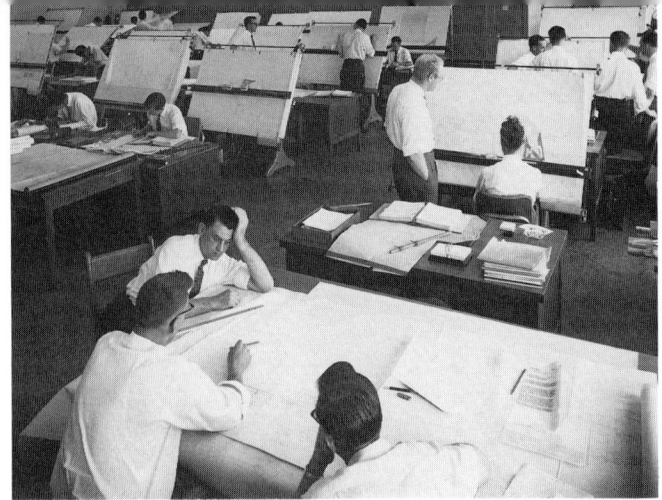

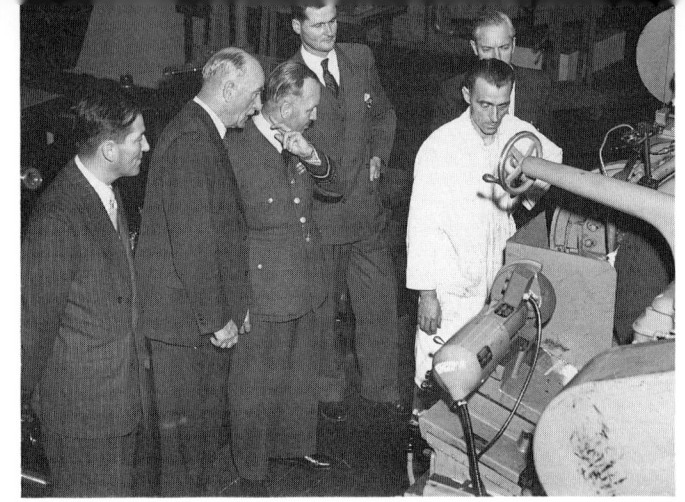

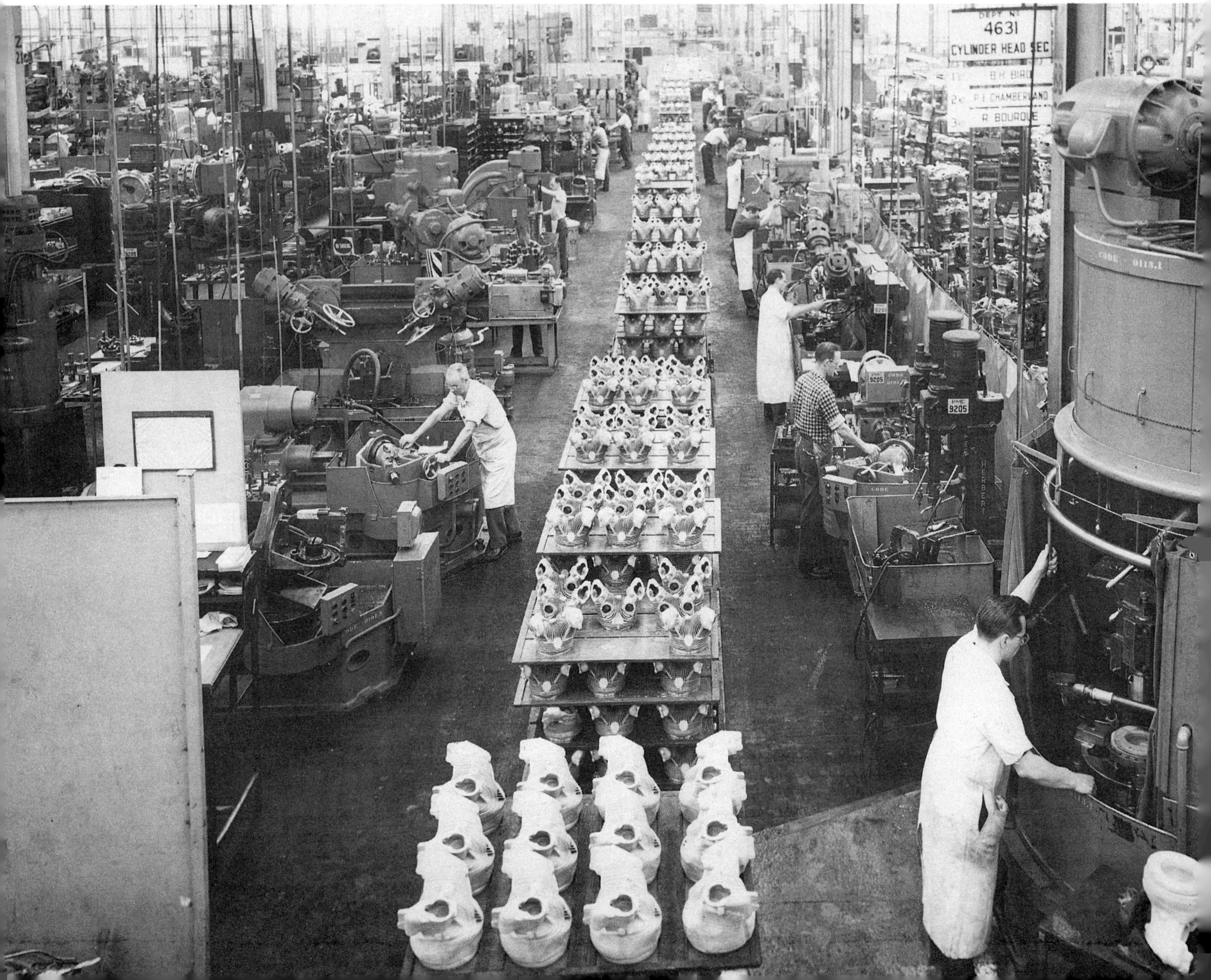

(Facing page, top left) **P&WC engineers and draftsmen involved in the Wasp program. (Top right) Ron Riley, James Young, John Drummond and Howard Spence accompany Air Marshal Wilf Curtis, Chief of the Air Staff of the RCAF, during a tour of the Wasp facility. (Bottom) Production in full swing in Plant 1—the cylinder head line. The floor is packed with machinery. (P&WC Archives, via J.W.R. Drummond, P&WC Archives)**

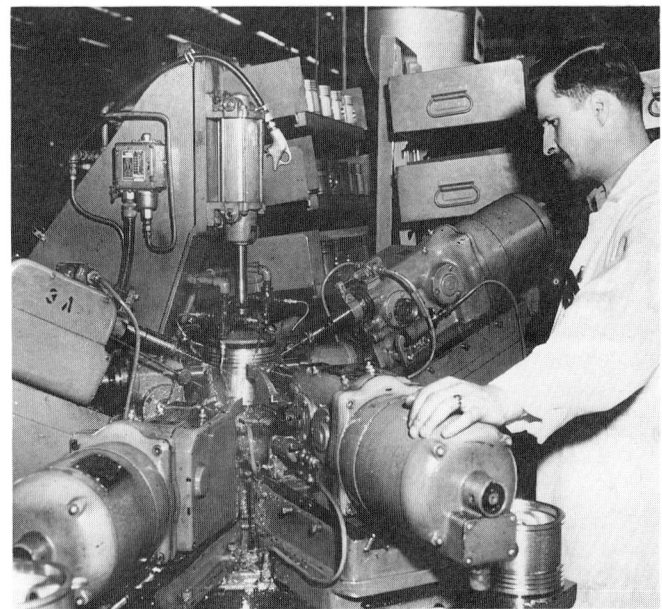

Fairchild's). When we moved into the plant, the new lab was all plaster and sawdust. The spectrograph had arrived from England, but I didn't dare take it out of the box until the lab was sterile. It took a lot of cleaning, but the lab finally opened, complete with industrial mats at the door and a prominent 'Please wipe your feet' sign.

"We routinely tested samples of metal. Typical were slugs of aluminum looking like hockey pucks. We had a chemical treatment to determine if they were the correct alloy. We made similar checks on finished components. One of the final stages in inspection was the interchangeability test. We would ship a Wasp to Hartford, disassemble it and interchange parts with one of Hartford's engines. This really put our quality control to the test."

The first Wasp was completed in December 1952, three months ahead of schedule. Sometime following its 150-hour qualification test (which had to be passed before delivery began), a hitch developed. One of the pushrods failed repeatedly. Eighteen of these slender, foot-long aluminum rods controlled the intake and exhaust valves on the nine cylinders. This was a problem which had not been encountered by P&WA. Examination didn't reveal any metallurgical or manufacturing flaw. Rick Stamm, who was in charge of the Wasp test cell, borrowed a high-speed camera from Hartford and cut a window in the pushrod cover. Contrasting paint highlighted the motion of the pushrod. The camera recorded vibrations in the rod which were traced to the four-bladed wooden test club propeller used to absorb the power of the engine. When a metal propeller was tried, the problem was solved.

The Harvard

The raison d'être for the R-1340 program was to provide engines for the Harvard trainer in Canada and its US Navy

(Top) Index drilling of oil holes in an R-1340 Wasp piston using a Kingsbury multiple drill. (Above) P&WC's material controls laboratory where incoming raw materials, purchased parts and assemblies were checked. The lab had five sections: chemical, spectrographic, radiographic, metallurgical and records. The chemical section analyzed metals, alloys, fuels and oils. (both P&WC Archives)

THE FIFTIES 93

equivalent, the SNJ. With the build-up of the Cold War in the late 1940s, training expanded in the RCAF, especially once it began training thousands of NATO pilots in the 1950s. At first, Harvard IIs were refurbished by Noorduyn at Cartierville, with P&WC overhauling the engines. An insufficient supply of Harvard IIs resulted in production of new Harvard IVs, of which 555 were built at Canadian Car in Fort William from 1951 to 1954.

Hopes of a big sale of R-1340s to the US Navy faded when the Navy decided to replace the SNJ with the T-28 Trojan powered by the Wright R-1820. However, P&WC was to find many other customers for the R-1340 through the 1950s. Besides overhauling thousands of them, engines were sold for numerous S-55s (including for export to places like India, these sales being handled through the United Aircraft Export Corporation of Hartford), hundreds of Otters, Kaman helicopters for the US Navy, and other users.

Standard Procedures

The Wasp program proved a great success. In large part this was due to carefully devised standard procedures. In these every detail of company operations was described as to how it should be carried out. From his early days as president, Ron Riley had insisted on standard procedures as the bible for daily P&WC activity and they remain so to this day.

The Other Work Agreement

The Wasp was the start of wider manufacturing at P&WC. Ron Riley summarized this: "From the beginning we recognized that a manufacturing operation could not be kept going on Wasp engines only, except in a time of emergency. As soon as Wasp production reached peak levels we therefore began to plan production for other parts. We also realized that the cost of Wasps to the Canadian government would be reduced if we could sell Wasps and other Pratt & Whitney parts to users throughout the world. We were then working under a manufacturing agreement with the Canadian government which provided that we produce engines and parts exclusively for it, and were paid the cost plus a small profit based on a target cost."

In August 1952, the "Other Work Agreement" came into force at P&WC under which the government was paid for

The first Wasp completed by P&WC receives final adjustments, then is run in the test cell. Gathered around it in the third photo are Philippe Nadeau, Charlie McGregor, Ted Smith, unknown, unknown, Frank Dando, unknown. (via R.H. Guthrie)

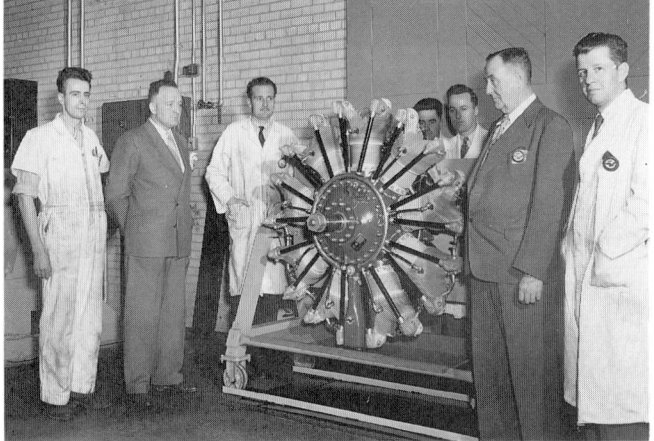

other work undertaken in the new plant, thus reducing its share of overhead. P&WC negotiated with Hartford for rights to supply Wasp spares to customers worldwide. The wisdom of this was shown when the Korean War ended in 1953. While there was no longer the rush to supply Wasps for the war, P&WC was able to keep activity up and the work force at a steady level. Hartford was phasing out production of spare parts for piston engines and focusing on turbine work, and in 1954 it began the transfer to P&WC of all tooling to make cylinders for the R-985, R-1340, R-1830 and R-2000. P&WC would be the exclusive source for new spares for these engines. The first (R-2000 cylinders) were delivered in February 1955.

Bob Losch joined P&WC in 1955 after graduating from McGill. He had five job offers but chose P&WC because of its year-long training program for junior engineers. This gave him experience in every department of the company: "Just as I finished my training, the transfer of the spare parts business from Hartford was building up. We were then responsible for the cost and the selling of these parts, and that entitled us to keep the profits. With those profits we later developed the PT6." The largest customer for new spares was the US Navy, which placed its first order in April 1954. Business in spares that had been $288,746 in 1954 grew to $2,133,167 in 1956 and $13,619,242 in 1959. The magnitude of the business is also illustrated by monthly production. For 1960, pistons were the largest item (average per month: 20,513).

During the busy spare parts era, John Drummond helped negotiate a valuable agreement with the US. He notes: "The basis for evaluating which parts were charged duty varied from district to district in the US. At times an urgently needed part sat for days in a customs office while people sorted through stacks of regulations. In conjunction with the export committee of the Aircraft Industries Association of Canada we approached Washington to get customs routes simplified. We were able to negotiate an agreement to use the Port of St. Albans, Vermont. This included an evaluation of engine and component parts which took into account US content and those items exempt from tariffs. This grew into an evaluation of our whole parts list. Also, customs laws said that the importer had to pay the tariffs, but we got an agreement that the exporter could do this, something quite significant in days before free trade was even discussed. Just as important was

Ron Riley, James Young, Dick Guthrie, Howard Spence, and Charlie McGregor touring the test cell. Ted Smith readies the engine for a run. A metal "club" propeller is in place. (via R.H. Guthrie)

R-1340 Wasp engines (600 hp) manufactured by P&WC

S1H1G, S3H1G	geared Otter engines
S1H1, S3H1	direct drive, mainly for Harvards
S1H2, S3H2	for helicopters

that we got assurance that shipments would not be opened at the border. This reduced the risk of parts being damaged in transit."

The Boys in Overhaul

When Ron Riley established P&WC's manufacturing division in 1951, James Young cautioned him, "Don't forget Jimmy Ross and his boys in the overhaul shop." They weren't forgotten—business for them multiplied and in 1952 was $7.75 million, more than double the 1951 figure. Most work was related to the Korean War—the RCAF was piling up training and transport hours. This meant all the more engine and propeller overhauls, and a great demand for spares, even though the RCAF was now placing overhaul work with other

A pair of Harvard 4s over BC terrain in July 1952. Experts will note that the closest to the camera has a Harvard 2 canopy. In the fifties and early sixties thousands of RCAF and NATO pilots trained on Harvards, learning to fly behind the Wasp. From Harvards, they progressed to multi-engine training on Beech Expeditors with Wasp Juniors. Countless others had trained on the Harvard a few years earlier in BCATP days. (Harry Tate via K.M. Molson)

Manufacturing Sales—Engines and Parts (Millions $)

Year	Government of Canada	Export
1953	$ 6.056	$ 1.273
1954	7.126	5.482
1955	9.661	4.260
1956	9.558	6.748
1957	7.370	16.009
1958	3,765	19.335
1959	3.109	23.336
	$46.645	$76.443

Total Sales P&WC 1953

Helicopters	$ 321,340
Helicopter Spares	463,227
Helicopter Servicing	122,331
Engines	536,245
Engine Spares	2,649,709
Engine Servicing	2,149,385
Propellers & Accessories	708,445
Propeller Spares	508,893
Propeller Servicing	894,142
	$8,353,717

(Facing page, above) Wasps in final assembly in Plant 1. (Below) New Harvard 4s being built by Canadian Car and Foundry at its Fort William plant in the mid-fifties. It was the urgent need for Harvards that spurred P&WC to undertake the Wasp program. CCF built 555 postwar Harvards. (P&WC Archives, via Canadian Aircraft Operator)

USAF aero-engine mechanics service an R-985 Wasp Junior in Korea in the early fifties. Later on, P&WC manufactured the spare parts for this and all other P&WA piston engine types.
(L. Bruce Best)

Wasp engine data from the P&WA *Operators Handbook*, January 1, 1941.

Wasp S3H1

Type	Air Cooled—Radial—9 Cylinder—Single Row of Cylinders	
Bore	5.75 inches	146 mm
Stroke	5.75 inches	146 mm
Total Displacement	1344 cu. inches	22.03 liters
Compression Ratio	6:1	6:1
Blower Ratio	10:1	10:1
Propeller Gear Ratio	Direct Drive	Direct Drive
Propeller Shaft Rotation (viewed from rear)	Clockwise	Clockwise
Propeller Shaft Size — SAE	No. 40	No. 40
Overall Diameter	51.75 inches	1314 mm
Overall Length	42.94 inches	1090 mm
Center of Gravity —		
Forward of Mounting Flange	7.91 inches	201 mm
Below Crankshaft Center Line	.38 inches	10 mm
Dry Weight	864 lbs.	392 kg.
Grade of Fuel — See Section entitled, "Fuel."		
Grade of Oil — See Section entitled, "Lubricating Oils."		

Accessory Drives	Type	Ratio to Crankshaft	Rotation Drive End
Generator	Int. 6 Spline (Rect.)	1.500:1	Clockwise
Starter	3 Tooth Jaw	1.000:1	Counterclockwise
Fuel Pump	Int. 11 Spline	1.000:1	Counterclockwise
Tachometer	Left Hand	.500:1	Counterclockwise
Tachometer	Right Hand	.500:1	Clockwise
Dual Vertical Auxiliary	Int. Hex.	1.000:1	Counterclockwise
Vacuum Pump	Int. 12 Spline	1.500:1	Clockwise
Propeller Governor	Int. 12 Spline	1.144:1	Clockwise

---

firms such as Standard Aero Engines in Winnipeg. Not only were these firms competitors of P&WC, but also customers, for Longueuil was the pipeline for their spares. P&WC was also responsible for ensuring that other companies met the rigorous Pratt & Whitney standards.

Besides military business, the company was busy in the early 1950s on the commercial front, and not just with old stand-bys such as TCA and CPA. An airlift was underway between Sept Îles and Knob Lake on the Quebec-Labrador frontier, where a vast deposit of iron ore was being opened up. Hollinger Ungava Transport had been formed in 1948 to carry men, equipment and supplies north from Sept Îles where a railroad was being built. HUT built up a fleet of DC-3s, augmented with Cansos and C-46s. Maintenance was done at Mont Joli on the St. Lawrence, and as the project developed, work poured into P&WC's overhaul shop. This work tapered off once the railroad was completed in 1954. Ron Riley noted: "Commercial engine overhaul work declined by about 40% by comparison with the previous year, due largely to the substantial completion of mining facilities in Northern Quebec and Labrador." He added optimistically that business was again on the upswing "in connection with the construction of radar warning systems in Northern Canada." These were the Mid Canada Line along the 55th parallel from Hopedale in Labrador to Dawson Creek, BC, and the Distant Early Warning Line across the Arctic. These projects put a huge fleet of transports to work, from Beavers to DC-4s and even larger types. Once again, the overhaul shop was swamped.

A glance through the 1955 operating committee reports gives a good overview of the busy times at P&WC 10 years after the war. In the overhaul shop on June 30 were 109 commercial engines and 340 from the military. There were contracts to service HS propellers for RCAF C-119s and

Engine Overhaul Activity, 1950

	QUANTITY		MAN HOURS PER ENGINE	
Engine	April	March	April	March
R-985	2	8	225.50	217.50
R-1340	16	18	232.50	233.25
R-1830 (carburetor)	3	1	398.00	418.25
R-1830 (no carburetor)	9	10	365.00	357.50

Leading P&WC men c. 1956: George Brooks (personnel), John Drummond (vice-president), Thor Stephenson (executive vice-president), George MacDonald (assistant to the president), Ronald Riley (president), Victor Tryon (finance), James Young (chairman), Dick Guthrie (engineering), Miles D. Beech (purchasing) and James Ross (secretary and overhaul manager). (P&WC Archives)

Neptune patrol bombers. There was a large spares order for US Army R-1340s for Otters. By mid-June, over 800 of the 1000 new Wasps had been delivered. R-2800s in for overhaul were being turned around in 30 days. An order had just been confirmed for six S-58s for the RCAF. The company was getting involved in the Avro Arrow and DHC Caribou, and the first R-1820 kits were awaited. Over 1800 people were at work in the offices and shops.

DEW Line Operator

One of the most famous names on the Distant Early Warning Line airlift was World-Wide Airways of Dorval. Owned by Don McVicar, a leading Ferry Command pilot, its DC-3s, C-46s and other types were familiar to anyone on the eastern side of the project. In 1988 McVicar, author of several aviation books, reminisced about his association with Pratt & Whitney engines: "A pilot tends to remember with a lot of fondness an engine which saved his life. The first time for me involved an R-1340. I was flying a Norseman on skis on an exploratory flight into the Eastern Arctic in 1942. I was on a short trip to a secret base called Crystal Two. The snow was sticky and we just had to have more than the recommended 600 hp for takeoff. With the throttle firewalled, I milked almost 700 hp from that Wasp and kept our expedition on schedule. I also flew the Marauder, powered by R-2800s, and the Canso with its R-1830s. The latter took a hell of a beating on our 20-hour ferry flights from Bermuda to Scotland.

"There was always a fierce rivalry between Pratt & Whitney and Wright, and in my opinion the engines produced by them during the war were equally reliable. That was evident in the Hudson, which was an excellent aircraft with either engine. In the 1950s, World-Wide used C-46s on the DEW Line. These had R-2800s—the improved 'C' version. Even at below-zero Arctic temperatures they performed faithfully (so long as they got lots of oil!). Later, we equipped with Super Constellations. They had the Wright R-3350 with their 'power recovery' system. That's when I wished I'd had DC-6s with R-2800s. In the long run, the unreliability and high maintenance costs of the Wrights drove World-Wide out of business. If we'd had better engines, we would have given TCA and Max Ward a hell of a run in the North Atlantic charter business. They tell me these days that you can dispatch a jet like a 767 and go to bed knowing that the phone won't ring from some horrible place in the middle of nowhere with

Freighters on the Hollinger-Ungava airlift. Seen at Sept Îles are a DC-3 (R-1830s) and a C-46 (R-2800s), both getting routine engine service. Mega projects of the fifties such as this, the Mid Canada Line and the DEW Line kept P&WC's overhaul shops busy. (NAM)

One of Don McVicar's Pratt-powered DC-4s at Malton in 1961. (Larry Milberry)

the brave captain on the line to report that *another* piston engine has failed."

Working with CPA

With TCA's swing from P&WA to Merlins in 1944, and later to Wright R-3350s (for the Super Constellation), Canadian Pacific Airlines became Longueuil's leading airline customer. CPA had come into being in May 1942 through the amalgamation under the CPR of 10 air services, all of which had been P&WC customers through the 1930s. The new airline had many types—77 aircraft in all, 39 with P&WA engines. Along with this *mélange*, however, came some of the top engine men in Canada, including Tommy Siers and Albert Hutt, who had begun on Liberties in the 1920s and had trained at Hartford on the Wasp before the formation of P&WC. Their rapport with Pratt & Whitney was so close that when Canadian Airways' Ju 52A had a failure with its Rolls-Royce engine, Siers was able to go to Hartford and use P&WA's shop to machine a new connecting rod. Later, Siers and his Canadian Airways team perfected the oil dilution system which revolutionized cold-weather engine operations. When accepting the 1940 McKee Trophy for this work, Siers thanked P&WC for its help on the project.

Converting CPA's big Bellanca from a Cyclone to a Hornet in 1943 was perhaps the first major project where P&WC and CPA collaborated. Next came CPA's purchase just after the war of a fleet of DC-3s from the US military. Retired CPA engineer Ferdie Vachon recalls this as a milestone in CPA's goal to standardize its fleet. The DC-3s had their share of trouble: "We had 14 DC-3s with the R-1830 and were suffering master rod bearing failures. Gene Schweitzer came out and observed our operation. He told us, 'You're coming in to land with wheels and flaps up and engines idling. That's what's ruining your engines.' His advice was to come in with power on, wheels and flaps down well back. 'Don't glide the aircraft in,' he said, 'and don't let the props windmill the engines.' We followed his advice, and that solved our problem."

In P&WC's 1948 summary of engine use by its customers,

The 1953 P&WC board of directors. In back are William P. Gwinn, G.M. Black, Jr., W.R. Robbins, H.M. Horner, James Ross and John Drummond. In front are G.M. Black, Sr., Ronald T. Riley, James Young, R.H. McMaster and H.G. Welsford.
(via J.W.R. Drummond)

it was noted of CPA: "Maintenance is satisfactory ... power plants are operated in accordance with good airline practice.... The only exception is on several of CPAL's routes in British Columbia. The extensive use of climb power is required, but there is no alternative, otherwise safety will be jeopardized.... Overhaul, maintenance and operations are being checked by P&WC as closely as possible during scheduled visits of a representative."

Soon after it began operating the Canadair C-4, CPA realized the high cost of running Merlin engines, which P&WC calculated at triple the R-2800 cost. The solution was to sell the C-4s to TCA and buy a fleet of DC-4s. Gene Schweitzer emphasized the need for P&WC to give CPA all the support available to keep the DC-4s serviceable: "The Pacific route will be flown entirely with DC-4/R-2000 equipment during 1952. This is a very important phase of their flying.... Undoubtedly, the importance of our service on spare parts, etc. is fully understood as an influencing factor for the acceptance of future P&WA equipment." One problem came up on the Pacific routes. Exhaust valves and cylinder heads were being damaged. The R-2000s were being run at excessive power settings because of a typographic error in the aircraft operating manual!

CPA's leap into international operations came in 1953 when it took delivery of the first of 20 DC-6s. These were Cadillacs of the airways, powered by R-2800s. Dick Ryan, a CPA vice-president at the time, recalls: "I think the DC-6B was one of the best piston airliners ever built, and its R-2800s gave wonderful service." The DC-6 earned money for CPA into the late 1960s. During the same era the company operated the Convair 240 and the Curtiss C-46, both powered by the R-2800, with support in the field and overhaul provided by Pratt & Whitney Canada.

Bread and Butter

Besides big customers such as TCA and RCAF, P&WC conducted business with smaller customers all over Canada. As always, it was the field service reps who kept tabs on what

(Above) One of CPA's stately Douglas DC-6s, R-2800s throttled back, lands at Prestwick after a transatlantic flight. The DC-6 was so beautifully engineered that, 30 years after its heyday, it is still in demand in such special markets as firefighting. Manufacture of piston spares at P&WC ceased in the mid-1970s. Frank Santo notes that the last piston engine (an R-2000) came through overhaul in 1979. (Wilf White)

Typical of P&WA-powered corporate aircraft in Canada in the 1950s were these two VIP beauties: CF-DJT, a DC-3 of the Robert Simpson Co. (previously owned by Avro Canada); and CF-MFL, with R-2800s, a Howard Super Ventura of Massey Ferguson. P&WC provided full support for Canada's corporate fleet. Beginning in 1959, such aircraft faced replacement by the early turboprops and corporate jets such as the Gulfstream and Jetstar. (Larry Milberry)

A view of the impressive Avro Jetliner during construction in 1948. With the Jetliner, Canada was on the brink of leading the world in jet transport technology, but politics scuttled the project. Plans had been in the works to offer the Jetliner with the P&WA J42, a licence-built Nene. (K.M. Molson Collection)

was happening. They went wherever there was an engine needing servicing or a chance of work. At all times, their job was simply to be as helpful as possible. On a visit to the RCAF in Vancouver in 1949, Gene Schweitzer analyzed a troublesome Wasp Junior in a Beech Expeditor, commenting, "The engine ... has not been overhauled for five years and blower bearing wear due to corrosion is a possible source of the trouble." A visit to Edmonton that summer gave another chance to help the air force: "The RCAF schedules the R-1340 overhaul period at 960 hours for the Norseman. This appears to be quite high for their type of operation."

An exciting market opened up as corporate aviation began to boom. Before the war, few Canadian companies had private aircraft. Now companies large and small began to realize that an aircraft could be an asset in profitability and growth. Imperial Oil, Shell Oil, Eaton's, Avro, Ontario Paper, Abitibi Paper and Canadian Breweries, to mention a few, had executive DC-3s. A flock of Grumman Mallards appeared. There were numerous Lodestars, especially among the oil companies in Calgary and Toronto. Except for a few types such as the de Havilland Dove, most of the larger corporate aircraft had P&WA engines. This new market gave the company experience in a field which was eventually to become its major focus.

The Commercial Jetliner

Shortly after the war the first jetliners appeared, beginning with the de Havilland Comet in England, which flew in July 1949. Two weeks later, the Avro Canada C.102 Jetliner flew, North America's first jet transport. The Comet went into production but soon faced disastrous problems. Though it had a long and safe career with Avro, the Jetliner, through lack of government support, was not produced. It was powered by four Rolls-Royce Derwents, but Avro realized that US operators would prefer American engines and P&WA was consulted. P&WC's study of subcontract work for the J48 had just been completed when the operating committee met on November 14, 1950. In part it recorded: "Avro would like one aircraft set of J42 engines on consignment, and an additional four aircraft sets by purchase, their order to be placed in the spring of 1951 for delivery in the late fall of the same year. P&WA has not agreed to supply any more than one aircraft set of J42s, as the model is going out of production. Avro believes that they cannot cope with the redesign of the aircraft to take advantage of the J48's increased performance by early 1952 when they hope to deliver aircraft to an airline operator for service testing." In the end TCA, which had originally been expected to equip with Jetliners, dropped its interest. The RCAF purchased two Comets, and Ottawa ordered Avro to dedicate its energies to the CF-100. The Jetliner prototype flew for a few years as the company "hack," then was scrapped.

The first Canadian airline to order jets was CPA. In March 1953 it took delivery of its first Comet, but the aircraft crashed at Karachi en route to Vancouver. A second Comet was not delivered. TCA became the first commercial operator of turbine aircraft in North America, receiving its first of 51 Viscounts (with Rolls-Royce Darts) in late 1954. CPA didn't enter the turbine age until 1958 when it bought the Bristol Britannia with Proteus engines. Eight were used through to 1966, replacing the DC-6B on international routes.

When US manufacturers introduced commercial jetliners, TCA and CPA opted for the Rolls-Royce Conway to power their DC-8s, the first of which were delivered in 1960. The Conway was the first commercial application of a turbofan engine. After accepting 11 Conway-powered DC-8s, TCA switched to the lighter and more fuel-efficient P&WA JT3D (first delivery January 1963) and CPA followed. Later jetliners like the Boeing 707, 727 and 737 and the Douglas DC-9 brought further business in sales, servicing and overhaul to P&WC.

Power for the Avro Arrow

In 1953 RCAF specifications were issued for a supersonic fighter. The contractor was Avro Canada, and the aircraft became the CF-105 Arrow. It first flew in March 1958 with a pair of P&WA J75s as an interim measure while Orenda completed work on its Iroquois. The J75 gave P&WC hands-on experience with jet engines. The initial RCAF procurement, as noted in the June 1955 operating committee report, was for 31 engines.

John N. "Tony" Clark was one of several P&WC men who worked on the J75/Arrow project: "It is a general principle that engine development takes longer than the airframe. In order to get a jump on flight testing the Arrow, Avro chose the J75, assuming that later on the Iroquois would take over and power production aircraft. I spent a year at Hartford learning how the J75 worked, how P&WA did an installation design, and engine performance calculations. I also went to Hamilton Standard to study the J75 fuel controls. Finally, I spent two months at Edwards Air Force Base in the California desert, where the USAF tested its new aircraft. At the time, Edwards was especially exciting, with several of the latest fighters being flown there. I worked under Lowell Ruby, the resident P&WA field engineer, and Bill Funk, the Hartford engineer assigned to follow the Avro program. I had a chance to work with them on the J75 installation in the Republic F-105 and gained plenty of first-hand experience, such as sitting with the test pilots and discussing their criticisms of the engine."

Security around the J75 was tight. Only a few from P&WC were directly involved, including Ron Riley, John Drummond, Mike Saunders, Bob Losch, Gene Schweitzer, Ken Sullivan, Bill Reynolds, Thor Stephenson and Tony Clark (who replaced tech rep John Fotheringham, who had passed away). J75 data flowed steadily from Hartford to the DDP in Montreal. From there someone from P&WC would carry it on a flight to Malton, where he would personally deliver it to Avro's design office. There was so much material that daily, even twice daily trips had to be made from Montreal.

While P&WA was providing flight-test engines to Avro, it had wider goals. Here was a chance to test the J75 twinned in one aircraft and to observe the results of installation and operations and make on-going improvements. Tony Clark adds: "The Arrow held promise as a lucrative project for us. We knew what Orenda was up against in pouring millions into a powerful new engine. Though we felt confident that Ottawa would keep the Arrow going, the Iroquois could be dropped at any moment. We had the J75 ready to fill the vacuum should that occur. Hartford had something even bigger coming up, in case Ottawa wanted a more advanced engine. That was the J58, rated at 32,500 pounds of thrust, much more powerful than the Iroquois. The J58 later powered the Mach 3 SR-71 Blackbird.'"

At its peak the Arrow program included 19 J75s. Only 11 were used and these logged 141 hours flying and 326 hours on ground runs. They performed beautifully on 66 flights, with nearly eight hours at supersonic speed. On November 11, 1958, Arrow No.202 flew at 1.95 Mach, its fastest ever. Bob

Jan Zurakowski flies the first CF-105 Arrow near Malton in 1958. At the time the world's leading fighter project, the Arrow was powered by a pair of P&WA J75s each rated at 17,200 pounds of thrust (24,500 in afterburner). The J75s peformed as expected on a program which saw five Arrows make 66 flights before the project was cancelled. (via Canadian Aircraft Operator)

Losch recalls: "The J75s we had were early models, and we had to do 15-20 modifications on them after they were delivered. I would carry these out or supervise Avro people. When ground runs were underway, with a pair of J75s putting out 30,000 pounds of thrust, the ground would shake. We had many roles to play regarding J75 installation—engineering, performance and flight test programming, attending debriefings with the pilots after test flights, post-flight engine inspection, etc. The program was extremely exciting and successful, and we hoped for big things from it."

In 1958 the government was wavering in its enthusiasm for the Arrow program. Money was being poured into it at a time when Ottawa was convincing itself that the future in defence would be in missiles, not manned fighters. In February 1959 it cancelled the Arrow outright. The existing aircraft were destroyed, and all that survived were the J75s and a few Iroquois. A P&WC memo of August 8, 1958, noted that the

An R-1820 is serviced on a Tracker at CFB Comox. After more than 30 years of service, the Tracker still operates in the Canadian Air Force on patrol duties. In 1988 two different versions of the Tracker were flown with PT6 engines. (Larry Milberry)

J75 "equals or exceeds the performance of the Iroquois except under supersonic cruise." The Iroquois had yet to fly other than in a B-47 testbed.

Engines for the Tracker

In 1954 the Royal Canadian Navy ordered 100 Grumman Tracker anti-submarine aircraft to be built under licence by de Havilland in Toronto. P&WC bid to install R-2000s in the Tracker, which normally had Wright R-1820s. The Navy stayed with the Wrights, but in April 1954 P&WC won the licence to build them in Canada. This was direct result of the company's great success in building Wasps. The timing for this project was excellent—it came just as the R-1340 program was winding down. P&WC would also do other business with the Tracker as it used Hamilton Standard propellers. The R-1820 150-hour qualification test was completed satisfactorily July 16-18, 1956.

The R-1820 gave P&WC further experience building engines and helped carry the company over from a period which emphasized military contracts into more commercially oriented times. John Drummond summarizes the program: "A contract was signed on July 26, 1955. The first engine was delivered in kit form from Wright that September [all subsequent R-1820s were fully manufactured at P&WC]. The engines were supplied on a cost plus basis, with a provision for depreciation of the machinery and equipment purchased by the company. The propellers were not the subject of a production agreement and were supplied at fixed prices established annually for each year's delivery." The last of the 300 R-1820s were completed in 1960.

Following production of the R-1820, P&WC sought to provide them for the Grumman Albatross which the RCAF was buying for search and rescue. The operating committee reported in June 1959 that it had secured this business and had orders for the first engines at $51,175 each. There were 10 Albatrosses, involving 27 sets of engines and HS propellers. Other R-1820 business during this period included overhauls for engines in RCAF and civil Piasecki H-21s used on the Mid Canada Line and in Sikorsky S-58s. Since it came from a different manufacturer, the R-1820 broadened P&WC's perspective. The company learned important lessons about engine specifications and was introduced to a valuable list of new suppliers.

Company Life in the Fifties

Ken Sullivan reminisces about his early days with Pratt & Whitney Canada: "My first job interview was in Toronto at the Royal York Hotel, where John Drummond was carrying out a search for talent. John had sent his trousers out for pressing when I arrived, so the interview was rather informal. I went down to Longueuil for my second interview. It was early spring, the time the St. Lawrence usually flooded low-lying areas. As I drove to the plant I noticed rowboats tied to lamp posts on St. Charles Street. My interview was in the 'headquarters' building—in reality, a dilapidated two-storey red brick building on Lorne Avenue. I noticed that the engineering department was mainly busy designing packing cases for shipping three-blade propellers. It seemed that nearly everyone I met was named either Perks or Smith. On my first day at P&WC I watched a company expeditor moving parts from the propeller shop to the overhaul shop about 100 yards away, using a child's sleigh. This was typical of P&WC's low-overhead approach to business. I joined the sales force, which was based in the red brick building. However, the

Wright R-1820s roaring, an RCAF Albatross takes off from the Strait of Georgia in this 1963 photo. Its engines were licence-made by P&WC. (DND PL123785)

Greek restaurant in downtown Longueuil was almost a second address. It had a large room with a planked floor. There was a wood stove over which Mrs. Simatos laboured while Mr. Simatos operated the cash register. Soup, a bacon and tomato sandwich, milk and dessert came to 55 cents. About 10:30 each morning, sales engineering (all three of us) would hold our daily meeting at the restaurant.

"Overhaul was still the company's bread and butter. The overhaul shop was a delightful place with its mezzanine providing an accessory overhaul area, a plating lab and an area to store disassembled engines. Tools were in short supply, and I can remember trying to borrow a wrench that had to be shared widely. One task in overhaul was to insert a valve guide into a cylinder. This was done by heating the cylinder and cooling the bronze valve guide. An ice cream cooler was used for this. It was also used to keep lunch perishables cool. The oven in which we heated cylinder heads also had secondary uses such as warming up pork and beans!

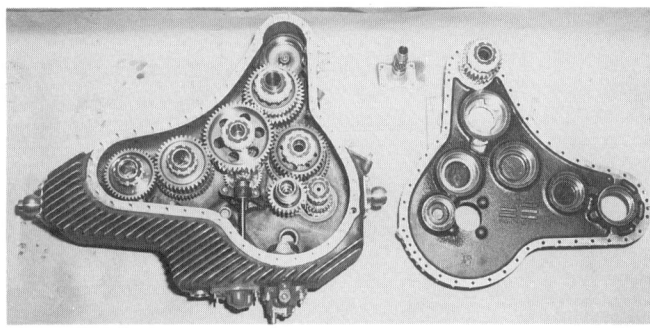

An Eland-powered Canadair CC-109 Cosmopolitan lands at Trenton in July 1961. Because of inherent technical difficulties, the Eland, supported in North America by P&WC, was replaced with an Allison engine. (Larry Milberry)

(Left) The CL-44 gearbox was a complex project for P&WC's young engineering department. (P&WC Archives)

"Across the yard were the test cells, blackened with oil from thousands of test runs. Behind was the shed where we stored wooden packing crates, completed engines and those awaiting overhaul. Several carpenters were always busy patching the crates. Engine cocooning was also carried out here. A spray gun spun a fibrous web around the engines to preserve them. Each layer was a different colour. One was pink, reminding me of candy floss. The last was silver. It took an artisan to produce a smoothly encased package, including its little clear plastic window through which one could see a humidity indicator and packets of silica gel desiccant. The box storage shed was P&WC's 'barn.' Lots of mysterious items were hidden away—an engine antiquarian's dream.

"One of my early sales trips was out to Wichita, home of Beech and Cessna. Beech was building the Model 18 with two R-985s. These engines hadn't been produced since the war, and we decided to offer them brand new ones. I arrived in Wichita just ahead of a blinding dust storm. I arranged a meeting with Leroy Bowery, Beech's procurement manager. It was a long taxi ride out to his office in a converted school. My interview was not so long. Bowery found the prospect of new R-985s interesting. I made my offer—1000 engines at $23,500 each. My interview ended within two minutes, and I was back on the street. Beech, it turned out, had all the engines it could use—good reconditioned war surplus R-985s at 10% of what P&WC wanted. To add to the enjoyment of my first visit to Wichita, the dust storm kept me stranded there for two more days. In the next thirty years I would return to Wichita about 300 times."

The CL-44 Gearbox

In the late 1950s Canadair had several programs on the go. It was still building Sabres and T-33s. Two new undertakings were the Argus ASW aircraft and its offshoot, a large turboprop transport, the CL-44. For the CL-44, Canadair required an accessories gearbox to operate such systems as cabin pressurization and hydraulics and P&WC was contracted to develop it. The gearbox would transmit 350 hp to six different

The Caribou demonstrator stranded in the Azores with one R-2000 unserviceable. In the second view, Mike Saunders of P&WC and Dave Fairbanks and Dave Kendrick of DHC prepare for the engine change. (A.W. Saunders)

accessory units mounted on it. "We had to design and develop the gearbox, procure the parts, build a prototype, build the test rigs and demonstrate that this was a proper piece of hardware," says Gordon Hardy. "We had never done this before, and the gearbox became the first product we ever certified."

With selection of a new engine project still in the offing, it was possible to focus on the gearbox without detracting from engine design studies. Heading engineering on the project was Hugh Langshur. Jim Rankin was the chief designer, and Joe Washburn ran the testing. The first unit, consisting of 320 parts, was tested in mid-October 1958. The design was approved and production commenced, with units delivered for 39 CL-44s. Through 1960, hopes ran high that an order for some 200 CL-44s for the USAF would materialize, but this was not to be. Nonetheless, the gearbox, which was the first major P&WC project conducted without Hartford's control, had allowed the design team to gain experience and brought welcomed earnings to P&WC.

The Napier Eland

In the late 1950s, General Dynamics transferred rights for the Convair Liner to its subsidiary, Canadair, which proposed to build it as a turboprop. Canadair chose the 3500-shp Napier Eland, a British engine, and converted two aircraft. An order was soon in hand for 10 aircraft for the RCAF. Napier would need a service/overhaul facility to support the Eland in North America, which P&WC agreed to provide. This led P&WC to build its first gas turbine test cell and $425,000 was set aside for this project in mid-1958.

The Eland did not leave sweet memories at P&WC. The RCAF, Quebecair and Allegheny Airlines, operators of Eland-powered Convairs, had perpetual trouble with their engines. Recalls Rick Stamm: "The Eland offered the ultimate in complexity. It taught us how *not* to build an engine!" In 1966 the RCAF converted its fleet to the Allison T56, while Allegheny re-converted to the R-2800, and later to the T56. P&WC's last connection with the Eland was taking the RCAF's engine inventory to the scrap yard!

The Caribou

Another program for P&WC in the 1950s was the de Havilland DHC-4 Caribou. In 1954 DHC had considered a design in the 13,000-pound gross weight category with two R-1340s. Next, a 22,000-pound design was looked at. Finally, thinking focused on a DC-3-sized STOL, i.e. in the 28,000-pound category. It was the US Army, one of DHC's most satisfied customers, that kicked the program off in 1957 by committing funds for five examples. P&WC was soon involved with the Caribou, which would use the R-2000 with Hamilton Standard propellers. The first engine arrived at P&WC from Kelly AFB, Texas, in April 1958, and thereafter engines were supplied by the US Army to P&WC for conversion and shipment to DHC.

Caribou sales got off to a slow start. Though the first three were delivered in September 1959, no large order had yet been received. The crash of the first Army Caribou didn't help, and layoffs were anticipated at DHC. Finally, following some impressive STOL flights before officials in Washington, the log jam was broken. In time, 293 were built, 165 for the US Army.

Through 1959-64, three Caribou 'round-the-world sales tours were completed. These were gruelling on aircraft and crew. The first lasted 221 days and included visits to 154 airports in 40 countries. There were 479 demo flights, and one engine was feathered 500 times! The tour had begun on a bit of a sour note when an engine failed on takeoff from Santa Maria in the Azores. A replacement R-2000 was flown out from Montreal, along with P&WC tech rep Mike Saunders. He, the three DHC crew and some local people completed the job, and within a week the Caribou set off. Over the next seven months, there was not so much as a hiccup from the R-2000s.

Achieving Objectives

By the end of 1958, Pratt & Whitney Canada had sales for the year from manufacturing of $24.5 million, only 15 per cent of which was Canadian government business. Most were for commercial export and the US military. In a 1960 review, John Drummond remarked: "The responsibility for the establishment of this manufacturing facility in Canada rested jointly with the Canadian government, United Aircraft Corporation and Canadian Pratt & Whitney Aircraft." He explained the government's role in terms of:

1. The R-1340 and R-1820 production contracts which enabled the company to train new employees and establish a manufacturing capability based on volume production.

2. Capital assistance contracts which let the company finance its plant while Ottawa provided machinery and tools.

3. The Other Work Agreement allowing the sale of parts at fixed prices while R-1340 and R-1820 business with the Crown was on a "cost plus" basis.

4. The "Canada-US Pact" which created a large export market.

The contributions of United Aircraft were:

1. Securing US government approval to transfer production to Canada.

2. Its transfer to Canada of major sales of spare parts, and waiving the usual royalty fee on parts for the US government.

3. Granting of a royalty-free licence to manufacture Hamilton Standard propeller components in Canada.

Drummond finished his review: "The fact that CP&WA retained more than 80 per cent of its earnings in the business during the 31 years of its operation is the most significant factor in the establishment of its manufacturing capability. Few subsidiaries of foreign companies have done as well."

Views from opposite directions of P&WC's Longueuil operation *circa* 1959. The photo on the facing page is north towards the St. Lawrence and Montreal East. Beyond is the Jacques Cartier Plant and the old Fairchild complex. At the bottom is the propeller overhaul shop with propeller packing boxes lining the fence. The smaller building beside it held offices. Just across the road is the "red brick building," then the old Walmsley plant. The large shed to the far right is the engine box storage with four test cells between it and the main plant. The large extension towards the river was built for armament manufacturing by Dominion Engineering during the war and is where Sea Kings were first built by P&WC. The Sea King space was later dedicated to spare parts manufacturing. (P&WC Archives)

The view above is looking away from the river. The one-storey building in the foreground is the cafeteria. The large building at centre right, beyond the rail line, was a wartime plant that made naval guns. These were test fired into the butts in the field behind. It was later home for P&WC engine overhaul. The street running in front of the main plant is Lorne Avenue (now d'Auvergne). (P&WC Archives)

Looking at the back side of Plant 2, with two helipads in the foreground. The first P&WC Sea Kings were assembled in the closest bay. They were simply rolled out the back doors for test flying. (P&WC via D.R. MacLean)

The Move into Turbines

At the beginning of the R-1340 program, P&WC comprised manufacturing, overhaul and supply, and accounting-treasury divisions. In February 1956 a new organization was introduced: manufacturing, engineering, purchasing, sales engineering and service, supply, works, accounting-treasury, and personnel. The company was soundly based, and in April Ron Riley announced that the Jacques Cartier plant (Plant 1) would be expanded (77,000 square feet) and new equipment and machinery added; office space would be increased by the addition of a second storey in the front of the plant (18,000 square feet). Helicopter overhaul at Plant 1 would move by August 15 to 25,000 square feet of rented space in the old Canadian Arsenals building adjacent to Plant 2 and a new helipad would be installed. But the real gem was that P&WC was to form a design and development group to work with existing power plants, accessories and helicopters. It was mentioned that the group would also conduct *original* design and development.

To illustrate how well the company was doing, in 1956 sales passed $29 million, compared with $3 million in 1950, and employment was up from 230 to more than 2000. Manufacturing was now diversified: the last new R-1340 was delivered in September, the R-1820 program was gathering steam and a range of new spare parts for P&WA engines was being added. Riley was careful to maintain a balance between this diversified product line and breaking new ground. Profits had to be sufficiently high before the company launched any original work.

In February 1957, Riley explained how the R-1820 program was "an essential government requirement and it offered us both more experience and more time to establish a commercial business.... We undertook to buy from the government all the machinery and equipment it had furnished for the Wasp engine production together with the substantial amount of additional machinery required for the new contract." Payment for machinery was spread over five years, and the shareholders put $1 million into the program. As a low-volume, marginal program, Riley's solution to the R-1820 was to increase volume with plant expansion. The piston engine spares business would remain lucrative for years to come.

In 1956, the first steps were taken to hire a design team of gas turbine specialists. Dick Guthrie recalls: "At every policy committee meeting we used to discuss what it would take and how many years would be required for us to become Canada's prime engine company. We were determined to reverse the picture as it existed in Canada with Orenda and Rolls-Royce the big names. Ron Riley and I looked at a variety of areas which could launch P&WC into new product development. In the end, we decided to focus on a small gas turbine engine. Ron asked me for recommendations about forming a team. I

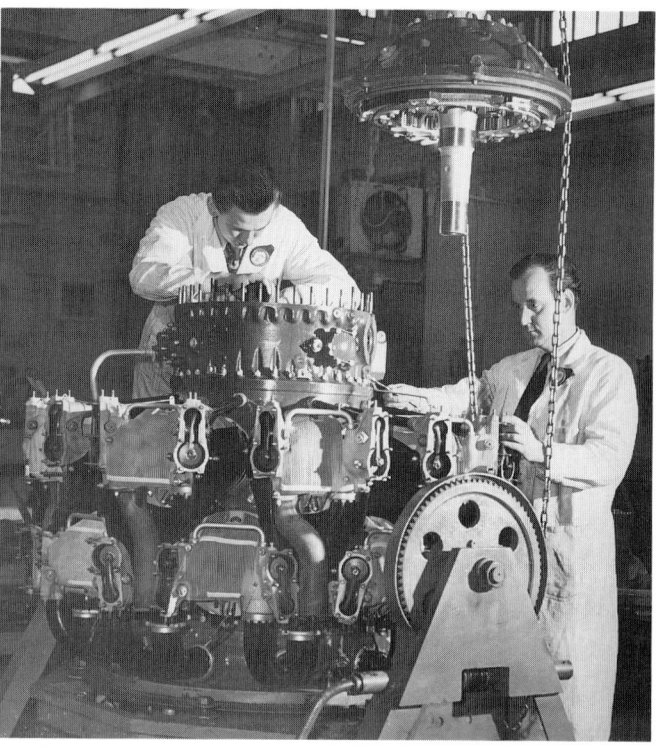

In the mid-to-late 1950s P&WC was beginning to study the gas turbine, but its interest in the piston engine continued as it added the complete line of P&WA piston spares, and piston overhaul was booming. Here Leonard McKaig hand-finishes an R-4360 crankshaft. Two overhaul technicians are seen installing the front section on an R-2800 Double Wasp. Finally, an R-2000 Twin Wasp is seen on a 250-hour test in early 1957. With funds earned from its piston activities, P&WC would soon launch its first big R&D venture, the PT6. (P&WC Archives)

thought it out, then told him that I needed 10 people, each with a particular skill. I wanted to put them down in Hartford for a year. Riley was agreeable, but drew the line at salaries. I could not have experienced men who would cost the company too much to hire. He told me, 'You can have $100,000 for the first year.'

"Now I had to hire 10 people to work for $6000 or $7000 a year. After seeking Hartford's advice, I began my search but found that the skills I wanted were not all available in Canada. Ads were placed in British newspapers. Gradually we started interviewing, but couldn't hire anyone until all the positions were covered by some satisfactory applicant—I had to have a balanced team."

P&WC's chief engineer in 1957 was Hugh Langshur: "I was surprised that we were allowed to enter the gas turbine business without being led by a 'big name.' The US industry had got off to a fast start using British and German expertise. We didn't offer the big money needed to attract such people." There were few job seekers in an industry that was booming. P&WC was a newcomer. Could it attract the right kind of people? There were only two sources of experienced talent in Canada—the NRC and Orenda. The NRC was rich in analytical expertise, but short on designers. Orenda had designers, but few would leave in the midst of its thriving Iroquois program. Elvie Smith, working in the NRC flight research section, was unaware that the company was hiring: "Doug Millar mentioned that he was going to Ottawa for an interview. I tagged along on my crutches, nursing a broken leg." Smith and Millar were two of six hired to start work on January 2, 1957. There were also John Vrana of the NRC, and Pete Peterson, Allan Newland and J.P. Beauregard of Orenda.

Smith had been working his way through the NRC engine lab, involved with several small teams and gaining an overview of engine technology. He had a longtime interest in aviation, dating from boyhood days on a farm in Eatonia, Saskatchewan. After attending the University of Saskatchewan, he went on to a masters program in engines at Purdue University. Smith joined the NRC in 1949. He studied state-of-the-art British engines and worked on an afterburner project using a Sabre 6.

Doug Millar had joined the NRC engine lab with a degree in metallurgy from MIT. John Vrana was a specialist in compressors and combustors. Gudmundur "Pete" Peterson had a BSc in mechanical engineering from the University of Manitoba in 1951. He was a performance engineer at Orenda from 1951 to 1956, then joined Lucas Rotax on gas turbine controls. Jack Beauregard had a masters in mechanical engineering from McGill in 1952. He took a job with Rolls-Royce in England and in 1954 went to the NRC in combustion chamber development. Next came a year with Orenda in thermodynamics. Following service in the RAF, Allan Newland had graduated in 1950 from Edinburgh University with a degree in mechanical engineering. He then attended the

Dick Guthrie and Ron Riley at a ceremony in December 1952 celebrating the completion of the first P&WC-built Wasp. Riley launched P&WC into manufacturing while Guthrie built up Wasp production and later the team that designed the JT12 and PT6. (P&WC Archives)

A portrait of Ron Riley done by Robin Watt in 1956. (via P&WC)

Institute of Technology in Cranfield, taking a masters in aircraft propulsion. Before joining P&WC he had been with Rolls-Royce, Sperry Gyroscope and Orenda. Pete Peterson recalls: "We were all excited about working at P&WC, but aware that it was a gamble. Dick Guthrie was the man there with the vision of what the program would be. Hugh Langshur had the contacts in the engineering department at Hartford."

By mid-June 1957 Guthrie had hired the rest of his team. Some had been recruited in Britain. Ken Elsworth remembers: "I saw an advertisement in the *Daily Telegraph*. In England, competition for promotion was intense, so I decided to apply to P&WC." When Elsworth arrived in Longueuil, he brought with him experience with Bristol, where he had worked on the Orpheus engine. Gordon Hardy had been in the RAF. He gained turbine experience with Blackburn from 1954 to 1957. The company had a licence agreement with Turbomeca of France, which was developing small gas turbines. Hardy rose to be a project engineer at Blackburn. He was examining the potential market in North America for small gas turbines when he noticed the P&WC ad and thought, "This is a glorious opportunity to get in on the start of a new program."

Also joining the P&WC team was Fred Glasspoole, who had been a draftsman with the company on the R-1340. Fernand C. Desrochers came to Longueuil from Orenda, where he had been from 1949 to 1955 working in systems, design and stress. He next worked for a hydraulics firm. Desrochers was a graduate of l'École Polytechnique in Montreal, and had earned the Ernest Marceau award for a thesis about the gas turbine. Also joining the team were Arthur Goss and Jim Rankin. Rankin was a gear specialist hired from Leyland Motors, which occupied the old Fairchild factory in Longueuil.

In the summer of 1957, the first US security passes were delivered for Guthrie's team. Engineers Beauregard, Desrochers, Millar, Newland, Peterson, Smith and Vrana journeyed to Hartford in June to work on design studies for a small engine. P&WA's specialists would assist as needed, but this was basically to be a Canadian effort. In August the other six men caught the train for Hartford. Gordon Hardy recalls: "We were called the 'dirty dozen.' Some of us were straight from England, so hadn't security clearance to enter the main plant. P&WC rented space in the Credit Union, just across from P&WA." Engineers from P&WA used to visit the P&WC office, and to keep those Canadians without clearances abreast of what was going on, lunchtime meetings were sometimes held under the trees outside the office. Hugh Langshur, who routinely travelled to Hartford on piston engine business, was the liaison between P&WA and the Canadian team.

Design Study "DS-3J"

In 1957 the RCAF announced its requirement for a small jet trainer. Canadair put forward its private venture, the CL-41, and this prompted P&WC to direct its gas turbine studies towards an engine suitable for Canadair. P&WC was heartened by optimistic forecasts of the market for small jet engines, and it was reasonable to expect support from Ottawa, given its generous aid to Orenda over the years. For 1957, P&WC set aside $340,000 for completion of a detailed design. Down in Hartford, things started happening, as Gordon Hardy relates: "We worked in the Credit Union as one team. Across at P&WA five other teams were putting down their own layouts for an engine. We all worked like beavers, an engine of 3000 pounds thrust being the target. Ken Elsworth and I produced a simple layout for a straight-through flow, axial compressor engine designated DS-3J. In the end, the six designs were submitted for the senior people at Hartford to review. It was the 'Credit Union Special' which won. We were elated and stamped 'Canadian Pratt & Whitney' on the drawings."

The RCAF would not select a trainer until the prototype CL-41 had flown. The aircraft was designed for a single engine in the fuselage. Besides the DS-3J, the Fairchild J83 and GE J85 were contenders, with the former being favoured initially. Guthrie and Langshur briefed Riley about matters in August 1957, stating that the J85 had three advantages: it was ahead of schedule and would be lighter and cheaper than the P&WC offering. They felt that their own engine, however, would be more rugged and easier to maintain and overhaul, and recommended designing in "significant performance advantages." They would work towards weight reduction by using light alloys where possible and try to reduce long-term costs by seeking other markets. At its September meeting, the operating committee recorded that P&WC should proceed on a jet engine (now designated DS-4J) slightly larger than the

J83 or J85, and noted, "The proposed engine would have applications in executive aircraft." Design assistance was assured from Hartford. Additional space was rented from the Credit Union and computer time booked with P&WA.

Security continued to complicate matters. Jack Beauregard was responsible for compressor design, and the initial three axial stages involved trans-sonic air flow. This was a classified area at P&WA. The computer calculated performance figures for the three stages, but results were deleted from the printouts for security reasons. There was also a question of patents and the export of P&WC-developed technology from the US. Drawings could not be brought into Canada either. People learned to live with such restrictions and the team made rapid progress.

In the fall of 1957 Canadair went ahead with the CL-41. P&WC, known around Canadair as "the people across the river," was now confident and laid plans for a gas turbine test cell. It set up an experimental engineering department under D.O. Blake and began recruiting staff. Cost estimates were prepared, with $292,000 recommended for 1957 and $1.7 million each for 1958 and 1959. These figures were based on the CL-41 flying by September 1959. By now, however, the true costs of an all-Canadian engine were beginning to hit home. The rapid schedule at Canadair would force P&WC to make immediate expenditures on facilities and new talent. Neither company, it was now apparent, could hope for any government R&D relief. Although sales at P&WC were 26% higher than in 1956, profits, due to overhead associated with growth, were 33% lower. It was not likely that the DS-4J could be financed from profits from the piston-engine spare parts business. Meanwhile, P&WA was itself becoming interested in the market prospects for a small jet engine.

In January 1958, Hartford took over the DS-4J. By month's end, the analytical and design contribution of the Canadian engineers was largely complete and by March they were back in Longueuil. That they had done their job well was shown when the prototype engine, henceforth known as the JT12, made its first run in a test cell at Hartford. Pete Peterson summed up the JT12 story: "We all felt disappointed when we lost control of the engine, but we didn't have the necessary facilities or manpower or the funds to pay for them. Today we think that it was the best thing that ever happened to P&WC."

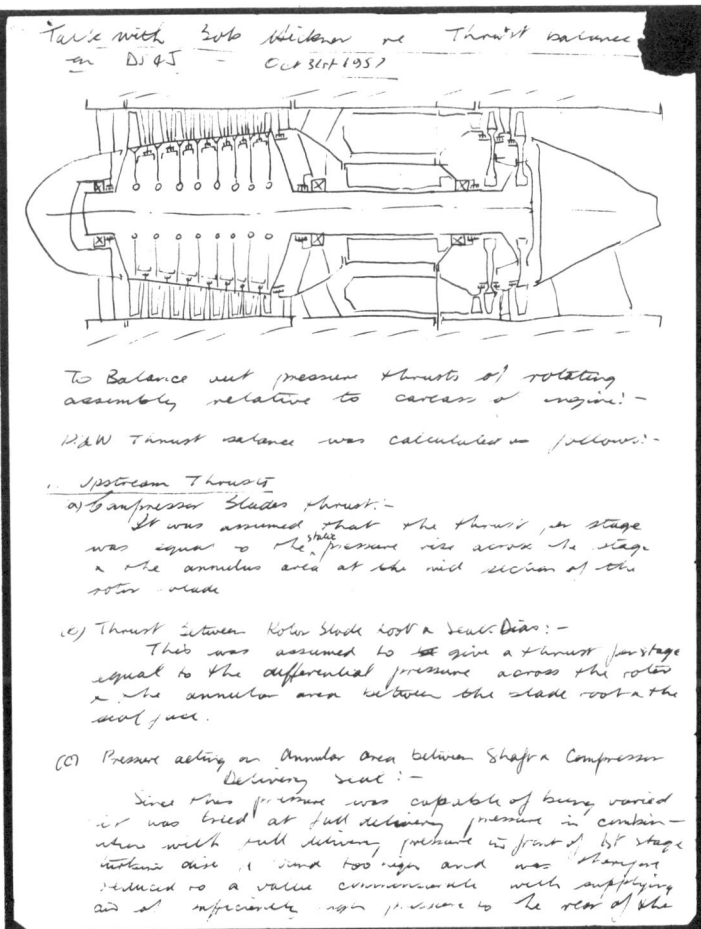

From rough sketches and quick notes such as these by Gordon Hardy arose the JT12 jet engine, of which over 2000 were built. Hardy refers to such early DS-4J sketches as "Leonardos."
(G. Hardy)

THE MOVE INTO TURBINES

Canadair's first CL-41 Tutor brochure announced the JT12 as its choice of powerplant. The 430-pound engine had a rated thrust of 2400 pounds. In spite of Canadair's enthusiasm over the JT12, a political decision resulted in the GE J85 being used in production Tutors. In the second photo, a JT12 is being fitted into the prototype Tutor as Bob Losch of P&WC supervises. (Canadair Ltd.)

Now we were free to move on and begin work on the PT6."

After test and development, the JT12 entered production at P&WA and was adopted for use in a variety of small aircraft. The first engine was delivered to Canadair in September 1959 and was run in the CL-41 on the 12th. It was damaged when foreign matter was ingested, setting things back several weeks. This may have been the occasion when Willy Krause of Canadair, while standing near the Tutor as it fired up, had a coin sucked out of his pocket and into the engine. Finally, on January 13, 1960, the CL-41 Tutor made its first flight. Canadair chief test pilot Bill Longhurst now assigned Ian McTavish as project pilot and the RCAF provided S/L Bob Hamilton as liaison pilot. This assured good pilot input into development. The Tutor was soon in high gear, and Canadair and the RCAF were pleased with the JT12. To CL-41 chief engineer Fred Phillips, the JT12 was ideal. It was conservatively designed, with, for example, large mass flow, low temperatures and rugged hardware. Early experience demonstrated ease and reliability in starting with no problems of excessive jet pipe temperatures. Air starts were always straightforward. Throttle controls were smooth and responsive. The initial pair of engines was priced by P&WC at $58,500 each.

Bob Losch and Mike Saunders were involved in the CL-41/JT12 program, and Losch recalls one incident reflecting on the ever-innovative Saunders: "Canadair had an engine run scheduled for the West German air force, but we were having troubles with the JT12, Mike located the problem in an oil pressure relief valve. We didn't have a spare, and, it being a weekend, couldn't get one from Hartford. Mike studied the part, noting that it looked like the relief valve from an R-2000. We drove down to the plant, and Mike had to climb the fence to get in, ripping his pants in the process. Inside, he found the part. It looked identical except for colour. Back we went to Canadair, installed the part, and ran the engine. Then Canadair came by with the visitors, and we made a successful demonstration. Before the airplane flew, we had the temporary valve replaced with the prescribed one."

After just 30 hours of flight evaluation, the Tutor won the trainer competition, beating out the American T-37, British Jet Provost and French Magister. In the end, however, the JT12 was not adopted. Although a successful European tour was flown, with several air forces expressing interest in the Tutor with the JT12, Ottawa chose the J85 engine, to be built under licence in Toronto by Orenda. This was considered a political decision, whereby the contract would be split between Montreal and Toronto. Placing the engine in Toronto gave Orenda work just when times there were slow following the Arrow debacle. In the end, 212 Tutors were built, and over 120 remain in service today. The main application for the P&WA JT12 (military designation J60) was in the four-engine Jetstar (162 built) and the Saberliner (over 500 built). These were mostly for corporate operators, the USAF and the USN. There were also successful industrial versions of the engine.

Further Design Studies

When it returned to Canada, the JT12 team began several design studies, but serious developments would have to await a market study. The military had been the first users of gas turbines, and in that market high costs took a back seat to performance over piston engines. The airlines were the next users, with their interests being improved reliability and durability, which would later translate into savings. Other markets had not yet emerged. A June 1958 P&WC market analysis identified various power ranges for turbine engines. The US Army was funding a 250-shp project. There appeared to be little action in the 500-1000-shp range in the US, but this was a prime market in Canada, where aircraft like the Beaver and Otter were plentiful. The 1000-2000-shp class engine was coming onto the US market with such types as the GE T58 for the S-61 helicopter. The Rolls-Royce Dart had a near-monopoly in the 2000-shp class, while the powerful Eland and Allison engines were appearing for mid-size airliners like the Convair 440. By late July 1958, P&WC had concluded that the most promising areas were in the 200-, 700- and 2200-shp range. It considered that the turboprop "appears to have considerable inherent capacity for increased power as the basic design is modified and refined during normal development."

Through September, an assessment was made of market potential for 250-500-shp engines. This was followed by a comparison of turboprop, turbofan and piston engines in single-engine transports and a study of small twin-engine transports. Ken Sullivan, Elvie Smith and P.J. Krones (P&WA) visited Piper, Beech and Cessna to get a picture of

A wide variety of aircraft used the Canadian-designed but US-built JT12, including the North American Saberliner and the four-engined Lockheed Jetstar. The USAF Saberliner is at Malton in the early 1960s. The Jetliner, a veteran of the Department of Transport, is now with the Atlantic Canada Aviation Museum in Halifax. It is seen parked in retirement beside the Museum's Voodoo (P&WA J57s). (Larry Milberry)

their needs. Piper had not yet done any design work for a gas-turbine-powered airplane and felt that in its range of small airplanes the added fuel consumption of turbines would make them unattractive. On the other hand, Beech was "turning to turbines with all possible speed." Beech was convinced that the future of light aircraft lay with turbines, especially turboprops, and was ready to install an engine as soon as it was available. An outgrowth of the Beech 50 Twin Bonanza was being planned which could be adapted to turboprops of about 450 shp. Beech was watching developments with a Beech 18 in France used as a test bed for Bastan turboprops.

Cessna was also keen about turbines. Like Piper and Beech, it had been visited by other makers of turbines, including Boeing, Allison and GE, and was impressed by Boeing's 300-hp T60, but would prefer a 400-hp engine for upcoming aircraft. Cessna already had jet experience with its T-37. First flown in 1954, hundreds had been delivered to the USAF. The T-37 had the Continental J69 of 920 pounds thrust.

Sullivan and Smith suggested that P&WC focus on a 450-shp engine with growth potential to 500 shp. Based on current aircraft wing loadings, this would provide direct operating costs comparable to piston engines of the same power, but at cruising speeds 50 mph faster. Cost was discussed—Allison was touting its 250-shp turbine at $4000 and talking of being able to twin it. P&WC therefore focused on a cost for its proposed engine of $9000. At the same time, P&WC agreed on the importance of working with Hamilton Standard to produce a propeller for the engine.

Which Configuration?

Choosing the most suitable configuration for its new engine (DS-10, later PT6) occupied much of P&WC's engineering effort in the final quarter of 1958. Market forecasts suggested which size of engine to build. Now the question was which configuration best suited the market and the goal of a low-cost, low-risk project. Reliability was the foremost consideration in finalizing configuration, followed by cost, fuel consumption, weight and maintenance. P&WC's lack of component test facilities dictated "the selection of relatively conservative aerodynamic and mechanical design parameters."

In October 1958 the decision was taken to go with a free-turbine rather than a fixed-shaft engine. There was still a lively debate pro and con these types. On the fixed-shaft engine, the gas generator and power turbine share a common shaft. On the free-turbine there are two units, one driving the compressor, one producing the power. The link between the two is not mechanical but it is made by the flow of hot gases through the engine. The fixed-shaft engine requires fewer parts, so is cheaper to develop. The free-turbine is more complex, hence costlier, but has such advantages as requiring less starting power and simpler fuel controls. The free-turbine eliminates clutch requirements in helicopters and makes easier the pairing of engines for more powerful installations. Fixed-wing aircraft could use an off-the-shelf propeller with a free-turbine instead of the costly tailor-made one required by a fixed-shaft engine. Aircraft manufacturers were attracted more by the free-turbine, which produced less drag in the case of a high-speed engine out. In such a circumstance, the free-wheeling propeller rotates only a small part of the engine. On a fixed-shaft turbine, propeller, compressor and turbine rotate as a unit. Should an engine fail, this could result in excessive drag and potential structural failure. Less beefing-up of aircraft structure would be needed with free turbines.

A gas generator operates at about 35,000 rpm vs 2500 for a piston engine, but its propeller turns at around 2000, requiring an rpm reduction of about 15:1. On a fixed-shaft engine the compressor and propeller operate at the same speed. Even at idle the compressor runs at a relatively high speed. With the propeller operating at the same speed the result is a very noisy combination. On a free-turbine, the compressor can be at high rpm with the propeller idling, making for a much quieter operation. In November 1958 Elvie Smith noted: "The general atmosphere is such that we assess it would likely be quite difficult to sell a single-shaft engine in competition with a free-shaft engine of similar size. The general predominance of free-turbine engines also bears significantly on the likelihood of suitable propellers being available." The PT6 was born as a free-turbine. This proved all-important to its success.

Next came the mechanical design of the engine. Having chosen the free-turbine, two configurations were considered. Previous free-turbines used concentric, or coaxial, shafting with the air inlet for the compressor at the front of the engine, just behind the propeller. The power turbine was at the back. A long shaft passed through the centre of the engine to transfer

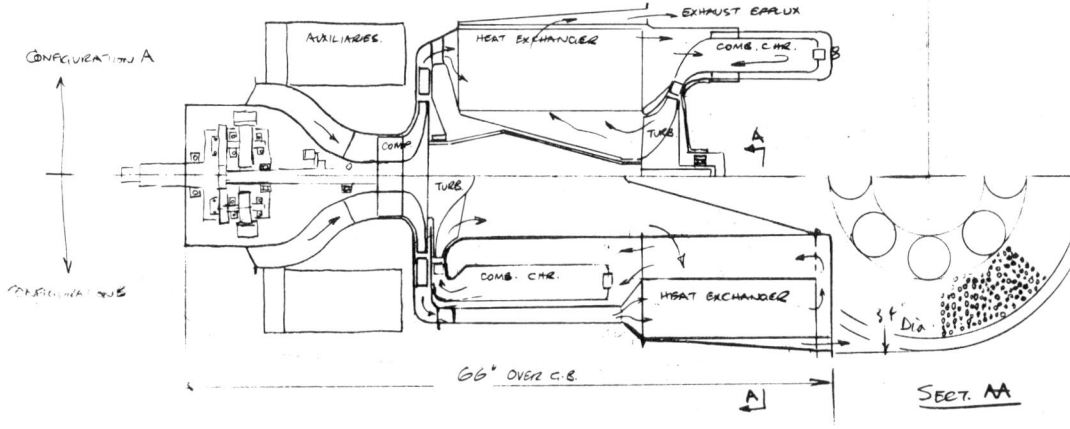

Another Leonardo. This shows Design Study 5/1 (March 27, 1958). DS-5 later evolved into DS-10. From such studies arose the PT6. (G. Hardy)

(Below) P&WC's engineering organization in the early days of the PT6.

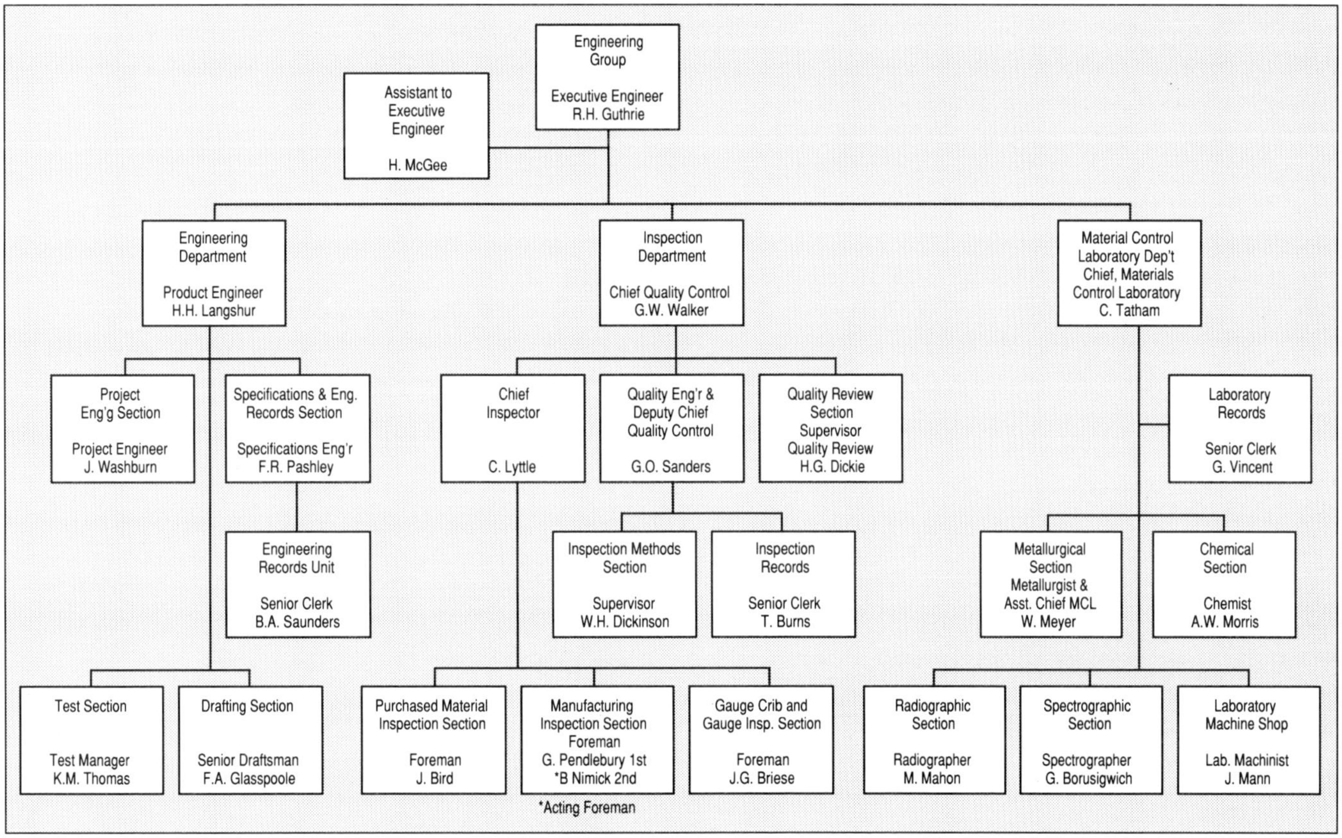

power from the rear to the front. This was straightforward for large engines, but not so with the space restrictions of a small one, where expensive and time consuming development would be required. Concentric shafting was quickly rejected.

The alternate was an "opposed-shaft" engine with the air inlet and compressor at the rear and the power turbine at the front of the engine, next to the reduction gearbox. The designers preferred this arrangement. It would have fewer features likely to require development. Components subject to replacement would be in the front of the engine, hence easier to service. For example, gearbox replacements could be made without removing the engine or accessories from the nacelle. Thus, in its final form, the new P&WC engine would be a free-turbine of a reversed-flow design having three units: gas generator assembly, power transmission assembly (free turbine plus propeller gearbox) and accessory gearbox assembly.

Specifications

On November 25, 1958, P&WC's marketing department at last had a product to tell its customers about and engineering had the design parameters on which to focus. The specifications were down on paper and stated: "The DS-10 is a 450 shp free-turbine turboprop and turboshaft engine suitable for fixed-wing, helicopter and VTOL aircraft. The size has been based on airplane design studies and surveys of the light aircraft manufacturers in Canada and the USA. The layout also makes feasible a high by-pass ratio turbofan engine (by-pass ratio 15:20). Preliminary studies suggest that such an engine would be a quite attractive powerplant for a light transport aircraft.

"The pressure ratio of 6:1 has been selected as being the highest value which, in a small engine, should be attainable without excessive design complication or development effort. The turbine inlet temperature has been set at a level consistent with reasonable turbine disk weights when conventional blade fixings are employed. If, as development proceeds, integrally cast turbine wheels prove feasible, considerably higher temperatures will be possible."

The Canadian team went to Hartford on December 1 to compare its design with one drafted by a team there. Wright Parkins, P&WA's engineering manager, critiqued both and gave the green light for the Canadian one. The immediate market would be general aviation and an emerging Canadian military requirement. The engine could be developed to 525 shp to compete for a rumoured US military helicopter requirement. Parkins saw nothing out of line with Canadian estimates of $4.4 million to carry four prototypes through a 50-hour test. Development facilities would cost another $2 million.

P&WC had an exciting road ahead, but it would be five years before delivery of the first production engines. The challenge for engineering would be great and, as Jack Beauregard recalled, "not infrequently clouded by serious technical problems." Marketing would have to work diligently to find customers if the program was to reach fruition. "Everyone understood that the engine program would evolve in response to the market," said Elvie Smith. Some financial assistance was later secured from Ottawa, but the burden lay with P&WC in re-investing profits and on continued world demand for P&WA piston spare parts made in Longueuil.

Allan Newland recalls the early PT6 days: "This was the first time we tried to put a gas turbine together. It's not surprising that we showed a great deal of inexperience in what we did. We had no history, no experience as a team, and only brought to the situation what background we had as individuals. This was a far cry from what would happen in a mature organization with a long history of design. Our inexperience did, however, have a positive aspect—we were uninhibited. We had no past failures. Besides, we had all the expertise in Hartford to draw on and were smart enough to know which questions to ask when consulting with people there. They keenly shared their knowledge and experience. They were first-class instructors. To say that this rubbed off on us would be an understatement."

Financing the Gas Turbine

In financing its new engine, P&WC subscribed to UAC's policy of viewing engineering and development costs as part of the price of staying in business. Its development staff was a fixture, whether it had a project to work on or not. Teams were not let go at the end of a project, then recreated when the next one came along. On the other hand, there was Orenda's way of covering development, having the government foot most of the bill and purchase most of the production, leaving the company dependent on a single customer.

There was, however, a role for Ottawa with the PT6. The Canadian market was too small to cover design, development and production costs yet allow for an acceptable selling price. Export markets would have to be developed, with the US being key. Canadian government support could be justified as there was a market appearing for PT6-type applications in utility and anti-submarine helicopters, STOL bushplanes and VTOL designs. Experience showed that military support was important in establishing a new engine's credibility in the eyes of commercial customers. P&WC was in a good position to be considered under the defence production sharing agreement with the US. The company asked Ottawa to add the PT6 to the agenda for discussion by the Production Sharing Committee, and added that some support from Ottawa was vital in establishing the common interest of the two governments. A request for support was made to the DDP in January 1959. P&WC asked for and received $1.2 million towards expenses for carrying the engine through 30 months to the 50-hour flight qualification stage. The contract with Ottawa was signed on March 16, 1960, for P&WC to provide four PT6A-2s for test purposes and four additional A-2s or B-2s with a model specification to be developed for each type. The aid was spread over fiscal 1959–60 and 1960–61. Other funds as announced in 1959 would be:

Capital expenses (P&WC)	$2.5 million
Design/development (P&WC)	5.2 million
Prototype engines (US and/or Canadian gov't.)	6.0 million
Product improvement (US and/or Canadian gov't.)	7.4 million

These expenditures would see the project through the 50-hour *and* 150-hour test phases.

With announcement of work on the PT6, P&WC was on a sound financial footing. The company reported that for 1958, "Sales at $34,601,126 were down about 1.5% from the previous year, but profit at $1,522,348 was 110% higher … inventories were reduced by $3,373,079 and bank loans were reduced by $5,222,000." However, P&WC was really in unknown territory with its new-technology engine. It ran a factory for piston parts and was not equipped to manufacture all turbine components.

R.D. Richmond, outgoing president of the CAI, invests Thor Stephenson with the chain of office as he takes over the presidency in 1956. Looking on is J.A.D. McCurdy, the first to fly a powered airplane in Canada. Richmond was at Canadair in this period as chief development engineer. Stephenson was with DDP. Stephenson became president of P&WC in 1959. Richmond, an aeronautical engineer from the University of Michigan, came to the company in 1960 as VP operations. In his earlier years he had been at the NRC, then was at Fairchild on the Husky design team. (Gazette Photo Service)

Twelve key men on the PT6 design team: Gordon Hardy, Jim Rankin, Fernand Desrochers, Fred Glasspoole, Ken Elsworth, Allan Newland, Pete Peterson, Hugh Langshur, Jack Beauregard, Elvie Smith, Dick Guthrie and Thor Stephenson. (PAC, Bruce Moss for *Weekend Magazine*)

Technicians in the experimental shop in Plant 2 assemble the gas generator of the very first PT6. (via G. Hardy)

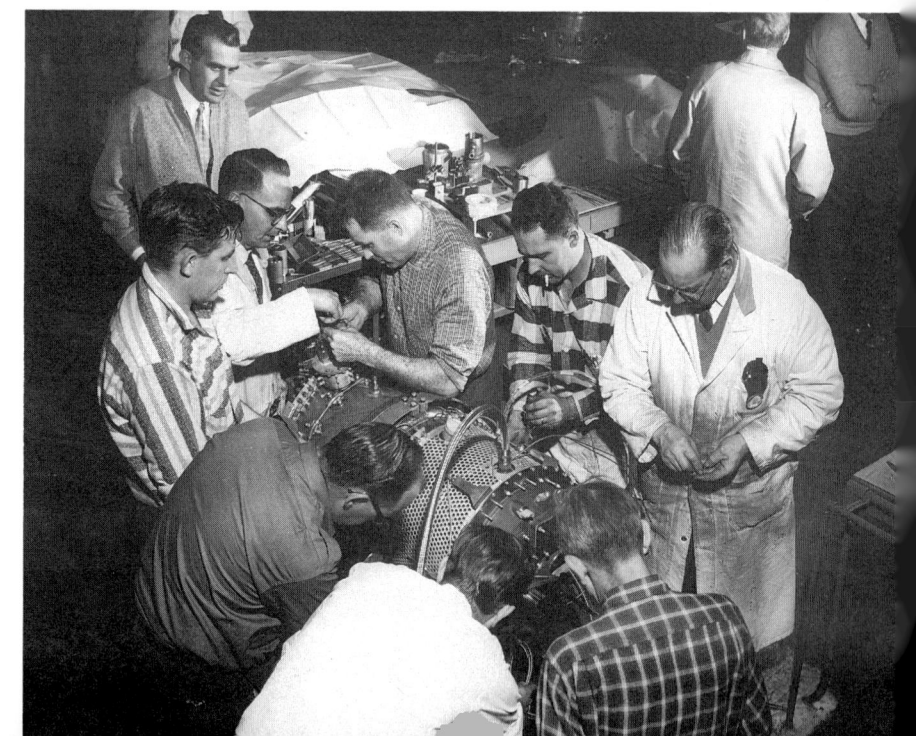

New Leadership

Ron Riley had seen P&WC grow from a small company to one of the leaders in Canada's aircraft industry. On January 1, 1948, at the age of 39 he became president, at which time James Young took over as chairman. Riley saw his dream of the company becoming a manufacturer come true. Ironically, just as his company was entering its most exciting years, his life was cut short. He had survived a heart attack in 1957 which left his health shaken. On a June day in 1959 he was lunching at the Mount Royal Club with several of the Canadian directors when he suffered a fatal heart attack.

Upon Riley's death, James Young (retired as of January 1, 1959, and now aged 76) became interim chairman. Thor E. Stephenson became president. He had been born in Winnipeg in 1920 and in 1942 earned an aeronautical engineering degree at the University of Toronto, then a masters from the California Institute of Technology in 1946. During and after the war he was at the NRC, where one of his projects was a glider used to study the tailless flying wing. In 1952 Stephenson joined the Department of Defence Production under C.D. Howe. He admired Howe's no-nonsense way of getting results and eventually became director of the DDP, overseeing contracts worth $400 million. He left in 1956 to join P&WC in sales. The prospect of working for Ron Riley drew him, as he wrote in 1983 to Harry Gray, chairman of United Technologies: "I went to P&WC because of my admiration for Ron Riley. For years I had dealt with all the senior executives in the country and Riley made the most sense."

Retired RCAF Wing Commander Frank Phripp knew Thor Stephenson in student and NRC days: "Thor and I first met at the University of Toronto where we studied aeronautical engineering under Professor Tommy Loudon. To Thor, learning was effortless, but what we most admired about him was his hockey ability. He played for our great Varsity Blues. In 1940 they regained hockey supremacy from the McGill Redmen who had held the trophy for 10 years. We met again after the war testing the National Research Council glider. Thor directed the testing and flew with me as observer on nearly half my flights. He was there when we were pushing to the risky limits where other such aircraft had suddenly pitched and tumbled with little hope of recovery. If he wasn't flying, Thor was in the control truck near the runway monitoring flights. Our test program concluded in September 1948 with the tailless being towed by a Dakota all the way from Namao, Alberta, to Arnprior, Ontario. Thor then reported on what further development was needed, but other priorities pushed our project aside and it was never completed."

Stephenson was to have a successful career with P&WC, leading it through the transition from manufacturing engines and parts as a branch-plant licensee to an integrated manufacturer of its own gas turbine designs for sale around the world. Stephenson was elected chairman of P&WC in November 1975, holding that position until retirement in April 1977. He passed away on June 3, 1983.

Thor Stephenson backed the move to get the PT6 into a Canadian or US military project. For the first half of 1959, sales efforts concentrated on the two governments. Though there was the $1.2 million offer of aid from Ottawa, that seemed to depend on P&WC arousing US interest. In May 1959 Stephenson wrote, "If applications cannot be developed [within a year] it is doubtful whether the project has any merit."

The message to marketing was clear—that year sales contacts were made all over the planet, 70 alone in the US. Six showed program potential, including Beech, Hiller and Republic, and 15 others were promising. In the US, there was interest in industrial use of the PT6 to power pumping equipment for the petroleum industry. There appeared to be no shortage of "interest," but before long this would have to be turned into a firm commitment. In December 1959 the US Army announced its requirement for the LOH—light observation helicopter—to replace existing light aircraft. This could mean thousands of engines for the eventual supplier. P&WC shifted its focus from the turboprop PT6 (for which there was not yet a customer) to the turboshaft for helicopter application.

Experimental

As the PT6 got underway, the experimental engineering department began working with the design requirements for engine test, build, inspection and instrumentation. Experimental had its roots with the DS-4J. One of its early tasks was to set up a gas turbine test cell. It was built specifically for large engines such as the Eland or the J57, which were used by the RCAF. Further cells would be added later to handle PT6 testing. There was also a small machine shop and engine assembly area where special parts and test equipment were fabricated, and where development engines and components were built up, stripped down and tested.

Compared to other departments, experimental was a flexible group, motivated by the needs of the engine program. Most of those involved were skilled mechanics and machinists who could turn out hand-made parts that looked like sculptures and met fine tolerances. Dennis Harding had been a machinist with the CNR, then worked with P&WC on production engines before joining experimental. He recalls: "We were a free-wheeling group and always doing something new. There was a lot of hand modification work. In the shop you could watch somebody adjusting the tip clearances of a turbine blade by using a magnifying glass and a small polishing wheel. We were all totally involved, and close to the action. We used to strip test engines down after a run, and an engineer would be looking over our shoulders making notes. Then we'd sit down over coffee and change the design. If a bracket was needed, you'd grab a piece of scrap stainless steel, measure what was needed, cut it out with hand-shears and trim the edges on a grinder. A designer would then take all the measurements and make up a blueprint."

Experimental had a number of rigs for independently testing engine components. Tests on certain ones were done in a spin pit without putting a complete engine at risk. A motor was used to accelerate components such as impellers, turbines and shafts to high rpm. Instrumentation tracked what was happening. If a failure occurred, the lead-lined pit contained any shrapnel-like fragments. A burner test rig was also built, and a water tunnel at the NRC engine lab in Ottawa was used. During this period, it was not unusual to look out a window at the plant and see a group of men huddled around a rig for testing combustor nozzles as flames leapt into the air!

The first two years of gas turbine work at P&WC had been spent on design studies and analytical research. Aside from the CL-44 gearbox, no work had actually taken place with gas turbine hardware. The PT6, though, did bring a build-up of staff and facilities for carrying the project through to a detailed design. By April 1958 details for the burner had been finalized, a test layout completed and a compresor design made. The first drawings were released for procurement in July. The test rig work began in August when the combustion chamber rig was lit for the first time. The engine was expected to run in November and pass the preliminary flight rating test in December. The first run would test 468 standard parts or two-thirds of the engine.

Design of the PT6 combustor, that part where compressed air and fuel are mixed and ignited to drive the turbines, was a challenge. John Vrana headed this work, and once design was complete, a clear plastic mock-up was tested in the NRC water tunnel. Water carrying small metal particles was run through the mock-up and a camera recorded the flow patterns. Eddies would indicate where, in the actual combustor, there might be a "hot spot." Temperatures were recorded to produce a profile of the gas temperatures at different places in the combustor. A range of pressure conditions, altitudes and atmospheric temperatures was simulated. Other tests focused on controlling the amount of smoke emitted by the combustor.

Rick Stamm, after several years working on piston engines, the Eland and the CL-44 gearbox, came to work on the combustor program: "Working 80-plus hours a week for two years became the norm. Things were such that few obstacles got in the way of making quick decisions when necessary. We were limited only by our imagination."

For a year, the combustor team struggled to turn its drawings into something tangible. First came a combustor with 42 intricate scoops made from sheet metal. Then one day Stamm sketched an alternate design, took it over to the experimental machine shop, and asked the foreman, Eddie Power, to make him a prototype. Just like that, the job was finished. The unit worked beautifully and became standard on the PT6.

In September 1959 the PT6 was thoroughly re-appraised. Aerodynamic and mechanical design was acceptable, but the engine was structurally lacking. Even before it ran, the design of the centrifugal compressor was suspect. This intricate component was cast in aluminum. Casting was relatively economic and produced the desired shape with a minimum of final machining, but it did not always result in parts that were metallurgically consistent. If one of these units burst in service, a catastrophe could result. The solution was to forge aluminum (later titanium) into a solid cylinder with precise physical qualities, but this required heavier machining to

Some of those involved in building the first PT6 pose around their "baby." From the lower left, clockwise, Barry Lees, Fernand Lauzier, Ed Power, Gilles Gélineau, Jim Harding, John Hunt, Gordon Hardy, Percy Hynes, Ben Judge, Jacques Lefebvre, Allen Titcomb, John Bennett, Rick Stamm and Fred Bates. The second photo shows Allan Newland, who was chief designer on the PT6 team, with the first engine. It is enmeshed in instrumentation wiring. (via G. Hardy, P&WC Archives)

The first complete PT6 runs in the test cell in February 1960. (P&WC Archives)

Chief test engineer on the PT6, Gordon Hardy, and test engineer Cyril Blizzard with the first engine. (P&WC Archives)

The complete PT6 compressor (three axial and one centrifugal stage, plus the compressor turbine assembly). The second photo shows damage caused in the test cell with the failure of a cast aluminum impeller. This led to use of forged impellers on production PT6s. Many lessons were learned in the test cells as the PT6 was being perfected. (P&WC Archives)

 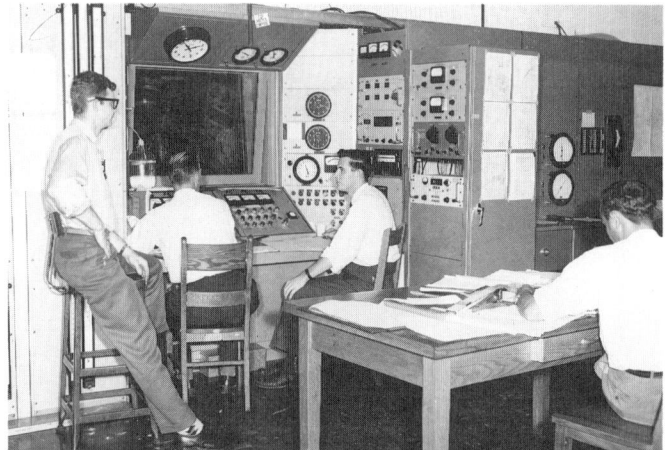

The PT6 undergoes icing trials in an NRC lab in Ottawa. (via G. Hardy)

Engineers monitor an engine run in one of P&WC's test cells. (P&WC Archives)

While P&WC was developing the PT6 and seeking markets, its main rival was Garrett AiResearch in the US. It developed its TPE331 beginning in 1959. It first flew in 1961 and was certified in 1965. The engine found many applications, for example in the Metro line of commuter aircraft. An early TPE331 is seen on its flying test bed—a Douglas A-26. (via Dusty Miller)

produce the end product. More metal would be machined off than remained in the impeller. This process was instituted, and to date not a single failure has been reported.

Another problem was that the engine was 20% above its design weight of 250 pounds. A part-by-part program was begun to reduce weight. The development engines did not attain design power and fuel consumption because of deviations from the drawings that had been accepted as expedient in the early days of the program. These occurrences are part of any development program and are corrected by introducing design changes and speeding up development testing of new components. The PT6 re-appraisal resulted in an engine called the Mk.2. Work continued on the Mk.1, and experience with it would be passed on to the new version.

As Allan Newland recalls, in moving from the Mk.1 to the Mk.2 there was a rapid maturing of engineering thought. The first designs were clumsy. In the move to the Mk.2 Newland notes: "We worked out the clumsiness, enhanced the aerodynamics and reduced the weight. But this was not without its headaches. The shafting for the Mk.2 gas generator developed uncontrolled vibration at relatively low rpm (in the 20,000 rpm range). On the Saturday following this discovery Ken Elsworth devised a dampened bearing mounting system for the thrust ball bearing. Other redesign was needed to the mounts holding the gas generator turbine vanes. These would collapse forward, severely damaging the rotor blades."

Everything was taken in stride, including management's decision to host a conference in Longueuil in November 1959, with senior aerospace people from Canada and the US to observe the first public running of the PT6. Preparations came down to the wire as the first test run slipped a week. The marketing people at P&WC borrowed UAC's DC-3 to shuttle the 65 guests to Montreal. The guests were being wined and dined at the Windsor Hotel in Montreal the evening of November 16 when Jack Beauregard came in to announce that the PT6 had been fired up for the first time. The next day, the gas generator, heart of the PT6, lacking only the reduction gearbox and power turbine, was run before the invited audience. The tiny engine, in a tangle of instrumentation tubing and wiring, looked out-of-place in the cavernous Eland test cell. With the announcement at the conference of federal aid for the program, PT6 credibility was running high.

Changes on the Industry Scene

Development of the PT6 coincided with big changes in the aviation industry. Election of John Diefenbaker's Conservative government in June 1957 marked the end of the C.D. Howe era. Little was heard hereafter of Howe, the energetic engineer-turned-politician who had guided TCA into existence, brought Canada's wartime industries into being and created a diversified postwar aviation industry. The new government was less inclined to view the latter as a national asset, vital to the defence structure. After examining cost factors in the Avro Arrow/Iroquois program, Diefenbaker simply axed it. Meanwhile, Avro had turned out the last of 692 CF-100s and Canadair the last of 1815 Sabres and 656 T-33s. Major programs such as these would have to be replaced for the industry to survive.

In the March 1959 issue of *Aircraft*, editor Robert Halford noted that the companies least hit by the termination of large contracts were those least tied to the Canadian defence market. He observed that having a variety of small customers was healthier than depending on one large one. He singled out P&WC, DHC and Canadair as likely to prosper where Avro failed. His figures showed that 77% of P&WC's business was in the civil market. Of company activity, 64.7% was manufacturing, 17% the sale of US-made parts and accessories, 12% engine repair/overhaul and 6.3% R&D.

From Orenda to P&WC

Long before Colin Wrong became vice president of engineering at P&WC, he worked at Orenda in Malton. Laid off in 1959, he headed downstream to Longueuil: "There had been 5000 people working on gas turbine engines at Orenda. The company was totally organized around military projects. Its Iroquois was the most advanced fighter aircraft engine of its time. I was mostly involved in design and development of its aerodynamic section. The aim of our group was to improve engine performance. Along with 15,000 others I was fired over the loudspeaker on February 20, 1959—Black Friday. From that day on, I swore that I would never again work for a company that had loudspeakers! I moved to P&WC in 1960, and was first of all surprised that everyone in the gas turbine organization was cost conscious. At Orenda, nobody had known what an engine cost! P&WC's intention was to get into the civilian market, and the competition was piston engines. Cost was critical.

"At Orenda, the government was paying much of the development costs. Our total engineering budget for 1959 was about $15 million. At P&WC the development costs were being paid for out of profits from the sale of spare piston engine parts. P&WC had inherited a standard price list from Hartford along with the spare parts lines, so it wasn't easy to increase parts prices to generate more profits, especially when our important customers were major airlines and the US military. We were aiming for a factory standard cost for the PT6 of $16,000. In other words, we wanted to keep the costs of material in the engine, inspection, and assembly (exclusive of R&D) to that amount. The target sale price was $25,000. The spread between these figures was to cover company overhead, the infrastructure to support the engine in the field, future R&D and profit.

(Above) A tiny PT6 beside a P&WA commercial jet engine. P&WC developed the specialty of dealing with extremely fine tolerances. Those required for the bigger engines were not so critical. In this photo the two men on the right are Tom Cutbill and Jack McPhee of P&WC. (P&WC Archives)

(Right) Following the PT6's 50-hour pre-flight rating test, engine components are being examined by Elvie Smith, LCMDR J. Barnes (USN), W/C W.R. Cole (RCAF) and C. Stim (US Army). (P&WC Archives)

"Many engineering differences made a comparison of the PT6 and the Iroquois especially interesting. Parts for the Iroquois were big. Inside the Iroquois we didn't have to worry about clearances the way we did with the PT6. The Iroquois didn't suffer a major loss in performance when clearances were .070", but the PT6 sure did. Its clearances were .0150" in order to give the performance we required.

"The demise of Orenda could not have been more perfectly timed for P&WC, which was in the midst of hiring overseas. Suddenly all these skilled people were out of work in Toronto. Included were strain gauge experts. Strain gauges are little transmitters used to measure vibration. They're glued to metal and wired to a power source and an electronic readout. As a metal surface deforms, it bends the strain gauge. This sends out a signal proportional to the deformation or strain. This is how we calculate the stress levels or vibration at a particular point in the engine. The whole strain gauge lab from Orenda moved to P&WC. Whereas we had been having quite a turn-over of personnel, the Orenda group came and stayed. The work was intricate and precise, and without the Orenda people our job would have been most difficult.

"At Orenda we had built axial compressors using British data. P&WC built axial compressors with data provided from East Hartford. It was fascinating to be immersed in a new technical culture and be able to discuss quite openly our different technical experiences. What had been problems for me at Orenda were not so at P&WC, and vice versa. To an engineer, any challenge is exciting. Aircraft engines are on the cutting edge. Even though the Iroquois had 25,000 pounds of thrust and the PT6 only 450 shp, the work at P&WC was not only exciting but fun."

On February 22, 1960, 14 months from full design launch, the complete PT6 with propeller went on test. Before the engine could enter production, more than 8000 hours of test cell running would be required. Countless tests of individual components would be made in rigs that precisely measured efficiencies and tortured the parts to destruction. The size of blading and high speed of shafts would call for ingenuity and the adaptation of laboratory techniques to test cell operations. Tubing the diameter of a pencil lead would be threaded through engine casings or down struts to measure the pressures within the engine without interfering with performance. Hair-like wires would carry minute impulses from miniaturized strain gauges cemented to areas of engine blading where components rotated at 37,000 rpm. The signals would have to be transmitted to the sensitive test cell measuring equipment.

In retrospect, development was hampered by the intricacies of the test equipment. It complicated matters in a program where the test engines had to be constantly assembled and disassembled. Much time was lost in correcting instrumentation problems. D.O. Blake commented, "With our surgical approach to instrumentation it is unfortunate that we have not developed an antibiotic to solve the engine problems."

The 1960 engineering budget had been estimated at $980,000, but before the end of the first quarter was revised to $1.7 million. Costs were escalating beyond proportion. On the credit side, it was noted that no catastrophic failures had occurred: "Not only does this speak well for the general soundness of the design, but for care exercised in setting speed and other limitations, and hours of running per build and the degree of inspection between builds." However, the approach was too cautious. Running on a single work shift, instead of uncovering problems rapidly, was masking them and delaying the inevitable. Down at P&WA there was a different approach to development—working three shifts with a philosophy of "make them and break them."

The Hartford Team

Through 1960 it became apparent that the costs of dealing with technical difficulties were far more than expected. By July, the Mark 2 engine was running but was beset by problems. Dick Guthrie was directed to draft a revised budget, looking for ways to reduce costs. The possibility of seeking help from Hartford in order to save the program was discussed. It could not be allowed to stagnate for a moment, not with applications pending for US Army and Navy programs.

In 1983 Thor Stephenson reminisced: "Wright Parkins was the auditor of what we were doing. We became great friends and he was very supportive. One day he came into the office and said, 'Your engineers are very intelligent, but are inexperienced and don't understand engine development. We should talk to Horner.' This we did at a meeting attended by Horner, Gwinn, Robbins and Mallet. It was decided that a team from Pratt, headed by Bruce Torell, would come to Longueuil for a short period, and this was done. After the meeting, Mallet cornered me and asked why I didn't oppose Parkins. I replied by saying that Parkins was correct."

Dick Richmond had come over from Canadair to be v-p operations at P&WC in 1960. He recommended Torell to lead the team, feeling that he would be acceptable to the people at Longueuil. Richmond, Stephenson and Torell were all natives of Winnipeg, although they did not know each other at the time. Torell was a graduate of the University of Minnesota who went to the NRC in 1942 to work in the engine lab (where he became friends with Stephenson and Richmond), and later joined Turbo Research and gained experience in the UK with Power Jets. This led him to the cold weather test cell in Winnipeg. In 1946 he left Canada for Hartford.

On January 2, 1961, the technical direction of the PT6 was

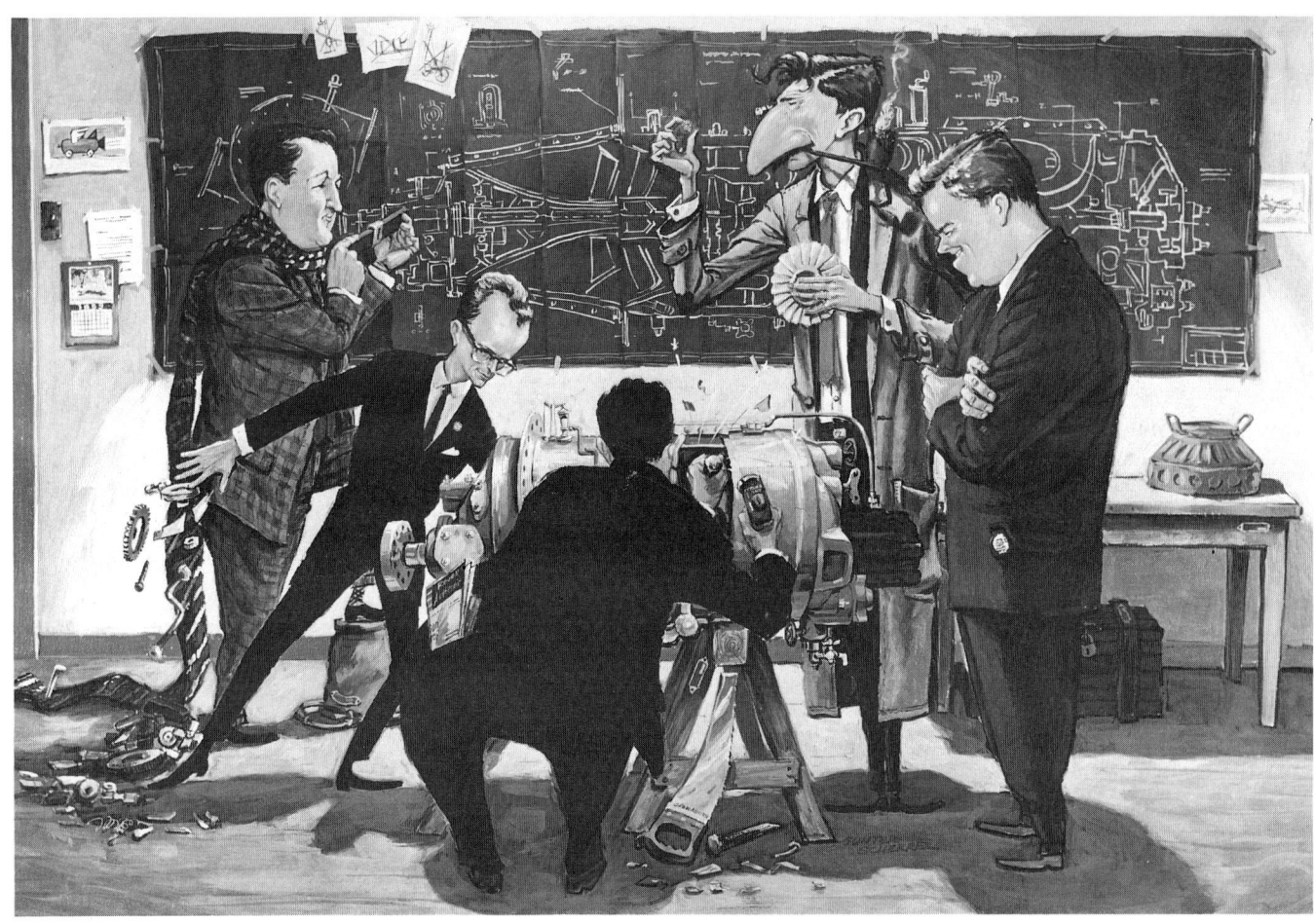

A caricature by Gunther Scherrer of PT6 personalities. From the left are Colin Wrong, Elvie Smith, Dick Guthrie, Hugh Langshur and Thor Stephenson. (P&WC Archives)

assumed for several months by a six-man P&WA team led by Torell, assisted by Art Nelson. Also involved were Ray Danahee, Paul Meyer, Nick Porto and Bob Spillane. Elvie Smith and Jack Beauregard represented P&WC. Smith recalls: "We learned how to develop engines from Bruce Torell. I had come out of a research background. The others were from design. But none of us at P&WC had ever run a major engine development program. Torell ran the program with an iron hand. Everyone would meet in his office at 8:00 a.m. with a report on the previous day's activities. They would decide what needed doing that day, then it was off to work. Torell quickly scrapped our single shift. From now on, development testing went on around the clock and cost was no object. We procured duplicate hardware to increase our inventory of components and thus reduced lead times."

Rick Stamm recalls of Torell, "When he was in town, he could be found in the plant at all hours. He was known to show up in the middle of the night wearing a raincoat over his pajamas." The Hartford team emphasized the importance of the project engineer in a development program. He was the one to orchestrate the specialists. Earlier management techniques gave way to central control and departments like experimental engineering surrendered their autonomy to answer directly to the project engineer.

In January 1961 the 50-hour test of the Mark 2 turboprop was completed at 385 shp. Icing tests at the NRC followed. The first runs of the complete Mark 2 turboprop and turboshaft engines were made in February. In March came a 50-hour test of the Mark 2 turboprop at the LOH rating of 450 shp. This cleared the way for the first in-flight running of the engine in the Beech 18 test bed that June, the official preliminary flight rating test (PFRT) in July, and the maiden flight that month of the Hiller Ten99 helicopter.

For Torell the PT6 program was an important step in his rise at East Hartford and in June 1971 he became president. Dick Richmond recalls that under Torell there were no financial restraints: "He was to be given whatever he needed to get the detailed configuration tied down. He knew that we had to spend money on alternative designs. If we had done things in sequence, it would have taken forever."

As the PT6 moved closer to production, management organization changed. Stephenson set up four divisions: sales and service under John Drummond, operations under Dick Richmond, engineering under Dick Guthrie and finance under Victor Tryon. Elvie Smith took over as engineering manager in October 1962. Dick Guthrie was named product planning manager, looking after long-term business opportunities. The PT6 was now safe, and P&WC had, in a crash course, learned the tried and true P&WA way of organizing a development program.

Years later, Stephenson added an interesting anecdote to the PT6 story: "The early days of the PT6 program were not encouraging, technically or sales-wise. As a result, James Young, P&WC's founder, and his great friend on the board, Hubert Welsford, went to Hartford to see Horner. They wanted the PT6 terminated and P&WC to revert to a sales and

P&WC's J57 test cell on the left, with the even larger Eland cell (in which the tiny PT6 made its first run!). These cells were built behind Plant 1. (P&WC Archives)

A line up of J57s (minus afterburners) at CFB Bagotville; then a view of the massive Voodoo. Note the length of its engines from intake to tailpipe. For nearly 25 years P&WC provided field service and overhaul for the Canadian Voodoos. One of the earliest P&WC tech reps on the J57 was Harry Mochulsky, and he was present at CFB North Bay in April 1987 when the last Voodoo was retired. The photo below shows jet engine overhaul in the old Navy building near Plant 2. For a period in the 1960s engines such as the PT6, J57, JT12 and Westinghouse J34 were coming through this facility. (Larry Milberry, P&WC Archives)

service organization. Horner rejected their pleas and the PT6 continued."

Further Spare Parts Work

While the Americans had sent their engineering experts to Canada for a few months, P&WC was saddled with paying the bills. Dick Richmond notes: "The PT6 was a real drain on finances. At one point we had well over the net worth of the company invested in the program. We wanted to get more piston engine parts out of East Hartford, and transfers continued until 1966 when we finalized the takeover of the whole line. This included the R-2800 and R-4360, the biggest of P&WA's piston engines." These came up in spite of fear that P&WC might not be able to handle Hartford's big engines since in-house production of the PT6 was well under way, as was the start-up of the Sea King program.

The demand for some parts was so great that the transfer and installation of some equipment had to be done on weekends so that productivity would not suffer. P&WA, which was forecasting a bulge in demand for military and commercial gas turbines, was happy to clear its own floor space. Richmond emphasizes: "These parts became vital in keeping our company viable so that we had the funds to develop the PT6. It was costly starting up the new lines of piston parts, but we knew that the investment would eventually pay off. Spare parts sales climbed to $70 million a year and were very profitable for us."

Piston Engine Spares Manufacturing at P&WC:
P&WA Engine Types

R-985 Wasp Junior	450 hp
R-1340 Wasp	550-600 hp
R-1830 Twin Wasp C	1200-1350 hp
R-2000 Twin Wasp D	1350-1450 hp
R-2800 Double Wasp	2000-2500 hp
R-4360 Wasp Major	2800-3800 hp

The J57 Years

With cancellation of the Arrow, Canada filled the air defence gap with the Bomarc missile and the McDonnell CF-101 Voodoo, the latter entering service in late 1961. One of the mightiest fighters, the Voodoo had two P&WA J57s, each putting out 15,000 pounds of thrust in afterburner. P&WC was contracted to service and overhaul the J57s and immediately hired service reps to handle field work. The early ones were Bill Armstrong (formerly of Orenda), Maurice "Chuck" De Maurivez (RCAF), Reg Hawke (Laurentian Air Service), Harry Mochulsky, Jim Nash (CPA) and Rod Turiff (Quebecair). These were based with the Voodoo squadrons and were P&WC's first service reps to be permanently stationed in the field (other than Russ McCormack, who was with CPA). In time, P&WC developed repair procedures for the J57 beyond those used by the USAF. Rick Stamm, first manager of the J57 program, maintained a dialogue with Tinker AFB, where the USAF had its own J57 overhaul base. In Canada the J57 began with a 750-hour TBO (time between overhaul) interval. When the last Voodoo left service in April 1987, the TBO had reached 1500 hours, exceptional for a large jet engine of 1950s vintage.

Harry Mochulsky had been a wartime flight engineer and had met Gene Schweitzer on the east coast in 1942. After the war he worked in the bush and on aerial survey around the world. In 1961, after he had finished a course at Rolls-Royce on the Conway and Dart turbines, he took a job with P&WC. Mochulsky notes: "Our job as service reps was to ensure daily engine reliability. After an aircraft had flown a mission, we would sit down with the pilot and maintenance crews to review any snags. It was a fascinating era both for the Air Force and for P&WC. One of the interesting aspects of servicing the J57 was checking the pigtails that fed fuel into the afterburner. You had to inspect them when the burner was lit. You can imagine how noisy that was! The J57 was the company's first in-plant involvement with a big jet engine, and it included construction of test cell No.7 at Plant 1. We had never been so closely involved with the J75s used in the Arrow." When not testing RCAF engines, P&WC's J57 test cell was used to test commercial P&WA engines in for overhaul and took overflow work from Hartford, including the 1000-hour endurance run for the JT3D turbofan.

James Young, retired P&WC president, looks on as Elvie Smith explains components on an early PT6. (P&WC Archives)

A Quarter Century Club dinner at the Windsor Hotel in the mid-sixties. The only woman in the club at the time was Betty MacDonald, seen in the checkered hat. (P&WC Archives)

There was always opportunity for a good time at P&WC and to recognize individual achievement. The Recreation Club and the Quarter Century Club have provided for the social side of life in the company over the years. In these photos from the fifties and sixties, typical functions are represented. In the photo above, James Young receives his 25-year pin from Frederick B. Rentschler, founder of P&WA, as Bill Gwinn looks on. (P&WC Archives)

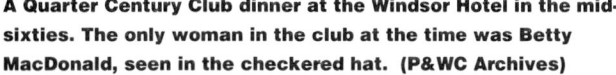

More Quarter Century Club activity. Enjoying themselves around the dinner table are Arthur Pond, John Drummond, Émile Quérel, Monique Bouthillier, Eddie Ober, Ted Harris, Ken Dawson, Harry Garton, and Jacques Gagnon. The group standing in the photo below shows Frank Santo, Sam Hill, Bob Reive, John Drummond, Gene Schweitzer and Bill Vertilneck. (P&WC Archives)

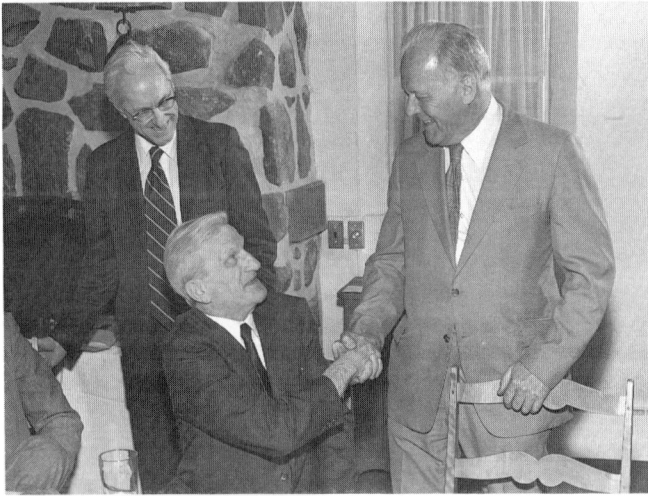

(Above) Gene Schweitzer and Ken Dawson congratulate George Rose on his retirement. One of his achievements was helping set up the company's first helicopter repair and overhaul shop.
(via E. Schweitzer)

Thor Stephenson, Harold Marr, Charles Charest, Bert Massé, Roger Trudeau, Leo Phelan and Eddie Ober around the table. Frank Santo and Jim Bujia are just behind Stephenson. Ted Harber is over Trudeau's shoulder. (P&WC Archives)

Flying the PT6

Early in PT6 development, P&WC began looking for a suitable test-bed aircraft. The DC-3 was favoured as it had the range and service ceiling but fitting an engine into its nose would take extensive re-stressing. The Beech Expeditor looked more promising. Aircraft HB109 was borrowed from the RCAF and converted at Downsview by de Havilland. On May 30, 1961, DHC test pilot R.H. "Bob" Fowler and John MacNeil of P&WC made the initial familiarization flight. The aircraft was found satisfactory and only 23 pounds of ballast was required to adjust the centre of gravity. A wide range of test flying was soon under way, including aircraft handling and performance, specific fuel consumption, propeller handling, propeller drag, air starting, accessory loading and noise/vibration levels. MacNeil took HB109 as high as 26,000 feet on occasion, to the amazement of air traffic controllers and airline pilots.

Flying the modified Beech was not always a delight, as described by MacNeil in a report of September 7, 1961. By then he had logged 26:50 hours and was able to comment: "It is very unstable longitudinally, particularly at the higher altitudes. It also has a rolling tendency about the longitudinal axis when high powers are selected on all three engines…. Care must be exercised at all times to be mindful of its shortcomings … most delays have been caused by instrumentation or propellers and governors (or both). I am very pleased with our engine operation to date. It starts quickly, both in the air and on the ground, and makes its thrust very obvious from the surface to 25,000 feet. The engineers and technicians who have made this possible have my humble respect and heartiest congratulations."

MacNeil carried out a series of in-flight propeller reversing tests in 1963. He described this as "quite interesting" and went on: "The aircraft is rather unstable in this configuration, and suffers from elevator buffet. As reverse power is applied the buffeting increases…. At this point [220 shp] a decision was made to stop further application of reverse power because elevator buffet was becoming too severe to manage, and the possibilities of flutter was a concern."

In 1963 John MacNeil took the test bed to Knob Lake in northern Quebec for winter trials. He departed Montreal with a temperature of +38°F and landed 6^1/$_2$ hours later at –21°F. This led him to comment, "Temperatures and snow fall at Knob Lake have been much worse than normal for this time of year. If the trend continues, we should get well below the –45° hoped for during our test operation." This program was interrupted when P&WC diverted the test bed to the US in connection with the US military competition for a new counter-insurgency (COIN) aircraft. MacNeil and flight test engineer John Hunt crisscrossed the US doing demonstrations for manufacturers with COIN proposals.

(Top) The P&WC Beech 18 test bed on a proving flight from Downsview in 1961. HB109 (later CF-ZWY-X) made 719 flights and logged 1068 hours testing many PT6 models and propellers. It was retired to l'École aéronautique in St. Hubert after its last flight on June 3, 1980. (Above) Port and starboard views of the first PT6 installation in the Beech 18. (DHC, P&WC Archives)

John Hunt, an early project engineer on the PT6, while visiting Flight Test in the late 1960s. Standing are Gord Linnen (pilot), Max Vejins, John MacNeil (chief pilot), Jack Beauregard, Dan Ketelson, Larry Samoil and Warren Connor. (via Don MacNeil)

(Below) P&WC men Juan Martinez, George Wilkinson, Bill Moore, an unknown pilot and Max Vejins during early work on the Beech 18 at DHC in Toronto. (Ken Douch)

Forty-five hours were flown showing off the PT6, with demos including maximum performance takeoffs (three engines), propeller feathering and air starts, and slam accelerations. Potential customers wanted assurance that the free-turbine engine would allow sufficient drag for a reasonably steep descent and landing. This was demonstrated to their satisfaction, but John MacNeil was not comfortable with the propeller reversing system, which had exhibited minor hesitation. Following a day of demonstrations at Lockheed at Burbank, California, he found the propeller reduction gearing overheated and about to seize. Consultation with Longueuil suggested that this was an isolated occurrence and a new gearbox was despatched, but the same problem developed on the next flight. Investigation showed that with the gas generator in low rpm, the reversing propeller was demanding more oil than the pumps could deliver. The solution was to change the size and location of the oil pumps and a diverter valve.

Hiller and the PT6

One of the earliest companies interested in the PT6 was Hiller of Palo Alto, California. It had been founded in 1944 by

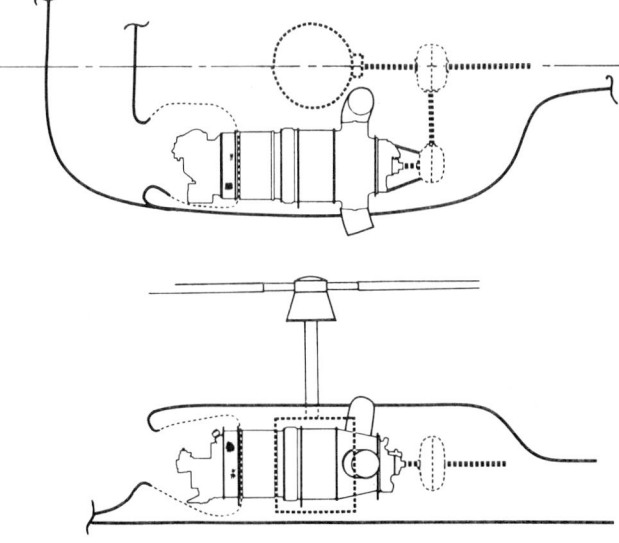

The first aircraft to fly solely under PT6 power was the Hiller Ten99 (July 1961). It is seen during winter trials in Canada. The schematic views below show the off-centre installation of the PT6B-9. (P&WC Archives)

Stanley Hiller, whose first major design was the Hiller 360 (UH-12) helicopter. In the 1950s he experimented with a small helicopter having ram-jets on the rotor blade tips. In 1959 he was briefed about the PT6. In February 1960, Ed Bolton, his manager, visited Longueuil for an update and Thor Stephenson paid a return visit to Hiller that April. Hiller was chasing a Canadian Army requirement for light helicopters and facing stiff competition from Bell. Part of Hiller's pitch was to sell piston UH-12s to Canada, then phase in the PT6. Army interest gave rise to a joint Hiller/P&WC proposal to the DDP for a flight-test program to carry the aircraft through FAA certification at a cost of about $500,000. This idea sat on the back burner for a while, but in 1963 a UH-12 was brought to Longueuil and fitted with a PT6. Ground runs were made to add to the knowledge of helicopter installations, but the government cancelled further development before flying began. Meanwhile, the Canadian Army had ordered 27 piston UH-12s for its artillery and armour units in Germany. In time, some civil UH-12s were converted to turbine power, but with Allison, not P&WC engines.

The US Marines were interested in an assault support helicopter (ASH) to carry a six-man squad. This would be larger than the LOH and beyond the scope of the Allison T63. After viewing the PT6 mock-up in Washington in March 1960, the USN asked P&WC for price, delivery and support information. Hiller entered the picture with a request to P&WC for an engine delivery date. The ASH encouraged use of the shaft version of the PT6. By December 1960 there was a formal USN request for the proposed ASH, asking for data about a 500-shp PT6 (military designation T74), funding estimates for a flight-test program, and what support P&WC would need to carry the engine to its 150-hour qualification tests. Hiller was now encouraged to built a prototype of its Ten99 based on a PT6. About the same size as contemporary machines, the Ten99 would carry twice the load in a roomy cabin.

In April 1961 P&WC shipped PT6 Mk.2 Serial 213 to Hiller for installation in its prototype. Stan Hiller realized that his engineers had erred with the installation, for the exhaust stack would blow hot gases right across one of the cabin entrances. This and other bugs were worked out, and the helicopter was due for its first run on May 19. Bob Losch was on the scene, but an early morning call from Bob Spillane put a hold on firing up the Ten99. Engineering suspected a faulty turbine blade. Russ McCormack and Fred Cowley, who were also at Hiller, removed the engine and took it apart. Cowley drove to Stanford University to weigh each blade on a laboratory scale. He could find no discrepancies. The program went ahead and the Ten99 flew in July, this being the first time a PT6 had powered an aircraft on its own.

After tests, the aircraft was shipped to Washington for military demonstrations, then trucked to Longueuil, where its engine was overhauled. A flight to Ottawa followed, and on December 11 the Ten99 was displayed to government onlookers. It was hoped that this would help clear the way for a funded PT6 flight test program. As it happened, the demonstration did not go well. As Fred Cowley recalls: "The helicopter went unserviceable for four days. We couldn't figure out how to fix the engine. All our other engines were for test cell experiments, and we had only a handful of fuel controls, which is what I needed. Each control was driven up to Ottawa for me to try. We only had one flight engine in the whole company at that time."

Once the Ten99 was back in operation, it flew to Longueuil and was parked between the guardhouse and the cafeteria at Plant 2 for a series of tie-down trials, to give P&WC further experience with helicopter installations. One problem studied was the recirculation of exhaust into the engine intake, caused by rotor wash. Following its Canadian sojourn, the Ten99 went home to California and is now part of the Hiller Museum in Redwood City.

Other Pioneers: Piasecki, Lockheed and Kaman

Frank Piasecki had been experimenting with helicopters since the 1930s, and his persistence resulted in a family of successful types including the H-21 of Mid Canada Line fame. Today, the Boeing CH-46 and the CH-47 Chinook are H-21 descendants. In the mid-1950s, Piasecki was studying such concepts as the VTOL Flying Jeep, which used ducted rotors. He then came up with a "compound" helicopter which included both rotary and fixed wing design features. There was a small wing, and a tail-mounted ducted propeller for forward propulsion and control. Designated the 16H Pathfinder, it was planned as a forerunner of a new breed of fast civil and military transports. First flight of the 16H powered by a PT6 was in March 1962. Unlike the Ten99, it would undergo many changes over the years as Piasecki explored its potential.

Fred Cowley worked with Piasecki and recalls the installation as straightforward, with a two-port exhaust: "The engine was only good for 25 hours between overhaul. It was mainly the turbine wheels needing work, so I would bring these back to Longueuil in my briefcase. Canada Customs had a hard time understanding what I was up to. And the old shoemaker on Chambly Road used to get regular business repairing the handle on my briefcase. I'd charge the repairs to my expense account."

The 16H's performance depended on what power the PT6 could deliver. With an uncowled engine and fixed gear, it reached 134 mph at 327 shp in early 1963. This was about the limit of the experimental PT6B-2. With an updated engine, it reached 173 mph in August 1964. P&WC was always available with field support, and Fred Cowley notes: "Frank Piasecki always tried to get me to stay another day in Philadelphia. He knew that I would work in the hangar as long as I was in town, so always tried to engineer a crisis of some kind at five o'clock so I would miss my flight to New York."

A Hiller UH-12 was fitted with a PT6 to give P&WC experience with helicopter engine installation. Hiller's dream of selling the Canadian Army a fleet of turbine UH-12s fell through. (P&WC Archives)

(Above) Fred Cowley was one of several P&WC engineers who spent much of the early 1960s on the road, supervising pioneer PT6 installations. (P&WC Archives)

(Left) Frank Piasecki's Model 16H during an early test flight with the PT6. The diagram below shows the engine installation. (P&WC Archives)

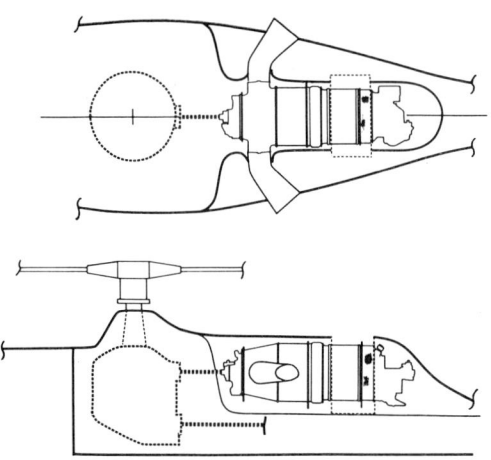

FLYING THE PT6 143

Spectacular for its day, the rigid-rotor Lockheed XH-51 flew at 242 mph on December 14, 1964. It was powered normally by a single PT6, but in this configuration also had a J60 (a military JT12) attached to a stub wing. Pilot Donald Segner is at the controls. (Lockheed California)

The 16H attracted the attention of the US Army, which had plans for a fast gunship, and under an Army contract the 16H grew. At first, thought was given to using three PT6s, but a single GE T58 was chosen instead. The airframe was enlarged and a 44-foot diameter H-21 rotor was used, plus a larger ducted propeller. That seemed to be the end of the P&WC and Piasecki association but years later, in 1988, Frank Piasecki announced his intention to revive the 16H for the executive helicopter market.

In the late 1950s Lockheed studied the mass-market possibilities for a simple flying machine. The old idea of a helicopter in everybody's driveway was briefly discussed but set aside in favour of experimental development. A helicopter was designed using a "rigid" rotor (a non-flexing, true rotating wing, as opposed to the standard flexing rotor blade). This concept offered lower weight and cost. A small helicopter (CL-475) was built to test the new rotor system, and in February 1962 the USN let a contract to build two high-performance research helicopters based on Lockheed's proposal. A target speed of 200 mph was desired, thus eliminating the piston engine. Lockheed selected the PT6.

Two years of testing the CL-475 had shown the rigid rotor to be inherently stable and the helicopter was praised for its manoeuvrability, ease of control, mechanical simplicity and speed. The new aircraft, the XH-51A Aerogyro, had one 500-shp PT6B aft of the transmission. There were two flush airscoops atop the fuselage and exhaust vents on the sides. The XH-51A first flew on November 2, 1962. Though a research project, Lockheed hoped that it might develop into a production machine and it made quite an impression when demonstrated to the Army at Fort Rucker, Alabama.

Lockheed was sold on the PT6 for the XH-51 and purchased two immediately. This warranted a high level of support from P&WC. Gordon Hardy, Fred Cowley and Bob Losch were all assigned to the project. Bill Andersen of P&WA (later with P&WC) was also involved and mentions: "The rigid rotor on the XH-51 was so stable that on the first flight the pilot was able to take his hands and feet off the controls and hover a few feet off the ground. The engine was generally good, but there were a few problems. While in hover, the rotor would cause the exhaust to go right up into the engine inlet. There was a considerable loss of power so we had to clean the compressor after each of the early flights. This is where we learned a lot about washing the PT6 compressor with water." Another problem was tackled by Gordon Hardy and the engineering staff. Bleed air valves had been installed in the engine to prevent compressor stalls. The valves opened rather abruptly to change the amount of air flowing through the engine. The resulting power change was enough to make the XH-51 hop around the ramp like a kangaroo! Hardy notes, "We redesigned the bleed air valves with a piccolo design that would open gradually."

The original XH-51A had a gross weight of 4100 pounds, a four-blade rigid rotor, and plenty of room for instrumentation. Its retractable gear reduced drag by 25%. Early on, it reached 174 mph. In 1964, NASA ordered one with three blades and seats for five. Lockheed built two further examples called the Model 286, which were almost identical to the XH-51A. The first was flown in June 1965 and certified by the FAA a year later. Lockheed would use the 286 to demonstrate the qualities of the rigid rotor and it was often seen at air shows performing loops and rolls.

Lockheed now received a further Army contract to modify an XH-51A for high-speed trials. A P&WA J60 (JT12) turbojet was added, attached to a 16-foot stub wing. The new version took to the air in September 1964 and in June 1967 it flew at 302.6 mph to become the world's fastest rotorcraft. With data gleaned from the XH-51, Lockheed won an Army design competition for an "advanced aerial fire support system" to replace various armed helicopters. This was the start of the AH-56A Cheyenne compound helicopter of which nearly 400 were ordered (but later cancelled).

The Model 286 outlived the program it had sparked. By 1968 the two demonstrators had logged over 1000 PT6 flight hours, mostly in VIP service, shuttling passengers between Lockheed's Burbank and Palmdale factories, and they had amazingly trouble-free careers before retirement in the early 1980s. Today, the original XH-51As, including the one with the J60, are on display at the US Army Aviation Museum at Fort Rucker. The two Model 286s were destroyed in a hangar fire at Shafter, California, on June 3, 1988.

Like Hiller and Piasecki, Charles H. Kaman formed his own helicopter company in the 1940s. A feature of his designs was overlapping twin rotors. Kaman's K-225 became the world's first turbine helicopter, flying in December 1951 with a Boeing engine. The USN purchased Kaman's R-1340-powered HOK and HUK types, and the USAF operated the H-43A and the turbine (Lycoming T53) H-43B rescue helicopter. Kaman then brought out a private-venture design, the K-1125, powered by twin 500-shp Boeing YT60s, which flew in August 1962 at Kaman's Bloomfield, Connecticut, plant. This was one of the first twin-engine medium helicopters.

Following the crash of the prototype K-1125, P&WC proposed that Kaman consider the PT6 in its replacement machine. Kaman agreed, as it was trying to win a USAF competition. With the dust hardly settled following the crash, PT6s Nos. 1003 and 1004 were shipped to Bloomfield, where they were installed, and the aircraft flew in April 1963, just eight weeks after the PT6 deal had been struck. The K-1125 project could not be salvaged, with or without PT6s, and after

The Kaman K-1125 was the first helicopter powered by twin PT6s. It and the various other early PT6 installations provided P&WC with vital knowledge at a time when the company was shooting up the experience learning curve. The accumulation of this knowledge led to the first practical applications of the PT6. (P&WC Archives)

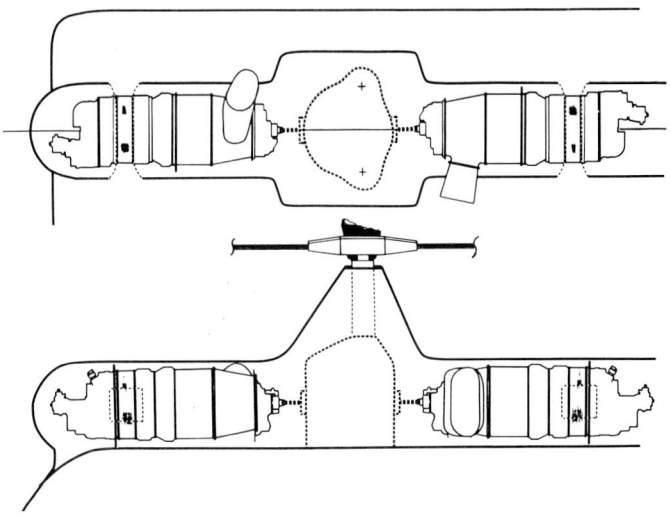

some test flying the venture was scrubbed. Kaman concentrated on developing the Seasprite for the USN and that type, powered by the T58, became a great success.

Losing the LOH

That the PT6 could win the LOH competition was unlikely. The Army was already funding the 250-hp Allison T63 turbine, which was in the power range for a light helicopter. The PT6 was, by comparison, too big, but P&WC decided to try anyway by reducing engine weight and improving performance and fuel consumption for lower power settings. At this time, Ken Sullivan and Tony Clark were the total marketing staff, and they sought to interest the 20 LOH airframe contenders in adopting the PT6. Sullivan and Clark were given some relief when Bob Losch joined their team.

Shortly after the PT6 had passed its 50-hour preliminary flight rating test, a milestone required by the US military, Clark and Losch took a PT6 mock-up to Washington. The military was interested and so were several manufacturers visiting Washington. "We took a large suite in one of the downtown hotels," Losch recalls. "Since there was not time to make other arrangements, we decided to make our presentation in the suite. We were wheeling the mock-up across the hotel lobby, getting the carpets all rolled up, while bell-hops were trying to tell us we couldn't take an engine through the hotel. Tony managed to divert their attention and we got onto the elevator. When we reached our suite we found that the mock-up would not fit through the door. We solved this by taking off part of the door jamb. We made quite a good impression with the people we briefed that evening."

Most agreed that should the T63 fail to live up to expectations or should the Army want more power, the PT6 was an alternative. When 12 contenders finally put their submissions on the table, only Republic specified the PT6 (for a re-engined Alouette) though all had looked closely at it. Bell, Hiller, Hughes, Kaman, Lockheed and Sikorsky all chose the T63. In May 1961 the US government announced Bell and Hiller as winners. There would be a "fly-off" between the two, and Bell finally won with the OH-58. A late arrival, the Hughes OH-6, also fared well and it too had the T63. The Army ordered large numbers of each. Civilian spin-offs from the Bell, Hiller and Hughes designs were soon in production.

Loss of the Army contract (and the later COIN program) was not as bad as it first appeared. P&WC would not have to contend with sophisticated US government requirements, which often involved drastic weight reduction and expensive use of leading-edge technology. The company could now revert to its focus on the commercial fixed-wing market. In the meantime, it had gained valuable marketing experience through the LOH competition.

Experimental Otter and Turbo Beaver

In the mid-1950s de Havilland and the Defence Research Board investigated the "outer limits" of STOL performance. An Otter was modified with redesigned tail, oversized flaps, and a rugged landing gear suitable for steep approaches and hard landings. A test program was flown under George Neal, then phase two was begun with installation of a J85 in the rear cabin. Thrust was directed out either side through adjustable nozzles. In this configuration Bob Fowler made the first flight in September 1961. Speeds as low as 48 mph were attained and landings made in less than 500 feet. As early as January 1959 DHC had been seeking information about the PT6. To study the effects of increased slipstream over the wings, the Otter was redesigned to take two piston engines. When P&WC heard of this, it proposed PT6s instead. John Orr of the DRB agreed to this so long as P&WC guaranteed to support the engines. The budget for the pistons was now used to lease a pair of PT6A-4s, and Fred Cowley became the liaison between P&WC and DHC. Bob Fowler flew the twin-engined Otter on May 7, 1963. The program, one of the most unusual in aviation, carried on until July 15, 1965, and much experimental data was accumulated. Throughout the flying program the PT6s performed reliably. As Bob Fowler put it, "They never missed a beat." The experimental Otter was, in Fowler's words, "not a Twin Otter by any stretch of the imagination…. It had flying characteristics all its own." The experimental Otter provided P&WC with another flying test bed and the opportunity to learn more about its product in operation. One suspects that it also had some bearing on DHC's later thinking about the forthcoming Twin Otter.

Engineering studies suggested a turbine-powered Beaver would be feasible, with design changes to the tail and a fuselage extension of 28 inches. The market was canvassed and

The DHC/DRB experimental Otter was the first twin PT6 installation to fly. It also had a J85 in the rear fuselage (note air intake on top of fuselage). This aircraft taught DHC and P&WC much about twin turbine fitment and operations. (DHC)

The prototype PT6-powered Turbo Beaver over Toronto Island Airport in 1964. CF-PSM served with DHC until 1988 when it was wrecked in an accident. (DHC)

FLYING THE PT6 147

the go-ahead given to make a prototype. The airplane was designated DHC-2 Mk.3 Turbo Beaver, and Bob Fowler made the first flight on December 31, 1963. (The PT6 received its civil certification on December 20, 1963 following 11,000 hours of test running and 1000 flight hours.) DHC's loyal customer, the Ontario Provincial Air Service, ordered 17 Turbo Beavers. The market, however, did not produce many further orders, and in 1968, after 60 aircraft were completed, the project was terminated. Though an excellent performer with its 550-shp PT6, the Turbo Beaver was unable to compete in a market where used Beavers and Otters were available at more affordable prices. De Havilland's hope of selling PT6 conversion kits to Beaver operators also failed to materialize, as few could afford the new engine to begin with. Small bush operators also balked at the high fuel consumption of the PT6 compared with the R-985. The Turbo Beaver did excellent work over the years and most remain in service today, some still with their original PT6A-6s, others with engines as modern as the -135.

Even by the spring of 1963, the fate of the PT6 was still a question mark. Millions had been spent on it but no major company had come forward with an order. To a large degree, P&WC's hopes were resting on what Beech would do with its proposal to build a PT6-powered aircraft. Dick Richmond relates of this difficult period: "Thor and I went skiing in the Eastern Townships just as we were making another unsolicited bid on the COIN program. We had stopped on a run and Thor said that he didn't know if we should keep the PT6 going. I realized that he was under a lot of pressure, but we had tried the engine in different aircraft, and with Garrett now on the scene, if we shelved our engine that would be the end of it. Before we continued down the slopes, Thor said, 'I'm going to Hartford to tell them that we've got to keep this thing going until we know what Beech has decided.'"

On December 11, 1962, the corporate name, Pratt & Whitney Aircraft Company, Limited, was changed to United Aircraft of Canada Limited to more accurately represent the widespread interests of the parent company (United Aircraft Corporation).

Beech and the PT6

Beech Aircraft was a pioneer in the US light aircraft industry. Founded in 1932 by Walter H. Beech, it produced such famous pre-war designs as the Staggerwing and the Model 18 twin. In 1944 Beech carried out its first design study of a turbine aircraft, but the company, like most others in aviation, was conservative—it moved cautiously with turbines and kept the Beech 18 in production for 32 years! In October 1958, it finally consulted with P&WC over size and configuration preferences for a small turbine. About six months later P&WC made its first PT6 presentation to Beech.

Beech now introduced its six-seat Model 65 Queen Air (military L-23F) powered by two 340-hp Lycomings. Allison was trying to interest Beech in a version with the Army-funded T63. The Navy-funded Boeing T60 was also vying for attention. At this time, Beech was buying 300-plus-horsepower Lycomings for about $7000. P&WC was talking $15,000 for a PT6 while Allison was offering the T63 for $4000. In early 1961 Beech and P&WC agreed to combine their products in a test program. The Army had just placed a follow-on L-23F order. To Beech this seemed a chance to demonstrate the type with turbines, in hope of selling the Army on converting L-23Fs, and bringing Beech closer to a commercial product.

At the National Business Aircraft Association convention at Tulsa in September 1961, Beech surprised P&WC by showing a cabin mock-up for a pressurized twin turbine aircraft, the Model 120, to seat six to eight. What hit P&WC was the plan to power the aircraft with Turbomeca Bastan engines. There was a growing French connection at Wichita. Beech was marketing the MS760 Paris executive jet and was cooperating with a number of programs to convert aircraft to French turbines. The Model 120 was aimed right at the corporate market P&WC was after.

Jim Lew of Beech and Thor Stephenson got to work promoting the PT6 L-23F to the Army. They presented the idea as a logical evolution of the small Army liaison aircraft. An experimental program would cover PT6 installation design, one airframe, two PT6s and Hartzell propellers. The package included a 100-hour flight test program. Beech and P&WC would contribute equipment, and Ottawa and Washington would help with financing. The Army, however, played a shrewd game—somehow it was able to convince Beech and P&WC to *give* it the L-23F with PT6s installed for one dollar! The deal was accepted and the project shot ahead. Now Beech

The first Beech PT6 installation was on the NU-8F, a modified Queen Air seen here in the shops at Wichita. (Beech Aircraft)

was involved in *two* turbine ventures—the corporate 120, and the L-23F. Frank Hedrick, Beech's executive VP, was the key man controlling these, and in late November 1962 he gave Ken Sullivan some good news—interest in the 120 was waning. A market for only 150 aircraft in that category was foreseen, and there were a dozen companies after it. The L-23F looked like a better deal, for there was no competition to worry about.

The turbine L-23F was designated NU-8F, and the rush was on to get it airborne before July 1963 lest it not get into the Army's upcoming budget. To accelerate things, Beech would take a second aircraft for FAA certification. The new aircraft

Mike Saunders of P&WC contributed greatly to the technical success of the PT6 and its acceptance by manufacturers such as Beech. With RAF Ferry Command during the war, he later flew in Argentina, then joined P&WC in 1952. He was working on a procurement matter for the PT6A-50 when he died suddenly in 1977. (Beech Aircraft)

Prototype PT6 Running Times, April 6, 1964

Aircraft	No. Engines	Engine Model	Total Time
Hiller 1099	1	B	85:15 hrs
Piasecki 16H	1	B	50:30
KV *Rimfakse*	1	B	302:05
LARC	1	B	78:00
Lockheed XH-51A	1	B	250:35
Kaman K1125	2	B	70:25
DHC Otter	2	A	201:10
DHC Beaver	1	A	62:55
Beech NU-8F	2	A	552:15
Beech King Air	2	A	105:00
Totals	14		1758:10

had a larger tail than the Beech 65, and added fuel for the thirsty turbines. Much day-to-day liaison between P&WC and Beech was handled by Dick McLachlan, an ex-Orenda marketing man. P&WC also sent some technical people to Wichita, the first being Bill Lee who was on loan from P&WA. He had been with the J57 team working on B-52s at Boeing's Wichita plant. In February 1963 he attended a PT6 course and recalls: "P&WC was a small organization. The comradeship was fantastic. I worked with everyone involved in the PT6 and remember thinking that the PT6 seemed so tiny compared to the J57 that you could put it under your arm."

Once back at Beech, Lee was joined by Mike Saunders, who took over training Beech personnel and briefing clients and distributors about the new engine. He was a natural for the job. As Lee relates: "Mike was a first-rate mechanic and a fabulous story teller. He did much to build Beech's confidence in the PT6, and Beech responded by opening its doors to P&WC." John Calhoun of Beech's design office notes how important the NU-8F was in the history of Beech and P&WC: "Frank Hedrick came in to see the prototype one day. He was very impressed by the neat appearance of the PT6 installation, and that's when he decided to back a commercial turboprop."

In May 1963 the NU-8F was going through its ground tests. One day Beech pilot Steve Tuttle was doing trials on the runway. Mike Saunders was in the right seat and Bill Lee was in the cabin. Racing down the runway, they went over a bump and inadvertently took to the air! A lot of folks from Beech and Cessna who were changing shift watched the event, but it didn't get into the history books. The official flight took place next day, with Tuttle and Jimmy Webber aboard.

On March 12, 1964, the NU-8F was delivered to the Army Aviation Test Board at Fort Rucker, Alabama. For six months it underwent rigorous testing. There were rough field tests at Fort Rucker, then the aircraft went to Texas for icing trials. The NU-8F would fly behind a Caribou which sprayed water from a boom to produce icing on the smaller plane. Cold-weather tests were made on the ground at Minot, South Dakota. Next it went to Arizona for hot-weather tests, and to China Lake, California, to punish the PT6s with desert sand ingestion. Hot and humid trials were done in Panama. The NU-8F was frequently commandeered by senior Army officers for use on "urgent business." It is said that it was probably

The NU-8F first flew in May 1963. Entered into with a certain amount of trepidation by Beech, this Queen Air/PT6 combination led to one of aviation's great success stories. The NU-8F was the beginning of the fabulous King Air line, powered by a series of PT6s. To date more than 3100 King Airs have been built. (Beech Aircraft)

flown by more generals than any other US Army aircraft, which appears to have paid off in the long run, for today all small fixed-wing Army turboprop aircraft use the PT-6. The NU-8F eventually became a training aid for army mechanics at Fort Eustis. In 1985 it went to the Army Aviation Museum at Fort Rucker (minus engines).

Enter the King Air

The NU-8F was at Colorado Springs on high altitude trials when, after 152 hours of flying, it was taken back by the Army for test bed work on a new infrared sensing system. It did not return to Beech, but as far as P&WC was concerned, it had more than done its job by winning over Beech to a light turbine-powered commercial twin. Beech ordered 29 PT6s. A market was forecast for about 10 aircraft per year. Beech was just finishing work on a pressurized version of the piston Queen Air for the 1964 market, but now changed direction and completed that model with PT6s. The aircraft was re-named the King Air 90 and officially announced on August 14, 1963. It would be a spacious pressurized twin available for delivery within a year. John Wilson became project engineer on the King Air. He admits that he had almost no knowledge of turbines: "At Beech we took a 'tinker toy' approach to new

Always a respected man to P&WC was J.C. "Jack" Charleson. One of Canada's first helicopter pilots, a senior inspector with the DOT and a key man with Okanagan Helicopters, he was later P&WC's representative in Ottawa. As an old friend of Walter and Olive Beech, Jack was close to discussions which led to the King Air. (P&WC Archives)

The NU-8F at Fort Rucker in mid-1964 during the first on-the-wing hot section inspection (HSI) of the PT6. (D.S. Miller)

Bill Lee and Mike Saunders with the NU-8F at Wichita. (P&WC Archives)

aircraft. With the Queen Air we mated a Twin Bonanza centre section, wings and tail to a larger cabin. Next, we took the PT6 installation that worked so well on the NU-8F and adapted it to a new pressurized fuselage. It was a bit more complex than that, but we sold management on the idea in quite simple terms."

Dick Richmond recalls that P&WC sold its first PT6s to Beech for $25,000 each but that parts for each engine purchased from vendors cost the company $21,000 since there was a very limited capability to make turbine parts: "We realized we couldn't live this way for long and although there were only 29 engines on order we decided to gamble and equip the factory to make many of the high-cost items such as gas generator and exhaust cases, turbine wheels, compressor discs, impellers and gears."

January 20, 1964, was the 28th anniversary of the first flight of the Beech 18, and on that day, before 3000 onlookers, the King Air took flight. Its début included a 280-mph flypast across the field. The aircraft entered its flight test program and Beech geared up for its sales campaign. It had a global sales network and kept close tabs on the ownership of over 100,000 light aircraft. The aircraft was certified on May 19, 1964, and made its first visit to Canada on June 11, winging in to the old Fairchild airstrip, where Frank Hedrick and other senior Beech men were met by Thor Stephenson. A month later P&WC took delivery of its own King Air. The first sale was made in September to Atlantic Aviation, which took aircraft No.6 (N770K) to use as a sales demonstrator. By the time of the November convention of the NBAA in Miami, there were eight King Airs flying and another batch on the production line. The fourth aircraft was already on a European tour, having flown non-stop from Gander to Paris on September 4, completing the flight in 9:50 hours. P&WC was optimistic about the tour, but didn't expect nearly the welcome it received— 27 King Airs were sold in three weeks to such customers as Volkswagen, Daimler-Benz and the Aga Khan.

P&WC in Wichita

With five years of work in experimental, Bill Vertilneck was a leading PT6 expert. When production began, he returned to the shop to train men in overhaul and assembly of engines. He was soon known for the way he could squeeze a few extra

Men from the PT6 shop gather around the first production PT6 being shipped to Beech on December 22, 1963. (P&WC Archives)

Inspection certificate for the first Beech engine.
(P&WC Archives)

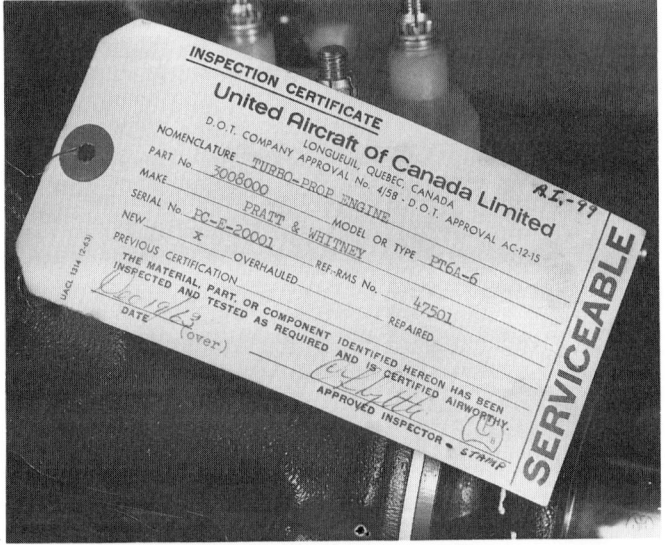

King Air No. 1 (N5690K) flies formation with the second prototype in May 1964. (Beech Aircraft)

(Below) One of the many challenges with the PT6 involved cold-start ignition. An ordinary sparkplug would not ignite kerosene. No suitable hot-surface device was available. John Saintsbury of P&WC experimented with the ordinary automobile cigarette lighter and found that its glow coil worked. Resistor tubes to regulate the coil, yet permit fast heat-up, were also needed and Saintsbury devised a practical means of combining these with the coil. Custom coils were made up and supplied to Champion, which produced the igniter plug shown here. It was mass-produced and is still in daily use. (P&WC Archives)

horsepower from a PT6 by making meticulous adjustments in turbine blade clearances. When a damaged PT6 needed some special care in Wichita, Vertilneck was called down: "I was free, so I went to Beech for a few days. I left two years later! P&WC put me to work checking engines arriving from Longueuil." Vertilneck helped train dozens of Beech employees. When a problem arose with engine icing, besides mods done by Beech to the nacelle and inlet, P&WC had work to do on the inlet guide vanes. It was assumed that this would be done at Longueuil and take $1^{1}/_{2}$ years. "That didn't make sense to me," said Vertilneck. "What did an engine know if it was being repaired in Wichita or Longueuil? I managed to convince the company to do the work in Wichita." He took five men from P&WC and a few from Hartford and showed them how to dismantle, modify and reassemble a PT6. Classes were in a local motel, and the modification line was set up in a disused air force hangar. Every King Air, except those in Europe, was ferried in for the mods. The program set up by Vertilneck proved effective and economic. In four months over 400 engines were modified, sometimes as many as four a day. Besides his main duties, Vertilneck did all the servicing on the PT6s in Mrs. Beech's VIP King Air.

While the King Air was the talk of the town, Beech had another PT6 project. In 1966, the AiResearch-powered Turbo Commander, Twin Otter and NU-8F were finalists in a US Army fly-off. The contest was narrowed to the latter two, and in the end Beech won based on price. In October 1966 it received a contract for 48 aircraft, now designated U-21, and powered by the PT6A-27. The first flew in March 1967 and over 200 were eventually delivered. Many remain in Army and National Guard service as general purpose transports.

Another product introduced by Beech was the *pressurized* Queen Air, the Model 88, filling the market between the unpressurized Queen Air and the King Air. Only 45 were sold in three years and in mid-production Beech realized it would be wiser to convert 88s on the line to King Airs. Beech made basic improvements in such areas as avionics, pressurization and interior styling for the King Air, but also looked to greater speed, range and payload. The King Air started with the 500-shp -6 engine, but the 550-shp -20 soon came along and with it Beech added propeller reversing (Model A90). With the C90 there was power enough to pressurize the cabin with bleed air and the separate Roots blower was dropped. The E90 was a C90 with 750-shp -34s. The 6-10 seat Model 100 (PT6A-28s) was stretched to accommodate 13 passengers. So it went, as a whole family of King Airs evolved. Throughout the process, P&WC and Beech had a friendly and informal relationship. At first it was P&WC that would approach Beech with a new version of the PT6, but then Beech began going to P&WC with particular power requirements. All the time, each company was moving from one proven stage to the next, an approach that kept costs in line and the King Air at a reasonable price, and customers became used to moving up whenever a new version came on the market.

Icing Problems

Beginning with tests at the NRC in 1961, P&WC studied the age-old problem of engine icing and how to deal with it in the PT6. The Beech 18 test bed was used to seek out icing conditions to see how these would affect engine performance. The FAA had recommended tests in the 6000-10,000-foot range where the worst icing seemed to occur. To keep ice clear of the engine inlet screen, an alcohol spray system was devised. An electrical system was tried but abandoned as it drew too much power. Further tests were made with a King Air seeking natural icing and doing trials behind a Caribou spraying water over it. Every indication was that a serviceable anti-icing system had been developed which met FAA expectations.

King Air No. 3 was delivered to P&WC in July 1964. Mrs. Beech is shown handing over the paper work to chief pilot John MacNeil. Looking on are Dwight C. Hornberger of Beech, and P&WC pilot Bob Hunter. The special registration, CF-UAC, represented the United Aircraft of Canada years at Longueuil. Another P&WC King Air was CF-PAW (Beech Aircraft)

Three key men on the PT6 at a social gathering in later PT6 years: Gordon Hardy from the design team, Ken Sullivan, who spearheaded the sales program, and Bill Vertilneck from servicing. Bill was known as "Mr. PT6" because of his knowledge and love of the PT6. (P&WC Archives)

(Top) A standard Beech L-23F of the US Army, many of which are still in service. (Gary Vincent)

(Above) One of the pressurized Model 88s with piston engines. Beech soon realized that this model was redundant once the PT6 won customer approval. (Sheldon Benner)

The seventh King Air and one of the earliest European examples. Originally owned in Leichtenstein, it is seen at Zurich in 1966. Over the years it was resold in West Germany, the US and Canada where, in 1988, it was C-GJBK. (Tim Martin)

The King Air had been in service for some months when the first of several engine problems was reported. Icing was suspected, but it was felt that the cause was improper use of the alcohol system. P&WC and Beech launched an investigation with Jack Beauregard in charge. He and Tom Gillespie (King Air program manager) attended the 1965 NBAA convention to publicly discuss the problem with operators and assure them that every effort was being made to solve it.

It became clear that the worst icing was occurring closer to 20,000 feet. A specially instrumented King Air with cameras in its nacelles ranged across the continent, searching for icing then deliberately flying into it. The cameras recorded icicles building up inside the inlet screen or on the inlet guide vanes, then flying into the engine, where damage was done to the first stage compressor. This research led to deletion of the vanes and to more rugged first stage blades. These "ice choppers" could handle any chunks entering the engine.

In the end, the answer to the icing problem was installation of an inertial particle separator. Air entering the compressor had to make a sharp turn around a flap. Such particles as ice or dirt could not make the turn and were redirected outside. A modification kit was made available for the nearly 200 King Airs in service in 1966, and new aircraft were modified on the line. The inertial air separator, patented by P&WC in 1965, has been installed ever since in the PT6.

Olive Ann Beech

When Walter Beech passed away, his wife, Olive Ann, became company president. J.C. "Jack" Charleson of Ottawa was a great friend of the Beeches and used to relate how, when the senior people in the company were having second thoughts about proceeding with the King Air, Mrs. Beech stepped in to order the program to go ahead. Tom Gillespie also remembers Mrs. Beech: "For her personal use, she preferred the Model 18, designed by her husband. She approved of the King Air, but that didn't mean she had to fly in it. I took her up for her first King Air ride and that changed her mind. She took the next aircraft off the line. Thereafter, Mrs. Beech would use various King Airs, and customers used to request these when buying their own. Mrs. Beech's King Air demonstrators became sought-after items! She started a tradition that pilots coming through the company training school would have

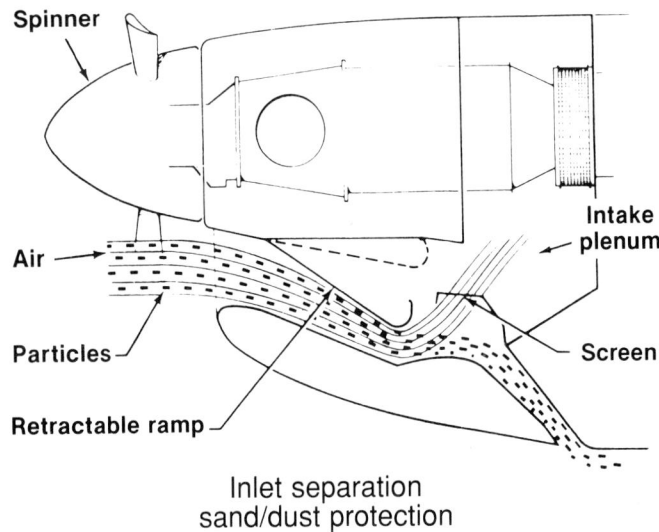

Inlet separation
sand/dust protection

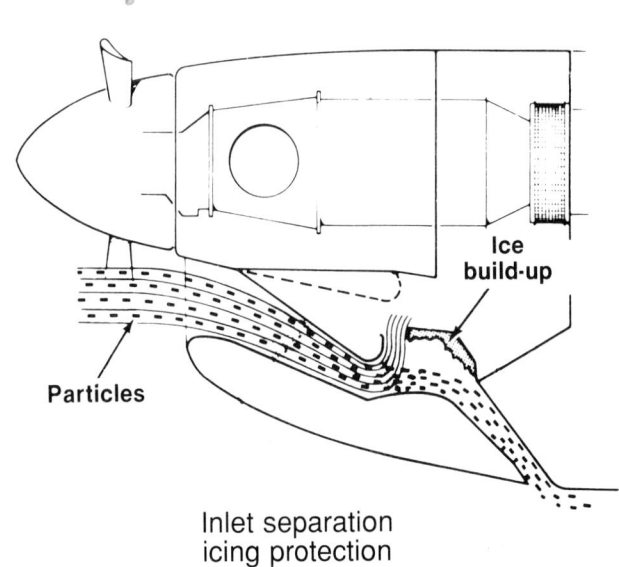

Inlet separation
icing protection

The solution to engine icing and other forms of damage in the PT6 was an inertial particle separator. It allowed ice to build up on a screen, break off and be dumped overboard. Foreign matter simply passed right by the intake plenum. (P&WC)

Guests at one of Mrs. Beech's luncheons in Wichita, April 14, 1966. Seated with her in front are Thor Stephenson, and Basil Petersen and Robert Hill (pilots for Northwest Refining). Standing are Dick McLachlan (P&WC), Tom Gillespie (Beech), Gene Schweitzer (P&WC), Leroy Bowery (Beech), Vic Emery (P&WC), Frank E. Hedrick (Beech) and Ken Sullivan (P&WC). (Beech Aircraft)

lunch with her on a Thursday. They would sit down at a special table with inlaid gold outlines of all the Beech aircraft, and Mrs. Beech would tell about each type. After lunch there would be a group photograph taken and the pilots would receive some souvenirs. If there was ever a successful marketing program, it was having all those pilots in for lunch with Mrs. Beech!"

Beech Turbo Trainer

In the 1950s, the Beech T-34A with a Continental 225-hp piston engine became the US Navy's basic trainer. But many pilots were going to fly turbine aircraft, leading to a project to make a turboprop T-34. Recalls P&WC's Dave Leslie: "The project became the closest thing to an engine competition we had met with Beech. It came just when we were having some reliability problems with the bushings in the PT6 gearbox. If they were sensitive in normal operations, Garrett wondered how we could make our engine reliable flying inverted in the T-34? The gearbox oil would mix with air and lose its lubrication properties. Our engineering people solved the problem of inverted flight, and Beech was convinced that the PT6 was the right engine for the T-34. We developed the PT6A-25 (derived from the -28 and derated from 715 shp to 400 shp) and the test program (first flight September 21, 1973) was so successful that an initial order for 116 aircraft was received from the Navy." First deliveries were made in 1977, and hundreds of T-34Cs have since been built for nations as widespread as Ecuador, Gabon and Indonesia.

The COIN Program

In the early 1960s the US military needed a light STOL observation and attack aircraft for counter-insurgency operations (COIN). The USAF sought to fulfil this role with conversions of aircraft such as the Cessna T-37, modified in 1963 into the AT-37. The Marines specified a new design with two small turboprops. The required aircraft would have a top speed of 275 mph and a single-engine ceiling of over 10,000 feet, be at home on a rough field, carry a sizeable weapons load and be easy to maintain. The wingspan would be limited by the need to operate from roadways.

A line-up of Beech T-34Cs at NAS Whiting Field, Florida, where the Navy trains all its new pilots. The 200 PT6-powered trainers at Whiting log a large percentage of USN annual flying hours. (Larry Milberry)

The PT6, as the military T74, would be ideal for the program and P&WC worked with aircraft manufacturers in 1962–1963 to assure a favourable place for the T74. When a formal invitation was issued in December 1963 to prepare proposals, P&WC despatched its test-bed Beech on the US industry tour. There were many contenders, including Convair and North American, most favouring the PT6, but in August 1964 the North American OV-10 Bronco (with two Garrett AiResearch engines) was picked. To date, AiResearch had provided its engines as auxiliary power units (APUs) in airliners, but they had not yet *powered* an airplane. Its proposed model TPE331 had only begun flight tests in 1961. Meanwhile, it was reported that while the Navy had favoured the North American entry, the Army and Air Force preferred the Convair Charger with PT6s. Normally, the Convair would have been out in the cold along with the other losers, but it had been pushing to come out with an airplane superior to North American's. Now, because of a fuss raised outside the Navy, Washington agreed to fund Convair's flight test program.

On the scene at Convair was Bill Andersen of P&WA:

The Convair Model 48 Charger flying with PT6s. Convair's last original design, the Charger lost the lucrative COIN competition to the Garrett-powered OV-10A. (General Dynamics)

"The Charger was being done in an old seaplane hangar on the far side of San Diego airport. The code name was LWT which stood for light weight transport, but the Convair guys called it 'last wild try.' Inside the hangar, on one side the team was designing the Charger. On the other side they were cutting metal and assembling the prototype. The engineers would gather around a drafting board and debate some issue. The program manager stayed out of the way in a corner office until the noise (or profanity) level reached a certain point. Then he'd come out, make a decision and it was done. There were no memos, committees or long inter-office communications. It was an efficient way of doing things. They went from an idea to an airplane in just six months."

The stubby-winged Charger derived much of its lift from propeller wash over the wing. Convair was enthusiastic about the PT6, which by now was in production and certified. The Charger was rolled out in late September 1964 and flew in November. It flew for over 400 hours but still lost to the Bronco. Convair, however, pushed on and the Charger was evaluated by the all-service flight test pilots. Finally it was lost in a crash. Convair's "last wild try" was just that—it never built another airplane.

P&WC played a small role in development of the OV-10 as the seventh aircraft had PT6s. It first flew on October 7, 1966, and there were test flights at North American's Columbus, Ohio, plant, then evaluation at Eglin AFB, Florida. These were successful, but it proved impossible for P&WC to dislodge Garrett from its firm position. Nearly 400 TPE331-powered OV-10s were eventually built for all four US services.

Engines for the Twin Otter

With its entry onto the STOL market, de Havilland Canada had explored other prospects, one being a twin-engine Otter with R-1340s, but each time the engineers worked out the numbers the results showed an airplane that could do no more work than a straight Otter. There was simply no suitable engine to make for a viable light twin STOL. Russ Bannock had taken over military sales at DHC soon after the Beaver appeared and made the first Beaver sale to the US Army. On a 1963 sales tour he visited Vietnam, where a large number of Beavers and Otters was operating. Company commanders told Bannock they would like a rugged Otter replacement with two small turbines and a tricycle landing gear. They were all-too-familiar with being shot at in their "single" Otters and longed for that extra engine for safety. A tricycle gear would make landings in a crosswind easier.

Bannock proposed the concept to his management committee. Phil Garratt had Dick Hiscocks put some performance data together and within a few days gave the go-ahead for five development aircraft. What made all the difference now was the availability of a compact, lightweight engine. De Havilland had been able to get some of the earliest experience with the PT6 and realized it was a winner. The design team aimed for an aircraft in the 11,000-pound gross weight category, against 8000 for the Otter. The prototype made its first flight on May 20, 1965, and was certified the following April. The airplane was simply dubbed the Twin Otter.

(Top) The first Twin Otter in Latin America, seen in 1967 on a short-term lease to VASP of Brazil. (Harry Mochulsky)

To DHC's chagrin, no US Army orders materialized for the Twin Otter, but sales were soon made in a new market. In the US the feeder, or commuter, market was opening up, and small airlines such as Pilgrim placed orders. In Canada, bush operators, including old standbys Wardair and the OPAS, placed early orders. Quebecair was another early operator, introducing three Twin Otters in 1968 on the Lower North Shore to replace DC-3s. Harry Mochulsky, who went to Sept Îles to help inaugurate service on April 28, found that the airline had superb training for its pilots and mechanics on the Twin Otter/PT6, describing it as "possibly the best witnessed to date." The Twin Otters began with a monthly flying rate of 200-250 hours each, and the service continued successfully for a number of years. One early problem dealt with was engine corrosion/erosion caused by the region's maritime environment and its dirt airstrips.

De Havilland emphasized Twin Otter international sales. A 1967 tour was to Brazil with pilot Grant Davidson, his mechanic, Ken Tilley, and Harry Mochulsky. VASP, the national airline, conducted a 20-day trial on some of its standard routes. Although Brazilians were impressed with the Twin Otter, few ever made their way to that country. The small Brazilian company CTA Engineering was developing a light twin design of its own and was anxious to have a look at the Twin Otter and the PT6 installation. The aircraft flew into CTA's base, São José dos Campos, where a close inspection was made by Major Ozilio da Silva and his personnel. CTA developed into Embraer, manufacturer of a wide family of P&WC-powered aircraft.

Every time a Twin Otter was sold, P&WC got a call to build two more PT6s (at $30,000 each). Production rose

A Peruvian Air Force Twin Otter takes off from the upper Amazon. The combined traditions of DHC and P&WC made for a utility transport with a thousand-and-one applications in less accessible regions. (Below) A Lewis Airlines "Seajet" Twin Otter 300 at Christiansted, St. Croix, in 1985. The pilot uses reverse thrust to position the aircraft at the dock. (E.H. Schweitzer, Matt Rodina)

(Top) Another early Twin Otter found its way to rugged Afghanistan and is seen undergoing an HSI. The Series 100 and 200 Twin Otters had the PT6A-20 of 550 shp. The Series 300 went to the -27 with 25% more power. This moved gross weight from 11,500 to 12,500 pounds. (Harry Mochulsky)

(Bottom) Twin Otters on floats have become mainstays in commuter and cargo work along the BC coast. In this scene Twin Otters are docked with Otters in Vancouver, June 1983. (Robin Brass)

steadily, reaching 10 Twin Otters per month. Fully equipped, these were going out the door for (Cdn.) $285,000. Production continued until 1988 with over 800 de Havilland Canada Twin Otters delivered.

DHC soon wanted improved performance for its Twin Otter to bolster sales in hot, mountainous regions. Beech needed more power and better specific fuel consumption to give the King Air higher cruise speeds and better range at altitude. This led P&WC to develop the PT6A-27, the first growth version from the -20, and the basis for all subsequent small PT6s. The -27 was launched in 1963, with designers looking for ways to provide more power while reducing manufacturing costs. A more efficient centrifugal compressor, and a pipe diffuser to increase mass air flow were developed. Advances in material technology resulted in turbine blades that withstood higher temperatures. Single-piece, cast vane rings were found to have better aerodynamics, and be cheaper to make than individual blades. The first PT6A-27 ran in early 1965 and was certified in December 1967. With the -27, the company realized that it would be difficult to cater to utility/commuter *and* corporate operators with a single PT6. Engines would have to be customized. The -27 best suited the former; while the -28 (a -27 modified for high altitude) was developed for a series of corporate types including the King Air E90 and the Brazilian Xingu.

PT6 Entrepreneurs: Ed West

Besides appealing to manufacturers such as de Havilland, the PT6 attracted that unusual breed of entrepreneur specializing in conversions. Ed West was a 1934 aeronautical engineering graduate from Stanford University. During the war he worked for Howard Hughes and after the war formed his own company. He purchased a stock of war-surplus turbo superchargers, components driven off the exhaust system of large piston engines to supply compressed air to the carburetor. West planned to insert a combustion chamber between the compressor and turbine to produce a small, inexpensive jet engine. Fortunately, his investment in these units was modest, for they expired quickly and dramatically in the course of his experiments. He turned to more conventional conversions, starting with a number of Catalinas. When the USAF disposed of hundreds of Beech 18s in the mid-1950s, West provided about

Fred Cowley (nearest the camera) sits in Ed West's prototype Beech 18 conversion during engine run-ups with the aircraft tethered. With a single fin, this was the Westwind IV. The Westwind I had the standard twin fins but nosewheel; the II was similar but had a stretched fuselage (17 passengers), and the III was a standard Beech 18 taildragger with PT6s ("a real money maker for your cargo and mail routes" stated the sales brochure). The Westwind was as fast as a King Air and dozens were completed, many of which are still operating. (P&WC Archives)

A Westwind III at Hamilton airport. Originally built in 1945 for the USN, it was converted to PT6s in April 1975 by Tucson Aircraft. Piston and turbine Beech 18s are still important in the air freight and courier business.
(Robert Finlayson)

This Westwind I was displayed at the September 1968 business aircraft show at Hamilton, Ontario. It had been converted from a standard RCAF Expeditor.
(Jack McNulty)

200 kits to bring them up to civilian standards. He then developed a series of Beech 18 mods, including wingtip extensions to increase gross weight and a single vertical tail.

In 1959 Tony Clark went to Los Angeles to make a presentation of the PT6 to aircraft companies in the area. West was impressed by the PT6 and its projected $15,000 price tag. In the summer of 1960 he visited Longueuil to tell marketing of his idea to fit PT6s on a Beech 18. He described a 275 mph aircraft that would sell for $75,000 and take over the market. P&WC sent its mock-up to West later that year so he could study the installation requirements. In May 1963 P&WC received its first order for production PT6s: two engines for Ed West—hardly enough to launch production, but an order was an order.

In the summer of 1964 West received his engines (PT6A-6s) and with the help of Fred Cowley began installation. The R-985s in his prototype had to be removed, new engine mounts and cowlings designed, a tricycle landing gear installed, tail trim adjustments made, executive interior fitted and so on. Pilot Wayne Black and Ed West flew the aircraft on November 2, 1964, and West immediately sought approval to ferry it to Miami for the National Business Aircraft Association convention. The "Westwind" breezed into Florida just four days after its first flight. As the engines shut down, Ken Sullivan was on the ramp to welcome Ed West with a champagne toast. Also introduced at the 1964 NBAA convention was the King Air. West announced that his conversion was faster; Beech teased that he had likely doctored his airspeed indicator! West tried to interest Beech in his concept, but got little encouragement. Beech's eyes were fixed on the flashy new King Air.

The Westwind was approved for PT6A-6, -20 and -27. Being a small entrepreneur, West had his problems: "We were underfunded so had to do our own sales tours, then hurry back

This PT6 Beech conversion was done up from a kit by the BC government at Victoria under Hugh Thomas, head of the provincial air service. It was referred to as "Hughie's Homebuilt". After years of service in BC it was purchased by Keewatin Air of Winnipeg which was operating it as lately as 1988. (Tim Martin)

Another Beech PT6 conversion (the Jobmaster) was designed by Aircraft Industries of St. Jean, Quebec, with conversion work done by Dee Howard in Texas. CF-RSV flew in 1966 with PT6A-20s but no sales were made. (Chuck Wolfe)

to the shop to do another conversion. Customers would bring their Beech 18s into Long Beach and leave them with us. Depending on what they wanted done, we would deliver a Westwind for $115,000–$185,000. One of our early customers was a tall chap from Birmingham. He had a Beech Super 18 which had more headroom than a King Air. He was delighted with the Westwind we delivered him at about half the cost of a King Air. We made money with the Westwind but ran out of customers with the recession in the late sixties. One of the first sectors hit was business aviation." After doing 22 Westwinds, West sold his assets to his old friend George Hamilton, who built more than 40 for use in corporate and commuter airline services, one of them a Vermont airline run by Andy Deeds, grandson of Colonel Deeds of P&WA.

Garrett also tapped the potential of the Beech 18, and many aircraft were converted to the TPE331. Aft placement of the exhaust easily distinguishes it from the reverse-flow PT6 with its exhaust at the front. (Larry Milberry)

FLYING THE PT6 165

Ed Swearingen's initial PT6 installation in his Merlin II. (P&WC Archives)

The Metro 3A (below) was certified in 1981 with the PT6A-45R, but no production has followed. (Harry Mochulsky)

(Below) Swearingen's PT6-powered Aero Commander in 1966. This version was not produced but many Garrett-powered Turbo Commanders were. (John Rose)

Swearingen Conversion

In April 1965 a new 6-8-seat executive aircraft took to the air with PT6s. This was the Merlin 2 built by Swearingen Aircraft of San Antonio, an enterprise of Ed Swearingen, whose specialty was modifying stock aircraft. The Merlin 2 evolved from an earlier conversion in which he had taken Beech 65 Queen Air wings and mated them to a new fuselage and a Twin Bonanza undercarriage. He added a pair of 400-hp Lycoming piston engines (vs 340 hp in the Queen Air). The result was an attractive combination, but Swearingen realized that his airframe was suited to small turbines. He happened to

(Above and below left) PT6 final assembly in the mid 1960s. Once assembled, each engine was test run then stripped down, inspected and rebuilt prior to acceptance and shipping. (P&WC Archives)

(Below right) Pratt men chat with a Canadian trade representative at the 1963 Paris Air Show, where P&WC first showed the PT6 in Europe. From the left, Ken Sullivan, Fred Cowley, Dusty Miller and John Drummond. (via D.S. Miller)

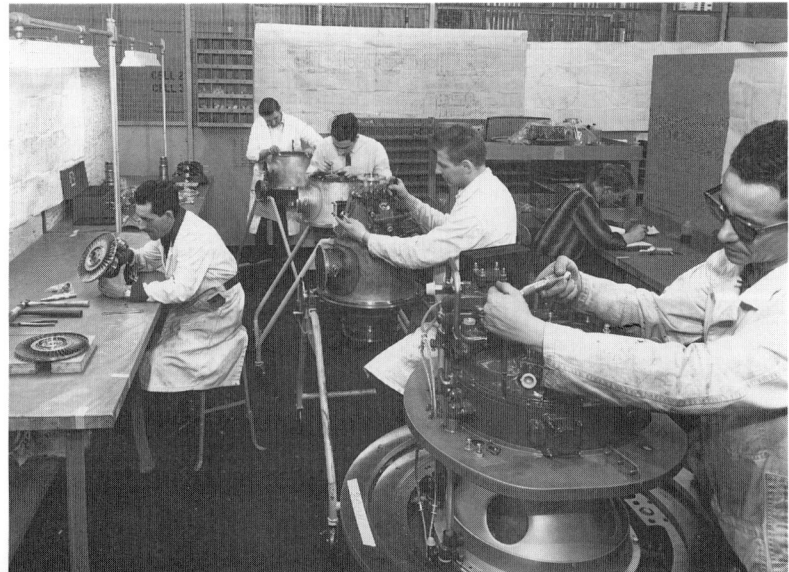

One of the famous US innovators was the late Jack Conroy (right). He used PT6s in his DC-3 trimotor. Here he looks over a PT6 with P&WC marketing man Chris Day. (P&WC)

meet Dick McLachlan, who had been visiting Dee Howard concerning the Beech 18 Jobmaster. No one at Longueuil had heard of Swearingen but within weeks he was sold on the PT6. He obtained a pair of PT6A-6s and the Merlin 2 was the result. It was certified in July 1966.

In June 1966 P&WC service rep Harry Mochulsky visited Swearingen's operation and reported that "one cannot help but be favourably impressed by the high quality, practical design and workmanship of the finished product." By this time Swearingen had also flown a PT6A-20-powered Turbo Commander. Mochulsky described the Merlin 2 as "one of the few aircraft designed with both aircrew and maintenance personnel in mind…. The fuselage is spacious and roomy, with ample cockpit room. All-round visibility could be classed as excellent, with extra large windows in the passenger compartment…. Engine installation can be considered excellent."

While the PT6 Turbo Commander did not enter production, the Merlin 2 became popular, but Swearingen, in a cash-flow squeeze, could not put money up front for engines. P&WC felt obliged to hold to standard terms offered to all its customers. Following completion of 36 Merlin 2s, Swearingen switched to the AiResearch TPE331 turbine. These were available as the maker was anxious to find applications for its new product. Swearingen was able to obtain engines on consignment. The resulting conversion was the Merlin 3, later stretched into the Merlin 4. Fairchild of San Antonio eventually took over the Swearingen line, focusing on the Merlin 4 (now known as the Metro) aimed at the commuter market. The line proved successful, and more than 400 aircraft have been completed.

The Canadian Connection

Carl Millard, founder of Millardair in Toronto, was good friends with many of the early PT6 entrepreneurs. In the early sixties he had talked to Peter Thompson and Johnny Brown of Aircraft Industries in St. Jean about a Beech 18 conversion. They were interested, and he introduced them to Dee Howard in San Antonio. Howard and Ed Swearingen had been engineers with Braniff Airlines in San Antonio, and when the war ended had gone into converting surplus military aircraft for the civil market. Howard won fame for his outstanding conversion of the Ventura bomber for corporate use (the Howard 500).

One rainy Sunday morning, Millard, Howard and Swearin-

A standard pre-war Grumman Goose with Wasp Juniors. Most surviving examples still use the piston engine. (Jack McNulty)

The first PT6 Goose conversion operated on the Alaska coast for a number of years. (via R. Losch)

gen met for breakfast in San Antonio. Millard describes the other two as "pure entertainment when they got together." Before long, Swearingen had taken his two friends out to see his latest innovation—the modified Queen Air, then just a wooden mock-up. Millard looked it over. It was an impressive effort, but he made two recommendations: replace the small windows with large "TV-screen" windows, and forget about the Lycoming pistons and go for PT6s. "Piston engines are gone as far as your type of airplane goes," he told Swearingen.

About a month later, Dee Howard came up to visit Carl Millard in Toronto. They flew down to Longueuil to talk to Thor Stephenson about their Beech 18 project. As they were chatting, the phone rang. They all got a good chuckle out of the conversation. It was Ed Swearingen on the line, giving his initial order for two PT6s to Thor! In the end, Swearingen's airplane (the Merlin II) turned out just as Carl Millard had suggested. As to the Aircraft Industries conversion, the Jobmaster, it flew successfully, but it took Howard about three years to get certification. By that time, Ed West's turbo Beech 18 had pretty well swept the market and the economy was in a tailspin. Only one other Jobmaster (CF-HOP) was ever made.

First Flights of Early PT6-Powered Aircraft

May 1961	P&WC Beech 18 Test Bed
July 1961	Hiller Ten99
March 1962	Piasecki 16H
November 1962	Lockheed XH-51A Aerogyro
April 1963	Kaman 1125
May 1963	DHC Experimental Otter
May 1963	Beech NU-8F
December 1963	DHC Turbo Beaver
January 1964	Beech King Air
April 1964	Fairchild/Pilatus Turbo Porter
June 1964	Helio Stallion
November 1964	Westwind Beech 18
November 1964	Convair Charger
December 1964	Northern Consolidated Turbo Mallard
December 1964	Potez P841
April 1965	Swearingen Merlin 2
May 1965	DHC Twin Otter
October 1965	Aircraft Industries Jobmaster
March 1966	Alaska Coastal Ellis Goose
April 1966	Pilatus Turbo Trainer
July 1966	Beech 99
July 1966	McKinnon Goose
October 1966	North American OV-10A

Strange but true—Northern Consolidated's piston/turbine Mallard at Victoria, BC. (via R. Losch)

Grumman Turbo Goose

One of the most enduring utility aircraft has been the Grumman Goose amphibian. It first flew as the Model G-21A in 1937 with two Wasp Juniors. The first Canadian examples (for the RCAF) appeared early in the war and most of these later flew in the bush with civil operators. In 1958, Angus "Mac" McKinnon, an Oregon businessman who had modified a number of smaller Grumman Widgeons, converted a Goose to four 340-hp Lycomings. Fitted with a plush interior, the Goose toured North America looking unsuccessfully for customers. Its payload was not improved, due mainly to the extra engines and a passenger cabin in the bow. The coming of the PT6, however, would launch the Goose on a new career.

Alaska Coastal Ellis Airlines was an air taxi operator serving logging and fishing villages from bases in Juneau and Ketchican with a fleet that in the mid-1960s included as many as 15 Goose. Company president Shell Simmons followed PT6 developments and had a Los Angeles firm design an installation for the Goose, believing that the new engine would enhance reliability along the rugged Alaska coast. Simmons specified a bare-bones conversion, the work being done in Victoria, BC, by Fairey Aviation. The aircraft flew in March 1966 and served for a number of years before being lost in a mishap.

Now McKinnon entered the PT6 Goose scene. He designed an all-new engine mount and introduced other mods. The engines were canted 4° outward, putting the rudder in the propeller slipstream to improve control. P&WC assisted in adapting the PT6 to the Goose, using experience gained with its *Rimfakse* motor launch in the early 1960s. PT6A-20s were installed and the aircraft was flown in July 1966 and certified in early 1967. Interest in Canada prompted McKinnon to open a shop at Victoria airport. One of his early customers was the BC government, which had a busy flight department that already included a Pacific Airmotive Beech 18 Tradewind conversion. Hugh Thomas, BC's chief air engineer, wanted a Goose with the PT6 so that the government could provide better service to coastal communities. The BC "Turbo Goose" with PT6A-20s (579 shp) was delivered in the spring of 1968. It had a new type certificate allowing a 12,500-pound all-up weight vs 8500 for a standard Goose. Empty weight was about 6500 pounds, leaving an exceptional payload. In June 1969 a Goose with the PT6A-27 (715 shp) was flown and certified. One example is CF-BCI, owned in BC by Crown Forest Products. Company pilot Keith McMann has reported: "The Turbo Goose is a great aircraft. In 1984 we upgraded ours to PT6A-28s, making it a 'G' model. With the earlier -20s we were restricted to 10,200 lb; now BCI weighs 12,500. We can

Fred Frakes' initial PT6A-27 Mallard conversion during roll-out festivities at Angwin Airport, Napa Co., California on September 5, 1969. Frakes flew the airplane in front of the large crowd, reaching 250 mph (vs 185 for Wasps). (F.C. Cowley)

fly with two crew, nine passengers, full fuel and 500 lb of baggage and still be under gross weight. We carry fuel for four hours and cruise at 180 knots compared to about 150 for a piston Goose." Nils Christensen, who worked for McKinnon in Victoria, claims about the Turbo Goose: "If you lost an engine, you could keep going and climb like a rocket. But I always thought the best thing about it was that the aircraft could take off from the water on one engine, something that was occasionally demonstrated."

In the late sixties, a used Goose sold for $25,000. PT6s were about $35,000 each. Full conversion, including new interior, avionics and retractable wingtip sponsons brought a McKinnon Goose to the $300,000 mark. At first there was a lively market, but that slumped with the economy about 1970. By the time things recovered, the Twin Otter on floats had monopolized the market. Only about 15 Goose conversions were completed, most for passenger and freight service along the BC/Alaska coasts.

Turbines for the Mallard

An early PT6 conversion involved Alaska's Northern Consolidated Airlines, which had a fleet of Grumman G-73 Mallards. NCA teamed up with Bob Lampson in Seattle to run a program at Victoria airport in 1964. The first unknown was

A Turbo Mallard of Virgin Islands Seaplane Shuttle at St. Croix in September 1987. (Matt Rodina)

what effect salt spray would have on the PT6. NCA decided to fit *one* PT6A-6 onto a Mallard. Test flying was assigned to NCA pilot John Waloka. Fred Cowley reported: "We raced up and down the runway doing taxi trials. On our final run I noticed John reach over and pull the gear up. We were flying! He then turned to me and said, 'Cowley, I've got my hands full with this 1340. Why don't you run the PT6?' We flew around for 20 minutes, each looking after the engine on his side of the aircraft. I was conned into this, but it was fun."

The Mallard was extensively tested and much useful knowledge was gained. NCA knew that the A-6 didn't have sufficient power and did not intend to use it in service. The test aircraft was later reconverted to pistons. One discovery during the program was that the PT6 could suck in great amounts of salt spray and still keep running.

In the 1940s J. Fred Frakes operated a charter airline in Alaska. He favoured Grumman amphibians and had Alaska's first Mallards. With these he could taxi onto a beach to unload cargo, but a problem with the Mallard was its small payload. Another was the tendency of the Wasp to be drowned by spray during takeoff. Frakes recalls that the Northern Consolidated PT6 Mallard trials proved "that water ingestion wouldn't kill the little devils." In 1967 Frakes met Gordon Simms of P&WC at San Francisco airport to discuss purchase of the PT6A-27. Simms was careful not to appear too enthusiastic, for Frakes' idea sounded a bit eccentric. He proposed a Mallard conversion, but the aircraft was over 12,500 pounds gross weight, placing it in the transport category. The conversion would be subject to exacting airworthiness standards. Climb and single-engine performance were vital, and P&WC had never certified the PT6 in this category. Nobody was sure that the A-27 was powerful enough to do the job.

Finally P&WC accepted Frakes' order. He hired an engineering firm to design the installation, then had his own men put together a mock-up of the nacelle. He showed this to Simms, knowing that it was not too impressive. Simms recalls: "Fred asked what I thought of the nacelle. To me it looked terrible, but I told him politely that it looked OK. At that he exploded, telling me that it was a piece of junk. That's when I realized he knew his stuff." Frakes took the engineering data he had paid good money for, stuffed it into a barrel and burned it. He realized that after all his years in aviation he was as qualified as anyone to do the work. He states: "I wanted enough horsepower to fly the Mallard under the Transport Category of Federal Aviation Regulations. The aircraft at 14,000 pounds would have to climb safely on one engine." Frakes needed a bit more power from the A-27 to meet this requirement. The intake had several vanes and a screen to protect against icing. Frakes suggested removing the screen and adding a second door to the inertial separator. He devised the appropriate mods and test flew the Mallard. He found that there was a net increase of 7% in power, enough to meet certification. His Mallard became the first aircraft in North America certified with PT6s and his two-door inertial separator became standard on other PT6 installations.

Only 59 Mallards had been built and by 1969 most were in corporate use. Frakes predicted a market for 17 conversions, but it took 20 years to see that fulfilled. He had earlier predicted that his first conversion would cost $320,000. "It took $325,000 and two years of work," he later said. His engines cost about $35,000 each, and his first machine had flown in September 1969. The earliest Turbo Mallards sold for $225,000. With the price of engines and labour today, Frakes estimates that the same conversion would be 10 times that! Many Turbo Mallards remain in use, some with Chalk's International Airlines in Miami, which serves the Caribbean. Nordair flew one in the Canadian North that was later used by

Even after the PT6 had entered production, the backbone of P&WC's financial stability for several years remained piston engine parts. Seen here is the R-2000 cylinder head and barrel assembly line. The largest engine for which parts were being made was the R-4360 which powered the huge Globemaster (left) still in use by the USAF in the stepped-up airlift for the war in Southeast Asia. P&WC sells piston parts to this day. Its last piston engine overhaul was completed in late 1980. (P&WC Archives, Larry Milberry)

FLYING THE PT6 173

Propeller overhaul also provided valuable revenue for P&WC through the early PT6 days. (P&WC Archives)

Canadair and in 1988 was based with a private owner in Edmonton. Overall, the Turbo Mallard proved itself a successful PT6 conversion. Fred Frakes gives Ken Sullivan of P&WC much of the credit and adds, "Ken was an important man for the end user. He understood the needs of little people like myself."

Frakes assisted P&WC on several later PT6 projects. Israel Aircraft Industries used an improved PT6 installation designed by Frakes for their Arava. So did Beech on a version of the King Air 100. In 1974 he began converting a 28-passenger Nord 262 airliner from Bastan engines to PT6A-45s. The new aircraft was christened Mohawk 298 and nine were sold to Allegheny. The Mohawk proved a money-maker with as few as 13 passengers aboard. The fleet was passed on to Ransome Airlines, and Fred Frakes notes: "Dawson Ransome says that the Mohawk is the aircraft that made his airline. It paid his way." Since renamed Pan Am Express, Ransome also flies such P&WC-powered aircraft as the Dash 7 and the ATR 42.

The Piston Engine Spare Parts Market

It was years before PT6 sales would rival those of piston spare parts produced by P&WC. The company foresaw the decline of this business, but through the sixties it was a vital part of operations. A 1965 P&WC market survey showed how big a market existed. From 1958 to 1964 commercial aviation had been growing at 13-14% globally and the airlines were converting from piston to turboprop and jet aircraft. P&WA piston-powered types were being replaced at a yearly average of 5.5% (11% for non-P&WA powered aircraft). Nonetheless, there was a $137 million spare parts business through 1964 for P&WA engines, of which P&WC's share was $42 million.

The R-985 was still the most widely used P&WA engine, 85% being in Beech 18s, Stearman biplanes and Beavers. There were some 3000 Beech 18s with R-985s worldwide, and they were still in production by Beech. There were over 1000 Stearmans (mostly sprayers) and 1600 Beavers, besides lesser types. The R-985 spares business was worth $7.3 million worldwide. Most R-1340s were in T-6s/Harvards, Otters and S-55s. There were 2701 Harvards, but the military was now replacing them. US military H-19s (S-55s) had turned to the R-1300, while the British were converting theirs to Gnome turbines. There was a 10% decline in S-55s each year. Most of the Otters were active with military services. These were expected to find their way into commercial markets. Agricultural conversions using the R-1340 were becoming more popular.

The survey projected good demand for R-1830 spares to at least 1970. About 95% of R-1830s were on DC-3s/C-47s, 3108 being active in 1965. While the US was phasing out C-47s, many were being transferred overseas through the Military Assistance Program. To the airlines, the DC-3 was still important because of its low per-mile cost, but turboprops were pushing it onto shorter, low-volume routes. Demand for new R-1830 parts was expected to grow as surplus stocks dried up.

Of R-2000s, 80% were on DC-4s/C-54s. Half the fleet was with the US military, and those aircraft were being pushed on account of a shortage of military transports. The C-54s were averaging 800 hours/year. Some 200 civil DC-4s were averaging 1500 hours. There were about 260 active Caribous, most busy in Vietnam and averaging 600 hours/year.

By far the greatest demand was for R-2800 spares, and demand was expected to hold until 1975. P&WC had added a few R-2800 spares in late 1961. The DC-6/C-118 was still competitive with contemporary turboprops. There were 592 in service in 1965, 141 with the US military. There were over 900 Convair Liners with R-2800s, about half with the US military. At least 290 Curtiss C-46s were flying, mostly as freighters. A variety of other types used R-2800s, as would the Canadair CL-215, which entered the picture in 1967.

The largest P&WA piston engines were the R-4360 which powered the Boeing C-97 (civil Stratocruiser) and KC-97 tanker, and the Douglas C-124 Globemaster. These were on the decline, but were being temporarily extended in the USAF. The only overhaul shops for the R-4360 were Aerodex in the US and Israel Aircraft Industries. In 1965 the R-4360 still supported a spare parts market of $33.8 million. The overhaul of one of these 3800-hp monsters took $15,000 in spares and 1000 hours of labour. As the 1960s progressed, it was obvious that there was a direct link between the world's use of P&WA piston engines and the profits that P&WC would have to direct into PT6 development and other projects that were planned. Thus did the piston engine become P&WC's ticket into the turbine world.

First flown in 1966, the Beech 99 began equipping commuter carriers in 1968 and remains active. This early example had -20s (550 shp). Later models used the -28 (680 shp), and a final version, the Commuter C99, moved up to the -34 (780 shp). (via Sheldon Benner)

A proof-of-concept Piper Navajo converted to PT6s is seen at Piper's Lockhaven, PA airport. Ed Swearingen did the engineering on Piper's behalf and the outcome was the Cheyenne series.
(via Fred Cowley)

High Times in the Industry

By 1965 manufacturers were making record sales, and 11,000 aircraft for the general aviation market were expected to be delivered, including 300 corporate turboprops and jets. Beech anticipated sales of $100 million, up 25% over 1964, with the King Air accounting for 40%. Several competitors were challenging Beech in the light turboprop field, including the Turbo Commander and Mitsubishi MU-2 with Garretts, and the Merlin II. Cessna introduced its flashy Model 411 twin piston. Likely the biggest rush, however, was with jets. The first Learjet was delivered in late 1964, closely followed by the Jet Commander. The Jetstar and Saberliner had preceded them and were still in demand. In August 1962 and May 1963 the British HS.125 and Dassault Falcon 20 had also flown. Other manufacturers were in the race.

The airline scene was booming in the sixties. Commercial jets had taken over in long haul, large turboprops in regional service. As the big airlines dropped shorter, less profitable routes, small commuter lines began appearing. These were soon looking for new equipment, as aircraft like the DC-3 became less economic on short runs with small loads. Typical of the smaller operators was NASA Commuter Airlines, which ordered three Twin Otters for service between the Manned Space Flight Centre, with its 1500-foot STOL strip, and Houston airport. Its fleet grew to a dozen Twin Otters. This type at first won most of the commuter aircraft sales, even though it had been developed for frontier operations.

Beech wanted its share of the new market and focused on a cheap, unpressurized entry. It had already sold a few Queen Airs as commuters, and its customers were so pleased that they came back asking for a larger version. This led to the Beech 99. It began as the original Queen Air 65, became the Model 80 with bigger Lycomings, then the Model 88 pressurized Queen Air. John Calhoun explains, "We then cut the fuselage and added about 80 inches. Queen Air firewall and nacelle sections were riveted onto the existing nacelles to reposition the engines forward for C of G reasons. It flew originally with piston engines but before long had PT6A-20s and an extended nose. This was a proof-of-concept prototype from which we developed the Beech 99."

The 99 was quickly underway as a 15-passenger commuter. It used an interesting production system: the fuselage jig for the Model 80 Queen Air was fitted with a screw jack. Whenever a 99 was needed the jig was lengthened appropriately. The earliest 99s on the market had 550-shp PT6A-20s and were underpowered. De-rated 650-shp -27s were installed to improve hot and high performance, and when the airframe was beefed up, the 99 used the full power of the -27s. The first 99 went to Commuter Airlines in May 1968, and 164 were delivered by late 1977 when production ceased (John Calhoun had originally estimated a market for 40). In 1980 Beech restarted production of the 99, this time using 780-shp -34s. Its decision had been prompted by deregulation in the US airline business and the great success of the EMB Bandeirante. A further 60 Beech 99s were built.

Piper Launches a Turboprop

Piper had been an enormous success building light aircraft since the 1930s, but it took P&WC years of effort to convince it to move into turbines. Dusty Miller helped with the first PT6 presentation at Piper's Lockhaven, Pennsylvania, plant in 1961. Thereafter, he visited Piper at least twice a year until a program was launched. At the presentation, Miller recalls W.T. Piper, Sr. saying: "We wish you luck developing the new PT6, but I must say that until an engine is available at a price and a fuel consumption at least matching today's piston engines, we are not interested." Piper finally entered the turbine picture by hanging two PT6A-20s on a PA-31 Navajo. First flight was in April 1967. There was some competition from AiResearch, but only until Howard Piper flew in the Metro IIB and disliked the noise level of its TPE331s. Piper finally seemed convinced of a turbine market, having seen the

The Norman NAC6 Fieldmaster sprays a crop at "zero feet." A British design, it uses the 750-shp PT6A-34AG. The engine is attached to the 520 Imperial gallon hopper, as are the wings and fuselage. (via P&WC)

(Above) Derek Emmerson, who laid much of the sales background to the PT6 Ag conversion market.
(P&WC Archives)

(Above right) A Grumman Ag-Cat converted to a PT6 by Fred Frakes for Air Rice.
(D. Emmerson)

Fred Ayres, whose conversions led to a family of PT6-powered Ag planes and a major production facility in Georgia.
(D. Emmerson)

success of the King Air, and having surpassed 500 sales of its own 6-8 seat piston Navajo. It now launched the PA-31T (with -28s, later renamed Cheyenne II), which first flew in August 1969. The following spring an order was placed for 70 PT6A-28s. Deliveries of these did not begin for over three years, as Piper waited for better economic times to return. The first Cheyenne was delivered in 1974. The PA-31T was followed by the Cheyenne I (-11s). The Cheyenne III (-41s), seating as many as 11, followed in 1977. Hundreds of PT-6-powered Cheyennes are in service.

P&WC and the "Ag" Market

Farmers had for many years been using aircraft to spray pesticides, herbicides and fertilizers on crops. At first, surplus aircraft such as the PT-17 (Stearman) were adapted. The occasional special design such as the Fletcher or the Grumman Ag-Cat (licence-built by Schweizer) came along, but in 1970 the piston engine, particularly the R-1340, was still king in the agricultural business. One day in 1974 Derek Emmerson met a customer in Longueuil who was up from the US to get an R-1340 crankshaft overhauled. He had walked into the plant with it and was anxious to get home where his "ag" airplane sat idle. To this time, P&WC hadn't devoted much effort to the ag market other than a Fletcher conversion in New

(Above) An Ayres Thrush Commander at work over fields near Salinas, California. (D. Emmerson)

One of the Sterner Aero Thrush Commanders at a Swedish base. The Sterner fleet specializes in aerial application of fertilizer over forests. This program has proven that such treatment greatly hastens forest growth. (Harry Mochulsky)

A Moroccan NEDS. The US State Department buys these sprayers for nearly $700,000 to supply to countries battling growers of such crops as marijuana and opium poppies. The two-seat NEDS has armour plating, self-sealing tanks, a quiet prop and advanced navigation gear. The second crew is primarily a spotter. (Ayres Corporation)

Ag operator Frank Livesay's special "Ag licence" on his 1978 Continental. (Frank Livesay)

A Schweizer Turbo Ag-Cat at work. (Schweizer Aircraft)

Mr. PT6, Bill Vertilneck, conducts a PT6 seminar with American Ag operators visiting Longueuil in February 1977. Bill passed away on December 3, 1988. (P&WC Archives)

Zealand, but this operator told Emmerson of the upcoming convention of the National Agricultural Aircraft Operators. In the next few days Emmerson put a PT6 display together and went down to the convention in Florida. He could hardly believe the results. Everyone wanted to talk about a PT6 conversion! One well known ag operator, Leland Snow, told him, "You people really should go after this market." P&WC realized what a valuable opportunity ag could be and began looking for its first customer.

The first step was to convince manufacturers to offer the PT6 on new production aircraft, but this proved difficult. Few could live with the cost of moving from $11,000 reconditioned R-1340s to PT6s which would increase selling prices by about $100,000. Grumman American, maker of the popular Ag-Cat, was not interested, but did suggest that P&WC do some door-knocking in the South where ag flying was integral with large-scale agriculture. This could give P&WC a better picture of what the market might accept. In May 1975 Derek Emmerson toured a number of operators and came home with a detailed understanding of the aerial spray industry. His focus was the 100-mile belt along the Gulf Coast of Texas and Louisiana where some 2.4 million acres of rice was under cultivation. The growers depended on aerial application of seed, herbicide, insecticide and fertilizer. Fields needed seven to eight applications per crop. There were some 400 aircraft dedicated to the task, all piston-powered types such as the Ag-Cat, Ag-Wagon and Pawnee. The largest had a 400-cubic-foot hopper which carried 2000 pounds.

Emmerson visited Lyon Air Service in Louisiana, which the year before had spread 20 million pounds over 176,816 acres. He spent several days following the spray operation around and talking to people at every level. As he put it, "We needed to get a feeling for the economics of ag operations, then pinpoint the benefits that might justify an operator moving to turbines. I bought myself a stopwatch and timed every phase of the application cycle." He visited Air Rice, headed by Bill Cardiff, who, along with his five brothers, had a large stake in local agriculture. Like other operators, Cardiff was concerned about problems with the R-1340. Failures were being reported, and within a few years this could leave a gap in the ability of companies like his to keep up with demand. After all, 85% of rice applications, from seeding on, was being done by air. Cardiff was delighted to hear of P&WC's idea for a PT6 ag plane, anticipating the dollar-benefits it would provide through larger payload, greater speed (i.e. faster turn-arounds), reduced pilot fatigue, and reduced maintenance and TBO expenses. He proposed that P&WC provide a PT6A-34 on loan. Fred Frakes was called in to handle the conversion.

The first Turbo-Cat was flying in the spring of 1976 and showing productivity 60% greater than piston planes. Soon Frakes was turning out conversions for customers anxious to go turbine. Grumman itself produced conversions in 1978 with -15AG and -34AG engines ("AG" signifying for agricultural flying, considerably more-rugged-than-usual operations, hence not for use on business or commuter aircraft).

Also anxious to enter the turbo scene was Fred Ayres, who had done many Ag-Cat conversions (from 220-hp Continentals to 450-hp R985s) at his Albany, Georgia, base. On the same field was Rockwell, builder of the popular Thrush ag plane with the R-1340. Rockwell was having trouble finding good Wasps and was considering switching to the R-1300, but operators, loyal to P&WA, were not pleased at this. Ayres spoke to Rockwell about doing a PT6-powered Thrush. They weren't interested, so he approached P&WC and made his own deal, hiring Serv Aero of Salinas, California, to do the conversion. His first Turbo Thrush flew in September 1975. Five kits were immediately sold to Sterner Aero of Sweden for forestry work.

Ayres faced an odd situation wherein Rockwell insisted that all Thrushes leaving its plant be fitted and certified with piston engines. The first 30 new Turbo Thrushes were retrofits

An Ayres Turbo Thrush takes on fertilizer during a quick turnaround. A Wasp-powered Ag plane is in the background. To this day, the Wasp and Wasp Junior still command the Ag market. Some R-1820s are also in use. The PT6 remains a costly luxury for most small operators. (via D. Emmerson)

of unsold R-1300-powered aircraft from Rockwell's inventory. Later, a token Wasp migrated back and forth across Albany airport to be installed on, then removed from each new Thrush for PT6 conversion. Notes Ayres: "We were doing so well with the PT6 that whenever we found an operator with enough capital to afford a conversion, we had a sure sale. We couldn't get airframes fast enough. The PT6 was right up front in our decision to buy the Thrush rights from Rockwell." Ayres Corp. took over the Albany plant in November 1977. Of 1978's production, there were 30 new aircraft with the PT6 ($184,000 each) and 12 conversion kits ($133,000 each). The Turbo Thrush evolved quickly. From overseas operators came a call for a second seat so loaders could be carried out to distant jobs and new pilots trained. Fuel capacity grew from 100 to 230 US gallons. The hopper was enlarged from 400 to 500 gallons. Weight-and-balance was corrected by filling the PT6 mount ring with lead shot. Ayres first offered the -34 engine, but in 1979 added the cheaper -11 and -15. By the late 1970s annual sales were 50 to 60 aircraft. The high times faded, however, with the recession of the early eighties, and such factors as the rising US dollar, high interest rates and farm surpluses.

In 1982, Ayres won a US State Department contract to produce a special Turbo Thrush for the drug war with two seats, armour plating, GNS (global navigation system) and power to climb out of any "hot" spot. At first, the 850-shp -41AG was used, but the 1376-shp -65R with five-blade Hartzell propeller was settled upon. These aircraft sold for $800,000 ($450,000 for the standard version with a -34AG). More than 30 anti-drug Turbo Thrushes have been built to spray illegal crops in Pakistan, Colombia, Burma, Belize and Mexico.

In 1980 Schweizer Aircraft of Elmira, NY, acquired manufacturing rights to the Grumman Ag-Cat line. It had earlier built under licence as many as 250 Ag-Cats per year. It now added the PT6, offering the -11AG, -15AG and -34AG, but found that price and long lead times for engine deliveries inhibited Turbo Ag-Cat sales. Although the occasional Turbo is made, most Ag-Cats sold today are still pistons. Schweizer's prices (US) for Ag-Cats in 1988 were: with Wasp Junior $127,500; Wasp $146,765; PT6 $373,000. Times between overhauls for the radials were 800-1000 hours, compared with 3500 hours or better for the PT6 (costs for the radials were less per overhaul). Most sales in 1988 were for export to such areas as West and North Africa, where one problem facing operators is the growing shortage of avgas.

Before it could capitalize on the benefits of lower weight, greater power and reliability, and longer TBOs, P&WC had had a number of matters to settle. Design had to consider the most rugged of operating conditions (flying off dusty, unprepared strips, constant g-loading, etc.). Engine component fatigue life had to be determined and safe operating limits set. A big advantage of the PT6 ag planes would be their ability to burn various fuel types. This was especially important with the energy crisis of the late 1970s. Cheap diesel fuel remained abundant in the farm industry, while avgas, used in the pistons, soared in price, and became scarce in parts of the world. Fuel alone drew ag operators to the PT6. Clean air was vital to operations, and in the dust- and chemical-laden environment in which the ag planes flew that meant special air cleaners to deal with 5000 cubic feet of air per minute coming through the engine. Self-cleaning vortex generator filters and replaceable barrier filters were devised for the job.

Engine Weights for Ag Planes

	PT6A-34AG	R-1340	R-1300
Basic engine (lb)	311	865	1050
Installed engine	664	1336	1659
Weight saving with PT6		672	995

Labour Troubles

The sixties and early seventies were troubled years for Quebec. Traditions were changing, in some cases being swept away. The once-supreme Catholic Church, for example, was losing its grip on its flock of over 5,000,000 Québecois. A spirit of independence was arising that included talk of a Quebec separate from Canada. There were big changes on the labour front, with unions flexing their muscles. Strikes seemed everywhere—among workers on the Baie James hydro development in the north, asbestos miners in the Eastern Townships, dock workers in Montreal and aircraft workers in Longueuil.

Beginning in the early 1950s, the United Auto Workers had been "testing the waters" among the non-unionized workers at

P&WC. The campaign grew each year until Local 510 was established in 1963 representing labour at P&WC. In August 1973 labour and management began negotiations pending expiry at year's end of an existing three-year contract. Except for a seven-week strike in 1967, there hadn't been any serious trouble between the UAW and P&WC. This year, the union was seeking some extra concessions including automatic dues payment by all hourly workers, not just those in the UAW; voluntary overtime; and an uncapped cost-of-living allowance. These were points on which the union would not budge. Not only would these be valuable concessions for Local 510, but winning them would set a precedent whereby members in Hartford could also benefit in their next negotiations. P&WC was as adamant about refusing such concessions as the union was in pressing for them. By the end of December negotiations collapsed, and a strike at Longueuil was unavoidable. George Elliot was assigned to manage the program to keep the plant running. On January 4, 1974, militant union workers disrupted activities in the plant and 21 were suspended. The plant gates were closed and those entering were screened. Three days later another group invaded the plant and harassed workers. This marked the beginning of what would become the longest strike in postwar Quebec.

Action on the picket line during the P&WC strike. Violence sometimes erupted during this difficult labour vs. management era at Pratt. (P&WC Archives)

While 2600 strikers took up their places on the picket lines, hundreds stayed at their desks and lathes. Clerical workers and engineers moved from the front office into the plant, adapted to shop routines and continued production. Hundreds of replacement workers ("scabs," to the strikers) were hired. Management was determined to keep P&WC operating. It was concerned about its hundreds of customers worldwide who were counting on a supply of engines and vital spares. In a separate move, production of the JT15 and PT6 was transferred to East Hartford and P&WC West Virginia. For the next year, all JT15 and 70% of PT6 production was carried out in the US, except for impellers and diffuser rings. By the end of 1974, 1133 engines had been delivered to customers compared with 744 in 1973. In the first half of 1975, the totals exceeded the figures for 1974, and a new production record was set. This kept aircraft rolling off production lines and provided little chance for competitors to dislodge P&WC from its market niche. However, it cost the company millions to keep up with production and it was fortunate in having a financially

A cartoonist summed up the P&WC strike with this masterpiece showing a badly bent eagle! (P&WC Archives)

sound parent in Hartford. Efficiency even improved—with engineers now working on the machine tools, many opportunities were spotted to improve work sheets and production instructions. Overall, however, the duplication of production in the US proved to be a losing proposition financially. Engine development suffered too, as all efforts were in keeping up PT6 production.

Meanwhile, the strikers were steadfast, and the more militant ones encouraged violence. By June there had been over 1000 cases reported of damage to employees' vehicles, 127 of damage to their homes, and 61 of physical assault. US activist Jane Fonda visited Quebec, telling university students: "It is inadmissible that a US multi-national corporation should try to break a Canadian union…. It is time to put this vital company under Canadian control." As the strike continued, workers began wearying, and many went back to their jobs. Negotiations continued, but the back-to-work settlement was elusive. The UAW wanted an immediate return to work. The company could not agree to this, citing responsibilities to the new employees and the need to organize the orderly return of machine tools from the US and the smooth resumption of production.

In early 1975 there was an uneasy mood among Quebec workers, especially those in the construction unions. The Cliche Commission, which looked into labour corruption, suggested placing four unions under government trusteeship until they could be cleaned up. The unions, under Louis Laberge, were furious. They felt that they had one powerful bargaining chip. Montreal was just gearing up for the 1976 Olympics, and the focus was on completion of sites for the games. The unions could pressure the government, perhaps into submission, by threatening to disrupt construction. On May 12, 10,000 workers stayed off the job at 40 construction sites. Thousands gathered for a huge demonstration, and this became the fuse that set off some of the country's worst labour violence. Some 2000 workers stormed the P&WC plant and 58 got inside. Armed with clubs, they took 10 hostages. A three-hour siege followed. Outside, police cars were overturned and set alight. Inside, over half a million dollars in damage was done to equipment. In the end, riot police stormed in and arrested 34.

The violence seemed only to make all parties more determined. The company would still not give in. The Quebec Federation of Labour called for a show of solidarity, asking its 330,000 members to stay off work. The government of Robert Bourassa described the QFL's action as "a kidnapping of Quebec's economic life." For its May 21 day of action, the QFL could muster but a third of its members. It finally backed off and ordered construction workers to return to work on the Olympic sites. Meanwhile, the government had appointed Gilles Laporte to make recommendations for a back-to-work settlement. He recommended that P&WC take all strikers back. The company was still hesitant about this, but on May 30 vice-president Elvie Smith announced: "We plan to begin rehiring as soon as we find exactly how many workers want to return…. I believe there should be a job available for each of them." Just a week before, the company had been offering to rehire just 250. The union, however, rejected any attempt to poll strikers about their willingness to return to work. It also insisted that the 34 arrested rioters be assured of their jobs.

Donald C. Lowe took over as P&WC president at the end of the strike. (P&WC Archives)

A King Air 200 lands in Switzerland. (Anton Huemann)

The strike dragged on. On June 2, the QFL warned of another round of trouble and talks again broke down.

A seven-member committee was now appointed, including prominent members of the clergy and well-known public figures like Claude Ryan, editor of *Le Devoir*. The committee would interview the remaining 976 strikers to determine a "head count." The results showed that 742 wanted to return to work. On July 23 the company offered to take back 300 right away, retain 100 more and put 250 on a waiting list. Again, the union was furious. Finally, on August 18 P&WC agreed to the government proposal. It would rehire 504 strikers within 12 weeks and the remainder by the end of February. Strike days would be counted into seniority, and an arbitrator would be appointed to determine whether the 34 rioters should be rehired. There would be no guarantee to the strikers about returning to previous jobs/pay rates. The union agreed to drop its demand for obligatory payment of dues for all workers. On August 26 the UAW voted 72% in favour of going back to work, and two days later Thor Stephenson and UAW director Robert Dean signed the papers officially ending the 19-month, 21-day strike. As so often is the case, the benefits of the strike were debatable. Now would come a period of re-establishing relations between labour and management. On the company's side, this began with the retirement of Thor Stephenson, who had been ill before the strike and whose health was worsening, and the selection of a new president, Donald C. Lowe. Lowe had a bachelor's degree from the University of Toronto and a master's in engineering from the University of Birmingham.

He began his career with General Motors of Canada and helped to build its operation at Ste-Thérèse, Quebec. In 22 years he rose through the ranks to head GM's Vauxhall subsidiary in the UK. He had nine plants and 75,000 workers but wanted to return to a North American setting. At P&WC he saw a similar "culture" to GM's—great reputation, product and human resources. Stephenson had brought the company through its demanding PT6 development years. Lowe would now guide it in becoming a fully integrated operation. Profits were still elusive—the costs of running the business were still beyond P&WC's financial base, something that Lowe was determined to change. In this he was encouraged by Harry Gray, the profit-minded head of United Technologies, whom Lowe viewed as one of the greatest US business leaders. There were some immediate problem areas he had to tackle, including the need to restore customer confidence shaken by the strike and to get the new JT15D through some rough technical times.

In May 1975, United Aircraft of Canada changed names again. The United Aircraft Corporation of Hartford had lately acquired Carrier (air conditioning) and Otis (elevators) and its name was changed to United Technologies to reflect its broadened scope. Carrier and Otis were established names in Canada and didn't need special representation, and Sikorsky had by this time repatriated P&WC's helicopter sales agency. It was decided to choose a name reflecting P&WC's primary activity and it became Pratt & Whitney Aircraft of/du Canada Ltd./Ltée. This bilingual name proved awkward in practice,

A Short 330 (PT6A-45B) taxis at Los Angeles. Hundreds of these commuter airline workhorses are in service, representing an ideal in airframe-engine combination. (John Wegg)

and in October 1982, the company adopted the simpler name Pratt & Whitney Canada Inc.

Growth of the PT6

From its beginning, the PT6 passed through countless changes. The first few installations were of -4s subject to factory inspection every 50 hours. A series of modifications, including a 13% increase in compressor air mass flow, resulted in the initial improvement—the PT6A-6, the first production model. By the late 1960s P&WC realized the market was ready for "growth" versions of the PT6. Beech wanted more power for an enlarged King Air, especially at high altitude. P&WC took advantage of available PT6T TwinPac technology, using its power section as the basis for the PT6A-40. Using the same internal dimensions, airflow and pressure ratio were increased. A second power turbine was added along with a redesigned reduction gearbox to transmit the higher power. Like the TwinPac, this engine would use directionally solidified turbine blades. The new engine for

Beech was the 800-shp -41, certified in October 1973. It had an output speed of 2000 rpm to reduce noise and vibration. As retired R&D vice-president at Beech, Leroy Clay, recalls, Beech had been aiming at a medium-sized aircraft since the early 1950s, when it jointly developed the still-born T-36 with Canadair: "The PT6A-41 allowed us to reach the goal we had had with the T-36. I was chief of preliminary design when we developed our Model 200 'Super King Air.' It evolved from two separate concepts, one from our military, the other from our civil marketing department. We brought the two together and came up with the 200."

The 200 resembled the 100 but had an added centre section and longer wing. The rear fuselage was beefed up to take the now-famous T-tail. What the military wanted from the 200 was an intelligence-gathering aircraft. This became the U-21D but opened the door for numerous military and civil versions. Dave Leslie of P&WC notes: "Beech was taking a major leap with the 200, spending far more than was its usual practice. The investment paid off and Beech found itself alone in a lucrative market." The old Beech rule had been to recoup investment with sale of the first 50 aircraft. Now it agreed to a break-even target of 200 units. The new aircraft flew in December 1972. More than 600 were sold through the 1970s, 200 in 1980 alone. To mid-1988, more than 1300 Beech 200s had been built with the -41 and -42, as well as 160 Beech 300s with the -60A.

In 1974 Beech pulled the PT6s from its prototype 200 and mounted a pair of JT15Ds over the wings. This was the FJ, or Fan Jet (unofficially, Funny Jet!), used by Beech to investigate small turbofan aircraft. The FJ could never compete with the Citation, but project engineer Jerry Hale notes, "Through it we got acquainted with turbofans and learned that installation was a lot easier when we didn't have to worry about propellers."

Short Bros. of Ireland was also looking for a new engine for its upcoming 330 commuter. The -45, certified in May 1975 (during the strike), was adapted to the 330's five-blade propellers. Aircraft like the Short 330 were the outcome of easing US airline regulations (beginning in 1969). Small scheduled carriers were recognized, and a dividing line between their aircraft and those of the big airlines was set at 12,500 lb (or 19 passengers). This opened the door for designs such as the Twin Otter and Beech. In 1972 the US relaxed

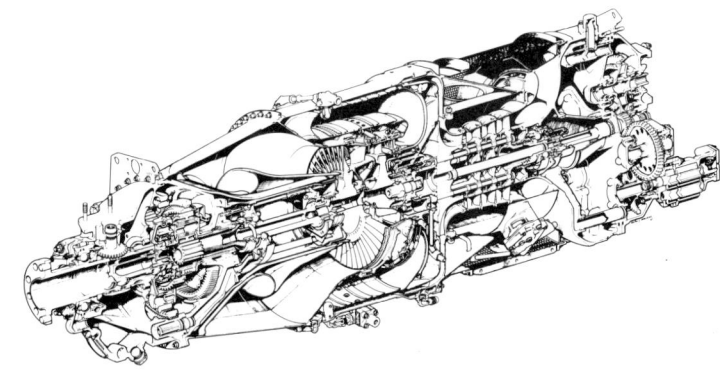

PT6A-20

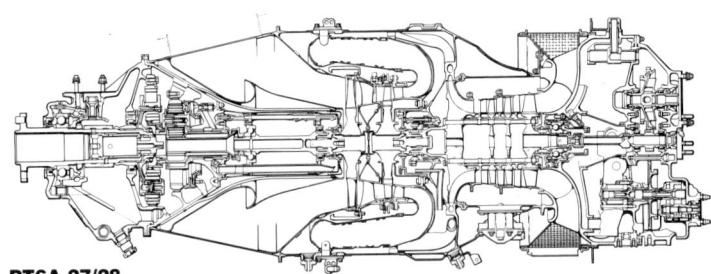

PT6A-27/28

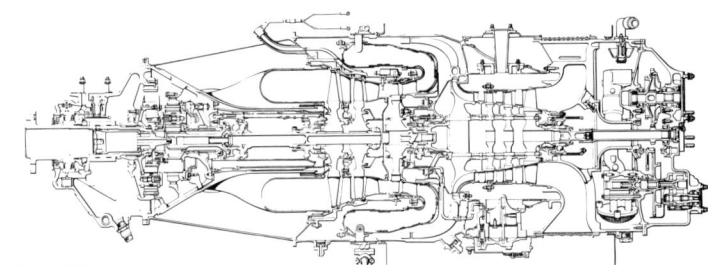

PT6A-61

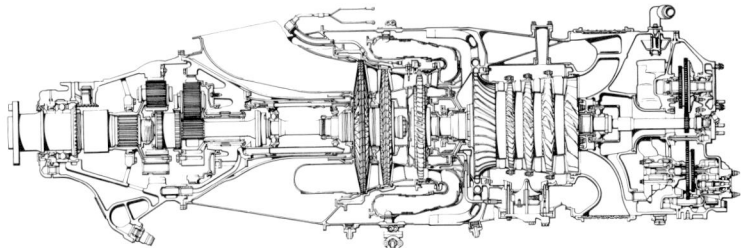

PT6A-65

Four versions of the PT6, shown in a variety of technical drawing styles. (P&WC)

De Havilland's Dash 7 demonstrator with its four PT6A-50s at Abbotsford on August 4, 1986. (Michael Macgowan)

A Short 360 gets airborne at Victoria, BC, in December 1985. The door was opened for this version of the Short series by the PT6A-65AR of 1424 shp. The latest Short 360 is the -300 with the PT6A-67R with increased maximum cruise speed of 218 knots, improved climb and hot and high performance, maximum takeoff weight of 27,100 lb, and reduced noise with its six-blade propellers.
(Michael Macgowan)

The Cox Turbo Otter at Abbotsford in the early 1980s. Several Otter turbine conversions have been proposed or tried, but none has so far caught on. (Larry Milberry)

regulations further, allowing 30-passenger aircraft for the smaller carriers. Short now brought out the 330, based on the earlier Skyvan. Time Air in Alberta became the first operator, launching service on August 24, 1976. P&WC stepped up its technical liaison to support development, test flying and certification of the larger commuters. The -45B (1198 shp) was introduced with an improved emergency rating (using water injection) to alleviate an engine-out situation on takeoff. The system, however, was heavy and complex. This led to the -45R ("R" for reserve), which did away with water injection, as it had a higher rating and a feature which automatically increased power in event of an engine failure on one side. US deregulation in 1978 removed the 30-seat rule for commuters, and aircraft like the 330 could grow further.

The PT6A-40 did not have a version for another new aircraft, the DHC Dash 7. This commuter program had begun in 1968. Besides having STOL characteristics, the Dash 7 would have to conform with a 75-db noise requirement. Part of the lift would come from prop-wash over the wing from four large, slow-turning propellers, meaning that P&WC would have to develop a reduction gearbox providing 1100 rpm. The engineering became complex and costly, and Ottawa was slow to come in with financial backing. For a time the project was close to shutting down. Derek Emmerson was the only P&WC man on it when things were especially low. Then, in July 1970, Ottawa stepped in with R&D funding, and DHC and P&WC pushed ahead. The new engine (PT6A-50) first ran in May 1973. For flight testing, the Beech 18 would be too slow and didn't have the altitude performance so an ex-Air Canada Viscount was purchased and modified to take a -50 in the nose. The first flight with a -50 on test was on May 10, 1974. In spite of the strike, work on the -50 continued and 11 prototypes were delivered to DHC to get the Dash 7 off the ground for the first time on March 27, 1975. Its new technology soon proved out and more than 100 Dash 7s were built by 1988, when production gave way to the more efficient Dash 8.

In late 1977 P&WC approved development of the PT6A-65 of about 1200 shp. This was in response to research indicating new growth in the commuter airline sector. Aircraft to carry more passengers further and faster than existing 19-seaters were foreseen. They would need power somewhere between the existing large PT6s and the PW100. The trick would be to get the extra power without making costly alterations. The answer was to use the PT6A-42 compressor with a fourth axial stage added onto the front; and introduce new airfoils. Airflow and pressure ratio increased while adding just 1.98 inches to overall length. After 23 months of development, the -65 was released to production. It ran for the first time a month later, and made its first flight in P&WC's testbed King Air 200 in mid-1980. The lead -65 customers were Short and Beech. Short needed the -65's power for its 36-passenger 360. Like the -45R in the smaller Short 330, the -65 was required by transport category certification rules to have an automatic power recovery system that would cut in should one engine fail on takeoff.

Beech wanted to get into the commuter market with a design based on the King Air 200. It went further and designed a new pressurized fuselage for its Beech 1900, which was aimed as a replacement for the Beech 99 or Twin Otter type of aircraft. This aircraft, Beech's reply to the Metroliner

Fred Frakes' Mohawk 298 commuter was derived from the Nord 262 with Bastan VICs (1080 shp). It uses the PT6A-45 (1180 shp). (P&WC Archives)

In 1987 Kelner Airways of Pickle Lake in northern Ontario became the first Canadian operator of the rugged Cessna Caravan bushplane. Here is one of its aircraft in a typical winter scene. (Robert S. Grant)

and Jetstream 31, first flew in September 1982. It was a battle keeping weight below 12,500 lb (above which the aircraft would be subject to FAR 25), but the introduction of a new regulation saved the day. It allowed operations with 19 passengers provided fuelled empty weight did not exceed 12,500 lb. Now the 1900 could take full advantage of the -65's power.

A parallel development was two new engines for higher performance business aircraft. The -60 and -61 were certified in 1982, three months after the -65. They were refinements of the -42, using new technology and gearboxes tailored to customer needs. Both gave added power at cruising altitudes. The -60 was adopted for the King Air 300. Its 1,700 rpm gearbox allowed reduced propeller speeds (and noise) while increasing performance. Piper chose the -61 of 2000 rpm in the uprated Cheyenne IIIA. The -62 with an inverted oil system (for aerobatics) was certified in February 1985 for use in the Pilatus PC-9 trainer.

In the late 1970s a number of operators, realizing that piston engines were losing favour around the world, particularly with shortages of avgas, became interested in turbine-powered light aircraft. Cessna considered modifying the Model 210 with a derated PT6. Russ Bannock at de Havilland was promoting a Beaver/Otter replacement. The Dash 7 was consuming all of DHC's energy, and when Bannock left in 1978, the utility plane project died. Elvie Smith was aware of the two concepts and encouraged Russ Meyer, chairman of Cessna, to contact Bannock and Dick Hiscocks, now retired from DHC, who had participated in all the postwar DHC designs. This resulted in Cessna solidifying its design for what is today the Model 208 Caravan. Equipped with a PT6A-114, it first flew in December 1982. The Caravan was quickly recognized as the ideal single-engine utility plane. It was especially attractive for small-package carriers such as Federal Express in the US. The improved 208B appeared in 1986 and FedEx built up its fleet to nearly 200 by 1988.

The service life of the PT6 had been steadily improving. In 1963 the average time-between-overhaul was 600 hours. By 1970 it had risen in some cases to over 5000. This was significant as the TBO was always the big factor in any operator's selection of an engine. Hourly operating costs dropped as TBOs increased. P&WC calculated an $18 per hour cost in 1963, less than $6 in 1970.

In 1974 P&WC had sales of $150 million and production of 1133 engines. By 1981 these figures were $650 million and 3319.

Changes in Production

In 1969, a little more than half of P&WC sales were licence products and the rest were its own, mainly the PT6, which by then was contributing to company profits. By 1974, on sales of $150 million, proprietary products had increased to 87%. For such a result, P&WC had had to spend years funding original research and design and reorganize for the shift from piston to turbine manufacturing. New machinery and processes had been introduced while continuing piston production and adding the Sea King helicopter. Responsible for these changes was Dick Richmond: "The three programs were interwoven. This required intricate manoeuvring since each division needed its floor space. It was quite a job managing and scheduling all the work going through the factory. We would move out one piece of piston engine equipment and drop in another for the PT6. We eventually moved the Sea King over to St. Hubert and used that space in Plant 2 for the piston lines. There was little commonality among the machines, but all divisions were drawing on many of the same skilled people.

"The only way to achieve cost control was by making a majority of the difficult parts in house. The initial production engines contained only a few parts made by ourselves and even though the actual volume would be small compared to making piston parts, the move in that direction was essential. Ross Brownridge led the initial study beginning in 1962. Later John Nicholson, Joe Coughlan, Charlie McGregor and Les Hunter also became involved determining cost-effectiveness and we expanded capacity to make the complex parts ourselves. As our market began growing with sales by the aircraft makers, we began inching up the engine price."

Parts production had originally been organized in lines, in part to make up for wartime shortages of skilled labour. Raw material entered the line at one end and progressed through many stages. To improve efficiency, in 1965 single-function tools were modified to perform more than one function. This

PT6 Engine Statistics, April 1977	
Engines delivered	11,600
Number of operators	1,700
Operating hours	26,500,000
Corporate TBOs (A-20, -27, -28)	3,500 hours
Commuter airline TBOs	8,600 hours
High time PT6A-20	18,053 hours
High time PT6A-27	16,681 hours
High time PT6A-28	8,163 hours
High time PT6A-34	5,912 hours

Total Flying Hours (Airlines), April 1977	
Twin Otter with PT6A-20	4,500,000 hours
Twin Otter with PT6A-27	2,300,000 hours
Beech 99 with PT6A-20	2,600,000 hours
Beech 99 with PT6A-27	1,000,000 hours
Beech 18 with PT6	265,000 hours
Total	10,665,000 hours

reduced the number of machines, shrank the size of the piston parts line, brought in the philosophy of a multi-skilled work force, and freed up space for PT6 tooling. Jim Ettritch was hired in 1966 to work in manufacturing: "We gradually increased our volume from 6 to 20 engines a month. We also studied what would be needed to boost production to 100 a month, but used to laugh at something that seemed so ridiculous at the time! Unlike piston engines, which had changed very little since the war, we had to plan PT6 production knowing that there would be a constant evolution in design. Even minute changes could bring improvements in performance and reliability."

While much piston engine tooling had been inherited from P&WA, many toolmakers had to be hired for the move to the PT6. The old radials were made of steel and aluminum, with heavy, rugged, easy-to-machine parts. The PT6 was much lighter, using nickel and titanium, which were tougher than steel and able to withstand higher temperatures and speeds. P&WC could only make a fraction of the 700-800 PT6 parts when production started, but expanded its range to include sheet metal parts, gears, compressors and turbine discs. New materials required changes in tools and skills. Tougher metals needed cutting tools made of tungsten carbide. Where parts were too hard to cut, they would have to be shaped by grinding. The first PT6 gas generators were made of 30 to 40 sheet metal parts and had seven levels of assembly. Skilled sheet metal men were needed, as were die-makers and welders with high-tech training.

Turbines brought a shift from line production to batch production, with machines used to do similar jobs on a variety of parts. One drawback of this new system was that no one had responsibility for the whole part. In a sense, tradesmen were no longer responsible for the end products, which were assemblies, so there was now greater emphasis on quality inspections after rather than during manufacturing, sometimes resulting in waste. Each different part required different tooling which put a premium on experienced mechanics with set-up skills. After exhausting the local supply of tradesmen, the company recruited from such sources as Orenda, the shipyards at Sorel, and overseas.

The introduction of the first numerically controlled machines (NCMs) in 1966 was a milestone. These went into Plant 2 to make PT6 gear cases. NCMs were complex yet flexible four-axis machining centres for processes like milling, drilling and tapping. They represented costly capital investment but were essential to maintain leadership in small engine development. The NCM approach was enhanced in the 1970s by introduction of CAD/CAM (computer-aided design/computer-aided manufacturing). Other innovations were the introduction of electron discharge milling (EDM) and electrical chemical milling (ECM), the former using a tool to eat through metal to make precise holes and grooves and the latter being the opposite of electroplating, with metal *removed* rather than deposited on a surface. ECM reduces the thickness of sheet metal parts beyond what is possible mechanically, which is important in making combustors. As PT6 component manufacture grew, time required to complete parts became less. In 1967 it took 135 hours to finish a gas generator, compared to 57 hours in 1975. At the same time, overall productivity grew while using fewer machine tools. In 1960, 2000 machine tools at P&WC produced 526,000 standard hours of work. By 1980, about 900 machine tools produced 1,297,000 hours.

Giving Rolls-Royce a Start

In March 1968, Lockheed and Rolls-Royce made a joint announcement—they would be a team in developing a large wide-body airliner. Initially there were design and financial setbacks for both partners, but the new airliner, the Lockheed L-1011 TriStar, was airborne by November 1970. One of its features was an auxiliary power unit (APU) to provide power for various systems while the aircraft was on the ground. Traditionally, such power had come from ground equipment, but airlines wanted greater flexibility. Hamilton Standard developed an APU based on the ST6 to provide electricity and compressed air (for example, for ground-starting the main engines, running the galleys and toilets, and keeping the TriStar air conditioned during waits on the tarmac). The APU was in the rear of the aircraft near the middle engine and its fuel and control systems, so there had to be maximum safety lest a failure in the ST6 turbine disc damage the propulsion engine. The APU was contained in a tight-fitting shield. To test this feature, an ST6 disc was scored in pie-shaped segments calculated to fail at the desired speeds, and an experiment was carried out observed by high-speed cameras.

A Lockheed L.1011 of the German operator LTU lands at Dorval in June 1988. Before an L.1011 can get off the ground, it needs the assistance of its little P&WC ST6. At right is a schematic of the L.1011 ST6L-73 installation. The ST6 is the Industrial and Marine version of the PT6. (Larry Milberry, Air Canada)

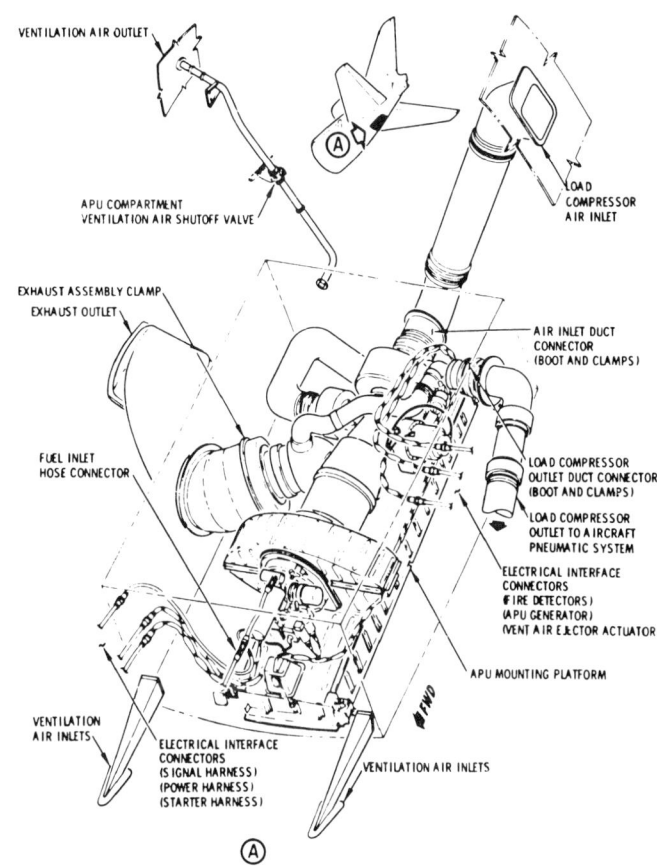

As predicted, the disc failed but no shrapnel penetrated the shield.

The APU is subject to long hours of operation. It is started long before the main engines and operates during flight and for extended periods after landing. Located at the back of the aircraft, it is subject to rough conditions, taking in such contaminants as corrosive hydraulic fluids, exhaust from the main engines, even bits of rubber from the tires, not to mention water, snow and ice. Some 355 ST6L-73s were delivered for the TriStar program and to date have accumulated over 6,000,000 operating hours. The TriStar's massive RB.211 engines—up to 50,000 pounds of thrust each in the 500 Series TriStar—can't do a thing until that tiny P&WC ST6 is first fired up!

Air Cushion Research

In the early 1970s, P&WC became involved in air cushion flight, working with Bell Textron in the US and de Havilland Canada on the ACLS (air cushion landing system), a research project to study air cushion technology in aircraft. At first, a small Lake amphibian was fitted with an inflatable air bladder under the fuselage to test the concept of operating an aircraft over land, sea or snow/ice in conditions that would be too rugged for an aircraft with a conventional landing gear. The inflated air bladder became the landing gear. For a larger test aircraft Bell fitted a DHC Buffalo with a 32 x 14-foot bladder, with air supplied by two P&WC ST6F-70s, one under each wing as part of an "air supply package" (ASP-10). Besides the engine, each ASP-10 had a two-stage axial flow fan, and an engine nacelle assembly including air intake, inertial separator, mount system, and fan discharge duct with vanes to direct fan air into the ACLS to inflate it.

Design of the ASP-10 began at P&WC in August 1971 and the first run of an ST6F-70 with fan took place in May 1972. All tests went smoothly, and two ASP-10s were forwarded to de Havilland Canada, where the Buffalo was fitted out. DHC

FLYING THE PT6 193

Views of the ACLS Buffalo showing the ASP-10 fit. The units were attached to the fuselage with the air intake inboard. The ST6s were in the bullet-shaped nose cone. In the air-to-air view the ASPs are inflating the large cushion. This program continued for seven years. No other ACLS flying programs have arisen since. (via Fred Cowley)

test pilot Bob Fowler made some initial flights, then the aircraft flew to Wright Patterson AFB in Ohio for trials with the USAF. All possible kinds of operations were carried out over a variety of terrain. It was found that in ground effect the Buffalo could operate over a range of obstacles as severe as a six-foot crater. Winter trials were flown at CFB Cold Lake. The ACLS Buffalo program concluded in the late 1970s and the aircraft was reconverted to standard configuration. There has been no production of ACLS aircraft, but the technology has had some application in the ACET vehicle.

The Bell Aerospace ACET (air cushion equipment transporter) during tests in the early 1980s at Grand Bend, Ontario. It used the ASPs from the ACLS Buffalo, once that program ended. The ACET is carrying an F-101 Voodoo (No.70308) to prove the concept of moving aircraft to runways when tarmac and taxiways have been bomb-damaged. The bullet-shaped ASPs each include an ST6. The ACET was later put to work at the USAF's aircraft disposal centre in Arizona. (Bell Aerospace Canada Textron)

Industrial and Marine

As activity with the PT6 increased in the sixties, opportunities for turbine engines outside aviation appeared boundless. The diversity of uses, along with the special engineering concepts involved, led to establishment of a separate office. On winning the order for propulsion systems for the RCN's new destroyers, this became the Industrial and Marine Division in January 1966. It was with the marine and, later, utility and gas pumping installations that "I&M" would be most successful. Dick Guthrie headed I&M and his staff included specialists in electrical, marine, mechanical, aerodynamics, hydraulics, vibration and noise. The goal was to supply complete power systems. In a ship, this would include everything from the controls on the bridge back to the propellers. The division included a quality assurance organization with tasks such as monitoring components received from outside suppliers. The worldwide network of PT6 service representatives was used to support the ST6, as the I&M version of the PT6 was designated.

The ST6 made an attractive industrial or marine engine. For marine use, it was adaptable to a variety of drive systems including propellers and water jets. It was light and compact, allowing for flexibility in installation and location within a hull, and the light weight reduced structural requirements in vessel design. Power output was smooth and free from oscillations common with reciprocating engines. This permited lower design factors for gear boxes, with savings in weight and cost. The ST6 used simple control systems. Starting was assured from -50°F to +120°F. Acceleration and

Gas flow schematic for the early ST6 and a complementary photo. The front of the engine is on the left, the screened air inlet on the right. (P&WC Archives)

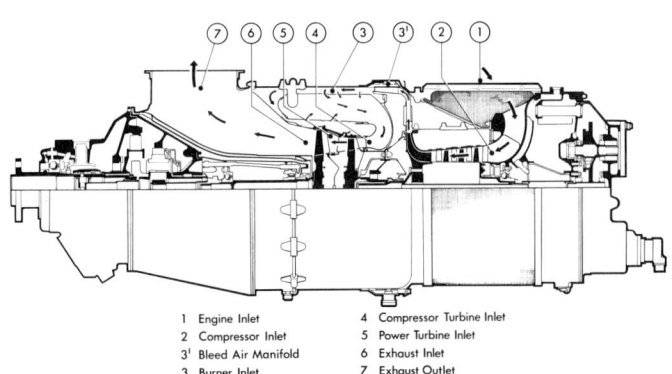

1 Engine Inlet
2 Compressor Inlet
3¹ Bleed Air Manifold
3 Burner Inlet
4 Compressor Turbine Inlet
5 Power Turbine Inlet
6 Exhaust Inlet
7 Exhaust Outlet

Rimfakse is shown with its ST6 throttled up, speeding over the waters of Chesapeake Bay. Data from the *Rimfakse* program was later applied to PT6 installations on aircraft such as the Turbo Goose and Turbo Beaver. *Rimfakse* had the first marine installation of the ST6. Heading *Rimfakse* operations was Reed Nesset, an ex-Norwegian Air Force Spitfire pilot hired on Ken Sullivan's recommendation as a sales engineer for non-aviation uses of the PT6. (P&WC Archives)

Silver Shark receives some finishing touches in the shipyard in Varraze, Italy, in 1967. The 53-foot luxury cruiser was fitted out for the Aga Khan to carry three guests and a crew of two.

Jim Wynne's ocean racing *Thunderbird* leaps through waves off Florida, its pair of ST6s at full power. (P&WC Archives)

deceleration were rapid. Kerosene or diesel fuel could be burned. For safety, the fuel control automatically prevented overheating during acceleration and limited maximum power. An intake screen prevented ingestion of foreign objects. The engine could withstand loads as high as 15g, which made it very resistant to damage from pounding in rough waters. Maintenance could be performed with simple tools and equipment; the engine was easy to split to inspect, repair or replace major components; and performance loss could be recovered by a simple compressor water-wash using a spray ring fitted to the engine.

The first marine installation was in a small Norwegian launch. Kongsberg Vapenfabrikk collaborated with P&WC in adapting the ST6 to a 34-foot launch, *Rimfakse*, to test a hydraulically actuated, controllable pitch, lightweight propulsion system. *Rimfakse* was first tested in 1962 and brought to the US and Canada. Reed Nesset was P&WC's man overseeing the project. Following 1000 hours of testing, including over 2000 engine starts and 200 crash stops, the US Navy Bureau of Ships approved the ST6 as a marine engine in October 1966. During this time, the compressor washing system and an air-assist system to improve atomization of heavy diesel fuel during starts were developed, becoming standard on marine and industrial applications of the ST6. *Rimfakse* provided several years of experience for P&WC and gained celebrity status as it cruised the waters around Expo 67 at Montreal.

The Aga Khan was looking at turbine power for a 106-foot yacht for cruising the Mediterranean. When his marine architect found that a suitable engine was not available, the Aga Khan opted for a half-scale cruiser powered by two PT6s. This was launched in Italy in 1967 and christened *Silver Shark*. P&WC oversaw engine installation and carried out performance checks. *Silver Shark* performed well, reaching 38.6 knots. With 4600 litres of fuel, it could run for 13 hours, or just under 400 nm, suitable for hitting the best resorts on "the Med," and was soon in service carrying the Aga Khan and his celebrity friends. Later in 1968, while the vessel was being ferried from Majorca to Sardinia, the two-man crew

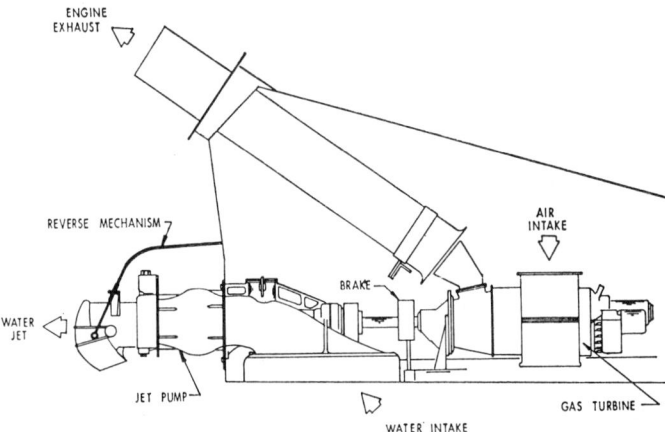

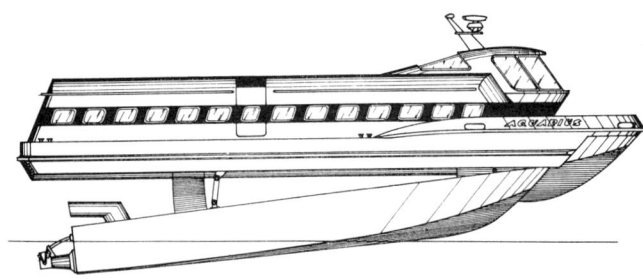

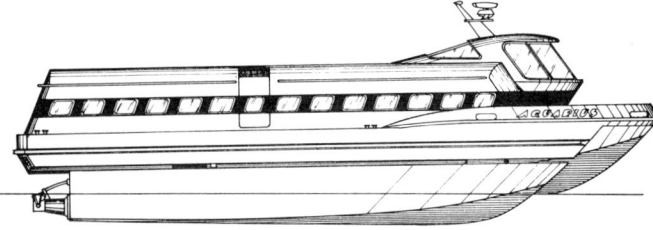

The *Aquarius* ST6 installation, and drawings of the vessel in a passenger version with skis raised and lowered.

rested overnight, letting the boat drift, and when they started up in the morning there were explosions. The crew abandoned ship, and *Silver Shark* burned and sank. It was determined that newly installed Ni-cad batteries had overheated to set off the events leading to loss of the vessel.

In 1966, ocean racing and the ST6 came together. Racing boat champion Jim Wynne designed a 32-foot boat called *Thunderbird* with two ST6s and a Chevrolet truck transmission providing the reversing gearbox to the propellers. On February 15, 1966 (the day it was launched), Wynne's boat clocked 68 mph. Four days later it travelled the 300 miles from Daytona Beach to Miami in six hours 48 minutes. On February 22, Wynne entered the Sam Griffith Memorial Race from Miami to Fort Lauderdale, across to Bimini and Cat Key, then back to the start. The day dawned with a stiff wind and rough seas. At 07:30 the race started with 31 entrants. Almost immediately, some had to drop out with damage. The heavy inboard engines in the conventionally powered boats were pounding the hulls to destruction. Few boats passed Fort Lauderdale and only *Thunderbird* and one other made Bimini.

The boats had been fitted with automatic overspeed limiters—when their screws came out of the water the engines would shut down to prevent overspeeding. *Thunderbird* had had its ST6s shut down five times as it leapt from crest to crest, but still came first, 2$\frac{1}{2}$ hours ahead of the only other finisher, and after a race of four hours 43 minutes. Because of its experimental status, *Thunderbird* did not qualify as the official winner of the race.

Several more marine ST6 projects now appeared. One was the 50-foot *Aquarius* with four ST6-70s, each with 510 hp. This was a trimaran with a central hull and two hydraulically operated "outriggers." The hull was a semi-planing feature to give an easier ride in rough water. Propulsion machinery and fuel tanks were placed in the outrigger "skis," remote from passengers. At speed, the skis would be lowered and the ST6s, driving powerful water pumps, would propel the vessel at 45 mph. *Aquarius* was built and tested in Florida. It had been aimed at the offshore oil trade to service drilling platforms, but the helicopter soon took over this market and *Aquarius* faded from the scene.

In November 1963 residents of Stamford, Connecticut, were surprised to see an amphibious vehicle driving through their streets and into the waters of Long Island Sound. This was the LARC (lighter amphibious resupply craft) designed by Consolidated Diesel and powered by a single ST6. LARC was experimental, but such projects proved valuable in the evolution of the P&WC engine. LARC was the first vehicular application of the ST6 and was used by the US Army to evaluate intake and exhaust systems which would allow operations through plunging surf without drowning the engine.

Another marine application was the 73-foot Alcan crew boat, *Nechako*, built to carry passengers and cargo for Alcan between its Kemano and Kitimat sites in northern British Columbia. It had a pair of ST6s and cruised at 20 knots.

The ST6 has been used in two programs as a driver for auxiliary power units. The first was in the hydrofoil HMCS *Bras d'Or*, which used the 390-hp ST6A-53 driving hydraulic pumps and a generator for ship's services and could be connected to the ship's propellers in an emergency to provide low-speed propulsion. The second use was in a Grumman hydrofoil for the Israeli navy. Main foil-borne power was supplied by two Allison 501s and APU power by two ST6K-77s driving hydraulic pumps and electric generators. Four K-77s were delivered to Grumman in 1980 to equip two vessels.

Hovercraft

A developing field in the 1960s was air cushion, or hovercraft, technology. ACVs (air cushion vehicles) of all sorts began appearing, from two-man units to giants plying the English Channel. ACVs could skim over land, ice or water on a cushion of air generated by engines and contained within a skirt surrounding the frame. Small ACVs often used converted automobile engines while larger ones used diesels or turbines. The first to adopt the ST6 was the British CC-7 Cushioncraft, but the only ST6-powered example to achieve limited production was developed by Bell Aerospace in Wheatfield, NY, and Grand Bend, Ontario. This was the 65-foot-long Voyageur built in 1971 and extensively tested in Canada. It first used GE LM100 engines, but from the second example ST6 TwinPacs were used. Voyageur No.1 was registered CH-BAC in Canada, and its early trials included demonstration work in the oil fields of the Mackenzie River delta, lightering trials on the Atlantic seaboard, and training the first cadre of US military ACV operators along the Florida Gulf Coast. A second

The LARC during trials with the ST6. (P&WC Archives)

Alcan's ST6-powered crew boat, *Nechako*, operated on BC's north coast in the 1960s. (P&WC Archives)

The Grumman-built Shimrit-class hydrofoil in Israeli service uses a pair of P&WC ST6s as APUs. (Grumman)

INDUSTRIAL AND MARINE 199

Voyageur was sold to Canada Coast Guard. From its base in Montreal it performed a variety of duties, including icebreaking and servicing navigation aids along the St. Lawrence as far down as Sept Îles. The Voyageur had a 22-ton payload and served from 1972 to 1988.

The real success story for the Bell ACV is the 26 units for the US Army, the last of which was delivered in 1986. Using the -76 TwinPac, these LACV-30s were developed from the Voyageur. Seventy-six feet long, they can carry a 30-ton payload at 35-40 mph, reach 62 mph with lighter loads and operate safely over obstacles as high as four feet. The Army uses its LACV-30s as lighters to unload supply ships anchored offshore. Lightering is normally done by slow-moving craft. It was the speed and manoeuvrability of the LACV-30s that sold the Army on the concept. In one exercise, two LACV-30s carried nearly twice as much cargo from ship to shore over a 24-hour period than any of the much larger conventional lightering craft used on the operation. The LACV-30s are based at Virginia Beach with the 8th and 331st Transport Companies.

The CC-7 Cushioncraft, powered by an ST6B-60 of 390 bhp. It was first run in April 1968. The 24'4"-long CC-7 carried two crew and six passengers at a top speed of 50 knots. Trials in the UK, Nigeria, James Bay, and, in this view, along the St. Lawrence, were successful, but no market opened up. (Cyril Blizzard)

In 1987 the Korean navy chose P&WC to supply two ST6T-76 lift fan drive systems to use in their lead ship of a planned class of fast ACV patrol craft. P&WC is also supplying large offset marine gearboxes, control systems and ancillary systems. The Koreans chose the 2500-hp Lycoming TF25 for propeller drive.

One special challenge with turbine-powered ACVs has been the conditions created as they fly along. They stir up clouds of dust, sand, snow or water, depending on the local geography, and much of this is ingested by the engines, restricting time between overhauls. An advanced air cleaning system was developed for the LACV-30 to provide clean, dry combustion air to the ST6T. Another problem was corrosion from salt. Careful washing and the use of special seals and coatings eventually brought this problem under control.

Power for the Navy: Bras d'Or and the DDHs

In 1963 the DDP awarded a contract to de Havilland Canada for a 240-ton anti-submarine hydrofoil for the RCN. Designated FHE-400 and later christened HMCS *Bras d'Or*, the ship would "fly" atop the water on wings or foils—a large main foil amidships and an adjustable bow foil. The hull was built by Marine Industries at Sorel, downstream from Montreal. At de Havilland, some 150 engineers and technicians were involved in design, and DHC built many components, including most of the foils.

(Above) A US Army LACV-30 speeds containers ashore at a base on the Atlantic seaboard. Another is seen crashing through heavy surf off California. (Above right) A beachhead showing a LACV-30 with 20-foot containers, offloading crane, forklift and flatbed trailers on artificial matting. A channel has been bulldozed for ease in beaching the ACV. (Textron Marine via Ron Helm)

(Right) A LACV-30 rescues a downed Beech L-23F from marshland near Langley, Virginia, demonstrating its natural versatility. (Textron International via Ron Helm)

HMCS *Bras d'Or* (top) flies on its foils above the waves off Halifax. The second photo shows a P&WA FT4A industrial/marine engine, similar to the one that drove *Bras d'Or*. The FT4A, rated at 25,000 hp for the DDH, is based on the widely used J75 aircraft engine. (P&WC Archives)

The heart of *Bras d'Or* was its main engine—a P&WA FT4A-2 rated at 22,000 horsepower. A marine version of the J75, it could lift the 151-foot ship out of the water and drive it forward at 60 knots. For normal cruising, a 1750-hp diesel engine was supplied. Auxiliary power was provided by a 390-hp ST6A-53 and an AiResearch turbine of 190 hp. Following setbacks caused mainly by a fire, *Bras d'Or* was taken to Halifax in 1968 for trials. A speed of 63 knots was recorded on one run. Even in 15-foot waves *Bras d'Or* seemed comfortable while foil-borne. Nonetheless, changing philosophies in Ottawa brought an end to this advanced naval project and it was mothballed after a brief series of tests.

In the 1950s and early 1960s Canada had a small fleet of modern steam-powered destroyers and frigates. First came the St. Laurent Class of the early 1950s, then the Restigouche and Mackenzie classes. With HMCS *Annapolis* and *Nipigon*, laid down in 1960, came the first RCN ASW ships built with helipads and hangars for Sea King helicopters. Finally came the Tribal Class destroyers, HMCS *Iroquois*, *Huron*, *Athabaskan* and *Algonquin* laid down at Sorel and Lauzon, Quebec in 1969. With these advanced "DDH280" helicopter destroyers came a revolution in technology. They were the

***Bras d'Or* in 1983. After years in mothballs on the Halifax waterfront, she was towed to the Bernier Museum near Quebec City. (CF Photo)**

West's first large naval vessels powered by gas turbines, each having a P&WA FT4A-2 of 25,000 hp.

The RCN had decided to enter the turbine age in July 1965. The choice was based on studies showing that gas turbines would give a ship faster "cold starts" (getting under way in 30 minutes vs. several hours for a steam-powered ship), increased speed and greater endurance between refuelling. They would free up space since the space required for gas turbine drives and ancillaries was 40% less than for steam turbines, boilers and ancillaries. Gas turbines required a smaller operating crew, produced less heat, vibration and noise (important in avoiding submarine detection), eliminated boiler cleaning and generally reduced operating and maintenance routines.

In November 1965 P&WC and five other prospective suppliers were asked to submit capability proposals for powering the DDHs. In January 1966, P&WC and a British firm were selected to submit formal proposals by June to supply power systems. While P&WC had no experience with large marine engines, the British bidder did and also had the advantage of lower labour costs. P&WC's proposal included two main gas turbines (FT4A-2s), two cruise gas turbines (FT12A-3s, each 3700 hp), two marine reduction gear boxes, one raft supporting the main propulsion machinery, two 150-foot propeller shafts, two controllable-pitch reversing propellers, a main lubricating system, a circulating sea water coolant system, two sets of engine intake/exhaust systems, a fuel

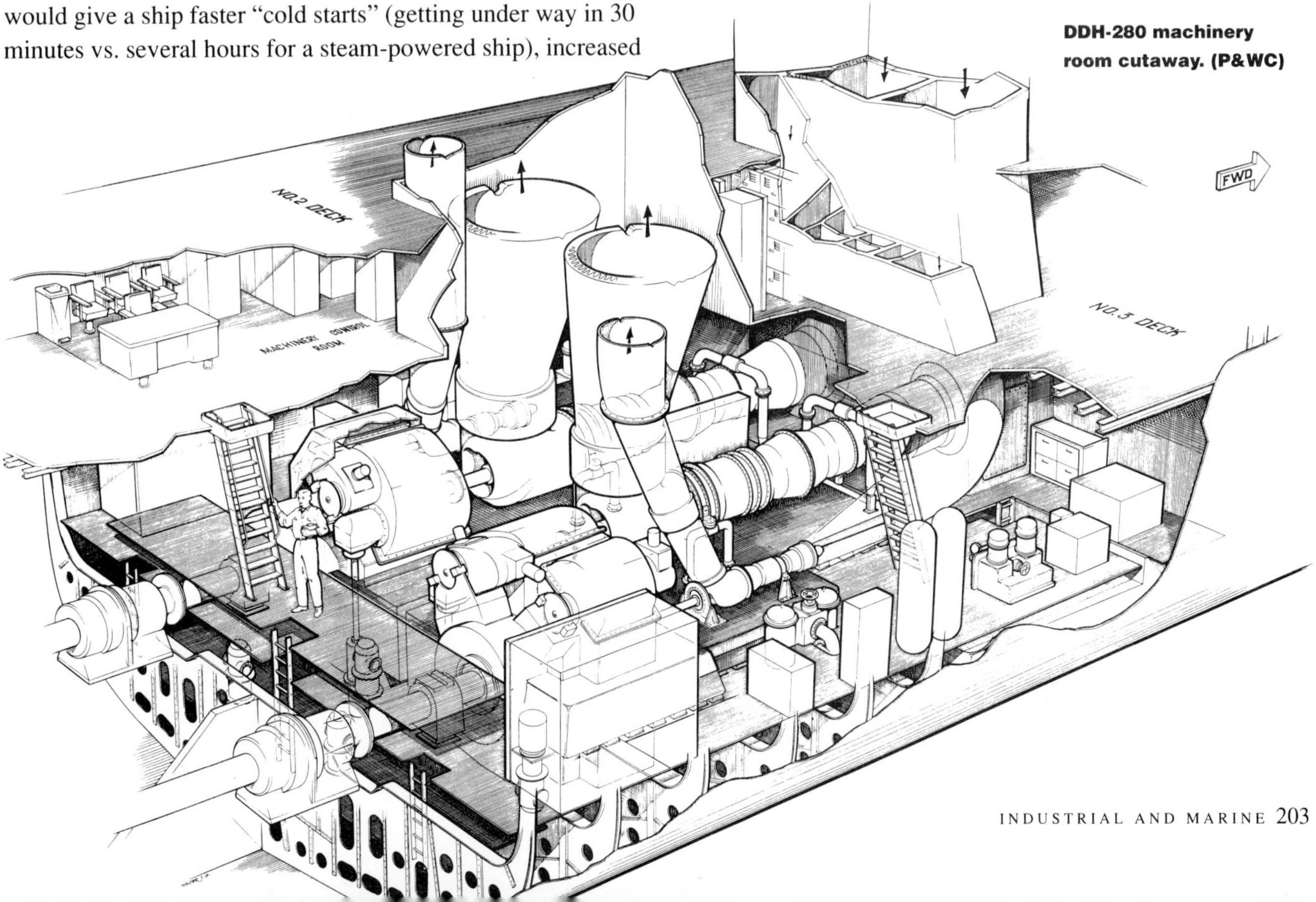

DDH-280 machinery room cutaway. (P&WC)

INDUSTRIAL AND MARINE 203

One of the DDH-280s, HMCS *Huron*, built by Marine Industries of Sorel and commissioned in December 1972. Currently being "Trumped," the DDHs were launched with two P&WA FT4A-2s and two FT12Hs. Under TRUMP, the FT4s are being overhauled, and the FT12s replaced with new Allison turbines. (CF Photo)

supply system, steering control system and other components. A fixed price for four sets of this equipment was quoted at $26.4 million. P&WC promised delivery of equipment for shore test for September 1968, and of equipment for the lead ship for January 21, 1969. There would be 28% Canadian content in the proposed package, a Canadian management program, equipment warranties and monthly progress reports.

On September 30, 1966, the I&M proposal was announced the winner. It had been judged technically superior to the British one, to have twice the Canadian content, to be cheaper, and to have a superior management plan. P&WC hired and transferred personnel to greatly increase the size of Industrial and Marine. The Crown's purchase order specified a limitation on expenditures of $35.5 million. The contract covered propulsion machinery, tools, drawings, manuals and spares to the value of $29.7 million, and allowed for shore test on a cost plus basis and an initial budget of $1.6 million. The shore tests, conducted at the USN full-scale test laboratories at Philadelphia, showed that the system was good, no major problems being encountered at the shipyards. The Crown soon

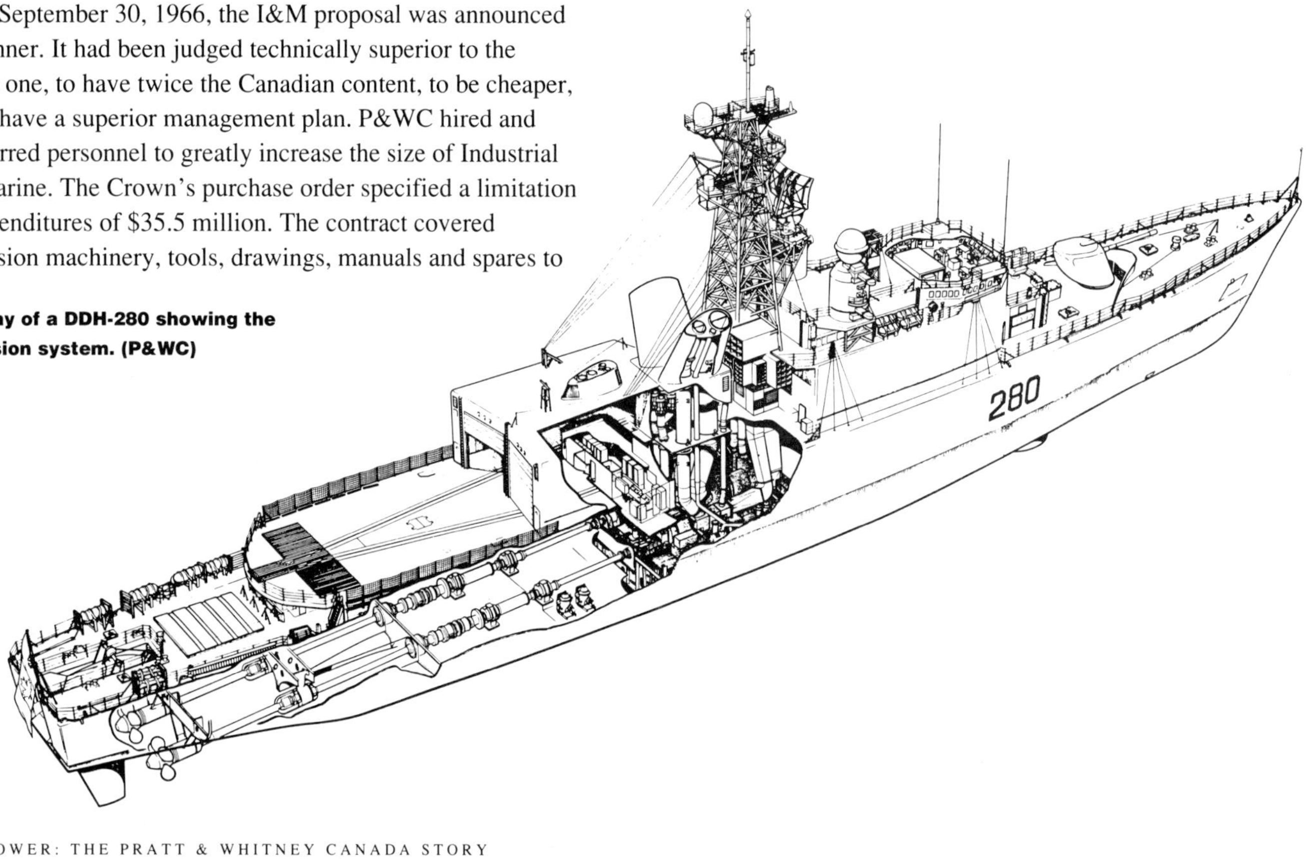

Cutaway of a DDH-280 showing the propulsion system. (P&WC)

realized that the P&WC contract was saving more money than expected, in spite of cost increases with design changes and shore tests. P&WC provided many improvements to the propulsion machinery beyond contract requirements and at no extra cost, for example improved combustion cans for the FT4 that greatly reduced smoke. While P&WC kept its costs under control, the overall budget for the DDH program soared from $192.7 million in September 1966 to $252 million in February 1971. The first DDH was due for acceptance in September 1970. Delays put this off to July 1972. The first ship commissioned was HMCS *Iroquois* built at Sorel. The last was commissioned in September 1972. Each had a double hangar to accommodate two Sea Kings and advanced ASW weapons and was defended by surface-to-air Sea Sparrow missiles.

The DDHs' FT4As each weigh seven tons and are 25 feet long. Their versatility permits a DDH to move from a dead stop to full speed in 80 seconds, or from full speed to a crash stop in 60 seconds, using reversible twin screws. DDHs have performed well for Maritime Command, all serving on the East Coast until the transfer to Victoria of HMCS *Huron* in 1987.

Ottawa announced a Tribal Class update and modernization program (TRUMP) in 1983, with Litton Systems Canada awarded the contract to define the modernization. Undertaking TRUMP marine engineering systems is I&M, and one of its tasks has been systems design for the use of space for the propulsion and electrical groups. A major change for the DDHs will be replacement of their FT12 cruise engines with 5000-hp Allison 570KFs. These fuel-efficient turbines are expected to save 52,000 barrels of fuel annually. The FT4s are being rebuilt. In preparing detailed project documentation for TRUMP, Litton spent $22.7 million. Nearly 20 years earlier, this phase had cost P&WC $1.4 million. TRUMP will ultimately cost $1.1 billion or more by the time the last ship is completed in 1991.

The contract to build six new Canadian patrol frigates (CPF) was announced in 1983. There were five contenders for this project, including P&WC (through its SCAN Marine division). The program was first discussed in 1976 and at its peak some years later involved 200 P&WC employees. After many delays, Ottawa announced a short list in July 1981—two contenders, including SCAN Marine—for the more detailed

Dick Guthrie (pointing) and a group of I&M and other P&WC staff (Leo Van Gelder, Donald Turcotte, Pat Ryan, Ed White, Chris Pascoe, George Robertson and Rod Houston) preview copies of the SCAN Marine Canadian patrol frigate proposal in December 1978. The CPF contract was later awarded to a rival consortium. (via R.H. Guthrie)

"definition" stage, and the contenders had submitted their proposals by October 1982. This stage at SCAN Marine involved spending some $25 million, which included ship design and schedules and budgets to carry out the whole program, as well as plans for support of the ships in service and ways to maximize Canadian industrial and economic benefits. In June 1983 the government announced that the CPF vessels (now called City Class) would be built by the Saint John Shipyard, with Versatile Vickers of Lauzon as major subcontractor. With this, SCAN Marine closed down. The first City Class ship was launched at Saint John in 1988 and the last is scheduled to be delivered in 1992.

The PT6 Goes to the Indy

"There's no way we can lose. Whatever the outcome of the race, the publicity is worth millions." So declared Andy Granatelli, president of STP, the maker of additives for automotive fuels and lubricants. STP had sponsored racing teams for several years at the Indianapolis 500, and the graphs Granatelli showed Thor Stephenson bore out his claims. With each race STP sales skyrocketed. Granatelli had been sponsoring the Lotus team, but for the 1967 "Indy" he proposed

Parnelli Jones in the 1967 ST6-powered Indy racing car. Had it not been for the failure of an inexpensive automotive transmission bearing near the end of the race, Jones would have been a sure winner. In spite of the speed of No.40, the crowd was not impressed—the ST6 lacked those all-important decibels of the traditional Indy cars. (via R.H. Guthrie)

ST6 installation in the Indy car. (P&WC Archives)

entering a PT6-powered car. He did not have to talk hard to convince Stephenson. The PT6 required few modifications to adapt it to racing, although the rules did require that the sponsor's products be used in the race. Happily, it was found that STP was an ideal lubricant for the PT6 fuel pump.

Driver Parnelli Jones had no trouble qualifying and looked unbeatable, at least for the first 492 miles of the 500-mile race, when a $6 transmission bearing failed and he was out of the race. Granatelli was back in 1968 with five turbine cars, only to face a major problem. The United States Automobile Club introduced a regulation limiting the size of the air intake on the turbine cars to 16 square inches. The PT6 cars needed 22. The object was to force out the turbine cars, which challenged the Indy tradition as a showcase for piston engines and tremendous noise.

Granatelli took the USAC to court, but a ruling could not be expected in time for the May 30 race, so P&WC went to work to bring the PT6 within the USAC guidelines. Gordon Hardy, project engineer for these engines, modified the four-stage compressor by eliminating the first two and reducing the diameter of the third. Now there was further trouble, as Parnelli Jones resigned, claiming the engine was underpowered and his chances of winning were remote. Even so, three turbine cars qualified, including one driven by Joe Leonard, who won the pole position. Early in the race a suspension failure claimed one of the cars, but the other two toyed with the field. Just $22\frac{1}{2}$ miles from the finish, the race was delayed. Leonard in car No. 60 was in the lead; turbo driver Art Pollard hung back from him. As the flag dropped to restart the race, both surged forward, then promptly lost power and left the race! Thor Stephenson and service engineer Fred Cowley had to exert the utmost tact to appease the frustrated drivers who saw their prize money go down the drain.

The trouble was a small extension driveshaft in each car, which failed simultaneously, probably because of misalignment of the shafts from the frequent installations and removals during the pre-race trials. The two cars were back on the racing circuit 10 days after the Indy. Now the USAC, realizing what a close call the turbine cars had been again, placed further limits on the intakes, effectively removing them from

The Nicholson Murdie "Utilizer" developed in Victoria. This wood chipping unit was powered by an ST6A-62 running the conveyor and chipper. A Cummings diesel drove a generator for electric/hydraulic systems. Weighing 40 tons, the chipper could handle spruce logs as large as 16 inches diameter and produce 12 tons of chips per hour. The self-loading arm fed logs in at the right to be debarked, chewed up, spit into the trailer on the left and trucked to the pulp mill. The Utilizer was evaluated in summer 1967 by J.D. Irving Company in New Brunswick. High cost of the ST6 led to abandonment of this project. (via Fred Cowley)

INDUSTRIAL AND MARINE

The B.C. government's turbine snow plough handling drifts in the Kootenay Pass. (P&WC Archives)

Airco of California adapted the ST6 as an auxiliary power unit. Fitted in a truck, it serviced American Airlines' early Boeing 747s. (Airco Cryogenics)

contention. Following the introduction of the PT6 to Indy racing, P&WC had many enquiries from racers but the high cost soon caused enthusiasm to wane. Andy Granatelli, however, loved the PT6 so much that he had one installed in his Corvette. With Indy cars and other non-aircraft experiments, P&WC was gaining experience with installation, operation and servicing of its engines.

Other Experiments

While P&WC engines were speeding cars around the Indy 500, a range of industrial uses was being tested for the ST6 series. At its research centre in Quebec, Domtar designed a "roadside chipper" to make woodchips from tree-length pulp wood right in the cutting area. The main advantage was in eliminating repeated handling of wood, mainly in loading, unloading and storage. The chipper had a 450-hp ST6 which provided power through a V-belt drive and also ran a generator supplying electric power for the attached barker and conveyor. Trees were fed into the barker, then the chipper, which ground them up and blew the chips into a van for transport to a pulp mill. The whole outfit was transportable on five pallets each 5 x 20 feet. Domtar tested the "jet-powered" chipper in the woods but the concept was not developed further. Accustomed to diesel or gasoline engines, crews did not readily adapt to the maintenance aspects of the turbine. This, combined with higher initial costs, led to curtailment of the venture.

Another project was an ST6-powered snowplough for the British Columbia Department of Highways, which had long been frustrated by limitations in running heavy ploughs up mountain roads. In July 1965 it replaced the 2000-pound, 250-hp diesel in one of its ploughs with a lightweight 370-shp ST6. The plough soon proved its worth on snow-clogged roads impassable to the diesel plough. One drawback was fuel consumption double that of the diesel. The ST6 plough operated in the Kootenay Pass region for several years.

Other truck conversions were for heavy Deutz and Dana road vehicles, John Deere (an earth scraper), and Komatsu and Nissan of Japan. None of these resulted in production, mainly because of economics. One venture that did lead to regular operations was the Halliburton Jet-Frac 4+4, placed in service in June 1966 in recovering crude oil. Each of the unit's four ST6A-62s drove a pump that forced slurry into underground fissures at gradually increasing pressures to force oil to the surface. A P&WC note of October 1969 stated of the Jet-Frac, "This unit has been operating without problems since hot section inspection. The unit is now operating in California and will be moved to New Mexico in the near future." Eventually prolonged use of the -62s caused them to fail and the project was abandoned. The operator felt they were not living up to expectations; the manufacturer felt the engines were being abused.

I&M entered another corner of the petroleum industry with compressors for pumping natural gas. In 1973 an FT4C-1 of 35,500 shp was installed for Alberta Gas Trunk Line, now Nova Corp. Hopes were high for pipelines from the Arctic feeding southern Canada and the US, and the energy industry expected a bonanza. Besides gas turbines for pumping stations along pipelines, Dome Petroleum was projecting massive ice breakers with FT4-type engines. With crude oil reaching $80 a barrel, such transportation costs were not a cause for concern, but the Arctic oil bubble burst and the grandiose schemes were

One of the heavy trucks experimentally fitted with the ST6, posed with a King Air. High initial cost was the main factor working against success of such conversions. (P&WC Archives)

The Halliburton JetFrac powered by four ST6A-62 industrial turbines. (P&WC Archives)

One of the stand-by power installations in Nova Scotia. Four FT4 turbines are in place. Fuel storage and power transmission lines can be seen. A detailed view of an industrial FT4 is also shown. (via R.H. Guthrie)

Hydro Quebec's La Citière generating station with a capacity of 280,000 kw uses eight FT4C-3Fs. (P&WC)

put aside. A fire in the early 1980s damaged the Nova station. I&M rebuilt it and it remains in service today.

A more successful oil industry application for turbines was adapting ST6s to gas pumping compressors. Sales of about 100 ST6Ls were made in oil and gas fields in such areas as Australia, Indonesia and the Middle East. One unit in Brunei has logged over 45,000 hours without overhaul. Running 24 hours a day, the ST6s are shut down only a few hours a year for routine oil and filter changes. STI Sulzer Systems of New York handles this specialized market.

Utility Projects

For utility projects P&WC packaged the whole installation, marketing a "turn-key" product. The first utility customer for I&M was in western Nova Scotia at Tusket, near Yarmouth. A hydro generating station had long served local power needs, but by the late sixties was no longer meeting demands. The provincial power commission was faced with expansion. It could bring in electricity with new power lines; it could construct a new oil- or coal-fired thermal generating station. These were costly prospects. A third option was to install an FT4 gas turbine plant, which would require a small site and have minimal impact on the environment. The gas turbine route was followed and the new 30-megawatt station was in production in October 1971. Its purpose was to provide "peaking" power—additional power available at times of peak usage, such as at the dinner hour or during a heat wave when air conditioners were running full blast or, in the case of Nova Scotia, in winter months with their heavy demands for light and heat. At other times, the gas turbines would be used as "synchronous condensers" to smooth out the flow of power in the regional grid system. Similar plants for peaking and standby electrical generating power were sold in New Brunswick and British Columbia. Nova Scotia followed on with six further units to fill a gap until steam thermal stations under construction came on line. Larger orders were received from Quebec and Ontario. In Quebec, La Citière was constructed at La Prairie south of Montreal. Powered by eight FT4s and commissioned in late 1978, it can provide equivalent power for a community of 200,000. La Citière was originally built for the peaking application and saved the province the cost of constructing additional power lines from James Bay or

I&M executives and board members tour one of the company's utility installations (FT-4s) in the 1970s: Elvie Smith, Ken Sullivan, Roy Crabtree, Victor Tryon, Michael L. Koerner, Marcel Bélanger, Herb Lank and A/V/M Frank R. Miller. (via R.H. Guthrie)

building a thermal generating station.

Since diesel fuel costs have risen and electrical energy conservation has been implemented, the FT4-powered stations have not been used so much for peaking as for smoothing out the flow of power in the regional grid systems. In the mid-eighties, La Citière was averaging a mere 20 hours per year for peaking and 8000 in the synchronous condensing mode. The same pattern has developed in other provinces. I&M has installed six Orenda stand-by engines at the Pickering B nuclear power station in Ontario. With the phase-out of FT4 production in the late 1970s, it teamed up with Rolls-Royce to install four industrial Olympus engines in the Ontario Hydro nuclear plant at Darlington.

The Turbo Train

The divisions of United Aircraft Corp. had traditionally functioned autonomously and successfully. In the early 1960s UAC decided to pool expertise from each division to explore new possibilities. It established its Corporate Systems Centre, one of whose focal points became America's rail system, which for years had been deteriorating. It was clear that rail service had lost the passenger carrying battle to the highways and the airways, but certain high density inter-urban routes held promise. Surely passengers on such routes as Boston-New York could be won over by fast, comfortable and reliable new trains. In 1962–63, discussions with the railroads and makers of locomotives and rolling stock led to agreement that a high-speed train with advanced propulsion (ST6) and suspension could be developed to operate profitably on existing rail infrastructures. In January 1965 the Corporate Systems Centre began design, most of the expertise being provided by Sikorsky. In 1966 the US Department of Transportation awarded a contract to UAC for two three-car trains to demonstrate that a train could travel at 160 mph on standard track, then to evaluate revenue service at speeds up to 125 mph. The first "Turbo Train" moved under its own power in Chicago (where it had been built by the Pullman Company) on May 11, 1967.

The Turbo Train was double-ended for quick turn-around. For the Canadian trains (built by the Montreal Locomotive Works and managed by P&WC's Helicopter Division) there were five intermediate cars between two powered dome cars at the ends. The MLW train was 430 feet long and could carry 302 passengers. Fully loaded it weighed 195 tons, compared with 750 tons for a standard train of equal capacity. The Turbo Train had cars of welded aluminum and a low-drag aerodynamic shape. The train's light weight meant that small gas turbines could power it. Four 455-shp P&WC ST6Bs were sufficient (six for the US trains), two each beneath the floor of the dome cars. Each engine weighed just 290 pounds and was 60 inches long and 19 in diameter. A fifth engine provided for such services as heat, electrical power and air conditioning.

INDUSTRIAL AND MARINE 211

The train had revolutionary suspension designed for a comfortable ride. The leading end of each dome car was carried on a two-axle truck, each axle driven by a shaft from its gearboxes and turbines. The use of a single-axle suspension system (one axle between two cars) meant reduced mechanical resistance. The cars were suspended by an "A" frame arrangement that provided banking on curves. Clamshell doors at each end permitted the coupling of two trains as required.

The engineer operated the train from a forward cab, a speed control handle and brake pedal being his main controls. He monitored the engines from a bank of lights. As there was a control cab at either end, turn-around was simply a matter of carrying the control handle to the opposite end of the train. All in all, the Turbo Train was something revolutionary for railroads steeped in the traditions of multiple axles and massive powerplants. They looked askance at the Turbo but also knew that they had to modernize or see passenger rail travel decline further.

While the US aimed at introducing the Turbo on the Boston-New York corridor, in Canada it was hoped to have it as a highlight of Expo 67. In May 1966 a contract was signed by UAC and Canadian National for five trains to be leased to CN and supported by P&WC and UAC. The first ST6s were delivered to CN in July 1966 for trials on a flat car. The first three P&WC men sent on course to CSC were Lucien Lavigueur, Laurier Comeau and Lynn Barlow. In time about 15 from the company were involved and many from CN. About 200 from United Technologies were temporarily assigned to Montreal. Deadlines came and went. Expo did not see the Turbo Train. It was January 1968 before the first train achieved its contract speed in the US of 170 mph. It was May before the first CN trials began, on the Montreal-Brockville sector. The summer was spent in training. On December 10 an inaugural press run was marred by a minor accident at a level crossing. The media seemed delighted at the accident and for years thereafter disparaged the Turbo Train. The "Turbo" was off to a poor start. Passenger service began on December 12, 1968, with a 3-hour 59-minute run from Montreal to Toronto (335 miles). It was hoped to build this service up and replace the popular "Rapido" diesel, but snag after snag began to plague the Turbo. After three weeks just one train was still serviceable. Insult was added to injury on January 30, 1969, when a switching engine piled into two stationary Turbos in Montreal causing extensive damage.

The designers had underestimated the severity of winters in Canada and the US Northeast. Snow clogged air conditioning and engine inlets. Water pipes were poorly insulated and froze solid (the quick fix was to wrap them in electrically heated tape). Besides cold weather problems, the electrical and control systems were overly complex and sensitive. A red light would illuminate, but it might be difficult to find out why. Sometimes the red light was a false alarm: it might have prompted an engine change but still came on when the new engine was run up. Because the train was a single long unit, a defect in a coach could not be easily solved by uncoupling and replacing it.

For several days in January and February 1970, one of the Turbos was sent to Capreol in northern Ontario for cold weather trials. Aboard were CN and P&WC/UAC personnel determined to pinpoint causes for the winter snags. After completing the trials, the train returned to Montreal where it and the others were torn down and modifications began. The trains were back in service in May 1970 following 85 modifications, but in August one was damaged by fire. Service continued until the following February, and during that period, 98% of all scheduled trips were completed. In February 1971 the trains were again pulled off the rails for further modifications. Meanwhile, there had been problems with the "Northeast Metroliner" as the American Turbo was called. Besides technical setbacks, the Metroliner was slower than expected, taking 3 hours 45 minutes to cover the Boston–New York route, only 35 minutes faster than the best run by a conventional train. Apparently few air travellers commuting along the corridor were won over. Otherwise, the Turbo faced such factors as the conservative railroad industry, a substandard infrastructure (countless level crossings, curves, etc.) and apathetic public response to passenger rail travel. In 1971 Amtrak, the US government passenger rail bureau, took over all US passenger service, and began looking elsewhere for trains. French equipment was ordered, and when further offshore purchases were made by Amtrak in 1975, UAC withdrew totally from the passenger train business.

The Turbo was back on the Montreal–Toronto run in December 1973. At last it was coming into its own and it gave

The sleek Turbo Train gathers speed as it pulls out of the Toronto yards and heads for Montreal in January 1976. (Larry Milberry)

excellent service. The previous accidents had caused a rearrangement of equipment—there were three nine-car trains. P&WC's representative at Toronto's Union Station was Alf Owen. He had been on the Sea King program earlier, but joined the Turbo team and moved his family to Oakville, near Toronto, in 1973: "We worked long days of 10-12 hours. Most days everything ran smoothly, and we got the morning train to Montreal away on schedule. I can't recall ever hearing any CN head-end crew complain about the Turbo. They all enjoyed it. Of course, as the crews were all paid by the mile, they especially appreciated the extra speed. The Turbo was fast once it got rolling, and on one stretch ran at 105 mph. One day we did a speed trial out of Toronto and recorded 140 mph. However, the Turbo didn't have the traction of a heavy diesel and was therefore slower to reach cruising speed. One day the Turbo was towed in. Its engine had quit before even reaching Kingston. CN began looking for the trouble. After several hours, I convinced them to shunt the train out of the station to where we could try to start up. I suspected something very basic. Sure enough, when the tanks were checked, they were bone dry. The train had left Montreal on partial tanks!" In 1976 Owen returned to Montreal. CN was by then capable of operating the trains, but always had P&WC to call on. As Owen's job had been so demanding—he was on call seven days a week—he had piled up extra holidays based on overtime and was able to take three months of well deserved time off. From December 10, 1973, to August 31, 1975, with three trains in service, CN recorded over 840,000 miles of revenue service, with only 10,000 miles lost to malfunctions. On 1384 Montreal-Toronto trips, 335,014 passengers were carried (averaging 242 passengers per trip). 615 Montreal-Ottawa trips were also made.

In 1975 *Business Week* reported that "after spending $100 million in the past nine years trying to develop trains able to whisk passengers between cities at speeds up to 300 mph, the Ford Administration is abandoning most of the program." CN purchased the remaining US trains. Supported by P&WC, these provided excellent service until the last trip was made on October 31, 1982. The year before, CN had opted for a new type of diesel train, the LRC (light, rapid and comfortable) from Bombardier in Montreal. The LRC had been in development for several years and came into service in 1981 on the Windsor-Toronto-Montreal-Quebec corridor. The sad fate of the Turbo was a trip to the scrap yard.

Many factors worked against the success of the Turbo. First came early development setbacks and accidents. Then, it was restricted to 90 mph (on average) on the old Montreal-Toronto track. Along the route there were local restrictions as in Dorion, on the outskirts of Montreal, where a 20-mph speed limit applied . The approaches to the stations in Montreal and Toronto were a confusing maze of tracks with complex switching and restricted speeds. The Turbo had to share the right-of-way with a variety of other trains, and there were over 100 level crossings on the route. There was even vandalism—rocks, bottles and other missiles were sometimes dropped from overpasses as the Turbo approached, shattering the bullet-proof windscreen. With such difficulties, the Turbo Train was not able to "show its stuff," unlike the famous Bullet Train in Japan, which ran on its own fenced right-of-way. As one person expressed it, "The Turbo Train was like

The Turbo lined up at Montreal beside a heavy diesel, and a vintage Northern steam locomotive. (CN)

an elegant thoroughbred sent to do the job of a draft horse."

To this day, the railroads are struggling to compete as passenger carriers. The steady inroads made by the airlines have tapped off more and more customers. The frequency and variety of flights on the Montreal-Toronto corridor, for example, make air travel more enticing, as do existing fares. In 1981, VIA Rail (formed in 1977) carried 8,009,000 passengers nationally, but this declined in 1987 to 5,865,000.

Specifications: UAC Turbo Train

POWER DOME CAR	
Weight	53 tons
Weight on drivers (2 axles)	34 tons
Overall length	73' 3"
Fuel capacity	1250 Imp. gallons
Fuel	Diesel
Range	1200 miles
Traction power	up to 5 ST6K771 engines
Electrical power	300 Kw at 0.8 power factor
Height above rail	12' 11"
Max. width	10' 5"

INTERMEDIATE CAR	
Weight	23 tons
Overall length	56' 10"
Height above rail	10' 11"

Overseas Inroads

Henry Potez of France started manufacturing aircraft during the First World War and his firm became an industry giant. In the late 1950s it launched a new project—a 24-seat pressurized commuter, the P840, with four Astazou turbines. The prototype flew in April 1961 but was beset by engine problems. Potez turned to the PT6 and the aircraft became the P841. In early 1964 a batch of PT6A-6s was shipped to Potez in Toulouse and soon the first order was received—from a German agent, Aero-Dienst, which had a sale in hand to the West German government. Besides all the work associated with a prototype, the requirements of Aero-Dienst's customer placed emphasis on grass field performance and the need for propeller reversing. The P841 toured extensively, and once all the bugs were out it was fitted with electronic gear. Later it was revealed that Europe's first PT6-powered aircraft was a German government spy plane, running up and down the East Bloc border gathering intelligence.

Only one other P841 was sold, to a West German department store chain for corporate use. Aside from the Grumman Gulfstream, it would be years before there was a turboprop to compare with the Potez. The P841 was ahead of its time—the aircraft was there, but not yet the commuter market that Potez had envisioned. A spinoff was the establishment at Aero-

A Potez P841 with PT6s is seen during takeoff. The 841 was the first four-engine application of the PT6. The only other one is the Dash 7. (via D.S. Miller)

Dienst of Europe's first PT6 service centre, which P&WC service rep Serge Page helped set up in 1965.

Pilatus and the Turbo Porter

The PT6 was aimed at a worldwide market, and the first aircraft it powered that flew into the remote corners of the globe was the PC-6B Turbo Porter designed by Pilatus of Stans, Switzerland. The modern-day Pilatus story begins in 1959 when the company flew a new STOL aircraft, the PC-6

One of the early PC-6Bs which went to Wien Alaska. (via J.W.R. Drummond)

Porter, with a piston Lycoming. In 1961 the Porter was adapted to an Astazou. About 50 Porters had been built by December 1963 when P&WC's Donald "Dusty" Miller arrived at Pilatus to present the 550-shp PT6A-6. Pilatus knew that North American operators would prefer a Pratt & Whitney engine, so was keen to listen.

Pilatus turned to market research. Sigurd Wein of Wein Alaska Airlines was excited about the PC-6B, and this encouraged the Swiss. In April 1964 Pilatus test pilot Guido Good took the PT6 Turbo Porter on its initial flight. Next day, Rolf Boehm ordered an oxygen bottle and checked out the cabin heating system by climbing to 30,500 feet. Suddenly everyone wanted a ride in the new airplane, and eight more flights were made that day, all full. In June, the PC-6B was taken by CL-44 to Fairchild Hiller at Hagerstown, Maryland. The US firm had distribution rights for the Porter in the Western Hemisphere and agreed to obtain a US supplemental type certificate for the PC-6B. The aircraft toured US military bases to drum up interest, and by year's end Fairchild had signed a licence agreement to manufacture PC-6Bs. It began by finishing Swiss-built airframes.

Wein Alaska placed the first order for the PC-6B, and Harry Mochulsky accompanied the first two on their delivery, leaving Fairchild on April 30, 1965, on a five-day venture. Snow and icing were encountered, but at no time was engine performance affected. Fuel consumption for both aircraft was monitored and found to be consistent. The flight gave Wein's pilots confidence in the new aircraft. At Fairbanks, Mochulsky helped ease the PT6 into service, as the aircraft began serving Alaska centres with only rudimentary airstrips. Part of their job was to feed into Wein's network, which included service to the rest of the US West Coast using aircraft as large as the Constellation. Many early adventures of PT6 service reps involved the Turbo Porter. Following the 1965 Paris Air Show, Pilatus sent a demonstrator on an Asian tour. Rolf Boehm and Émile Wick flew the aircraft. Wick would later deliver many Turbo Porters, often on the lengthy route to the East. He regarded these trips as routine, in spite of overflying the shark infested waters of the Indian Ocean. As he put it, "Uncle Pratt's vacuum cleaner never let me down." There were visits to remote locations, some in New Guinea, where the Pilatus gang met their first head hunters! In the end, the tour resulted in good sales, including to the Australian army, which took 19 aircraft.

In Canada veteran aviator Barney Rawson and ex-USAF Col. C.E. Fitzwater toured a PC-6B across the North in 1965. At one point, their propeller was damaged, and when they reached Montreal an inspection was in order. The aircraft tied up on the Back River, and Bill Vertilneck was called in to help. His wife, Grace, was an experienced P&WC worker and Bill enlisted her help. They removed the power section, lifted it onto the back seat of their car, and drove it to Longueuil, where it was disassembled, inspected and built up: "We drove back across Montreal to the Turbo Porter. Grace was putting the bolts on the C-flange when the pilot arrived on the dock. He stopped as if in shock at the sight of Grace then asked, 'Who the hell is working on my airplane!' 'My wife,' I replied, to which he answered, 'Don't you realize I have to fly across the ocean in that plane?' In the end, we had a meal and nice chat. The pilot was relieved when he heard that Grace had been working on engines for years."

Mountains and Deserts

Air Ventures of Nepal was an early user of the PC-6B, flying on relief work near the Tibet border. The company was served by P&WC rep Serge Page, who had transferred from experi-

Some of the Turbo Porter's welcoming party in New Guinea! (via D.S. Miller)

(Top) Ladi Marmol sprays Sudanese cotton fields with the Turbo Porter. This was the first Ag application of the PC6.
(Harry Mochulsky)

(Above) Field service rep Harry Mochulsky with a gang of young locals during his stay in Sudan. No matter how remote the region, P&WC has always maintained field service wherever needed.
(P&WC Archives)

mental engineering to product support to enjoy a little travelling. He recalls how on December 4, 1965, Gene Schweitzer interviewed him in the morning and he was on his way to Potez in France that night! Somehow his three-week assignment stretched to a year, much of it in Nepal. Arriving there, he found that nobody knew about turbine engines. He had to train the mechanic and fly whenever he could with the pilot. Page states: "Nepal was primitive in those years, including the transportation system. We used to land on river beds or mountainsides and bring sick people back to Katmandu for medical attention. It was a real adventure. As tech rep I used to do jobs that are handled today by entire departments."

In 1965 came the first opportunity to try the Turbo Porter as a spray aircraft. Harry Mochulsky went to Geneva to join Ladi Marmol of ADS (Aerial), a British firm, and fly with him in a Turbo Porter to Sudan. He was to see how the engine bore up in the heat and dust of the desert and carry out a 150-hour hot-end inspection. The aircraft operated in the Gezira Plain, where cotton attracted swarms of insects. ADS was responsible for spraying 126,000 acres, and timing was critical. Spraying was from dawn to dusk, making short flights with three- or four-minute turn-arounds to replenish the chemicals in a 200-gallon cabin tank. The engine was shut down every 10 trips, or roughly every $2^1/_2$ hours. Spraying was done at cotton height and 110 mph. After a day's work several small birds could be found lodged in the engine inlet screen, but the PT6 kept running. Another hazard came from "the locals" who tossed rocks at the low-flying aircraft, angry because it

Swiss-registered PC-6Bs during the rice spraying contract in Indonesia in 1969. (Harry Mochulsky)

sometimes scattered their flocks. One day Marmol had an engine failure, but just as he was about to land, the PT6 restarted. This was something new. He and Mochulsky tried to repeat the episode and after some effort they had another failure followed by a restart. Advice from Longueuil would be needed but telephoning from Khartoum was hopeless. Mochulsky flew to Cairo to make his call. Mike Saunders had him ship back the fuel control, and it was found to be clogged with dirt. This led to developing a system for cleaning the fuel control in the field.

The PC-6B was a success as a sprayer. The pilot felt secure behind the PT6 and didn't have to keep one eye peeled for a spot to make a forced landing should his engine quit. There was little pilot fatigue compared to piston-powered sprayers. The PT6's ample power meant that the throttle didn't have to be constantly jockeyed in turns. Marmol's main complaint was that it was just too hot in Sudan. For this he had his own solution. At times he could be found sipping beer in a desert canteen run by an American construction company, his Turbo Porter parked outside the door. In the end, ADS did not purchase a PC-6B. The entry price was simply too high compared to equipping with piston aircraft such as the Piper Pawnee. A few years later, however, a Sudanese organization purchased several PC-6Bs, and Pilatus set up a company in association with Ciba-Geigy to spray Indonesian rice fields.

In 1966 the Turbo Porter was certified with the PT6A-20 and kits were marketed to convert from the Astazou. The 1967 production scene at Stans showed about half the PC-6s coming off the line with PT6s. That December the -27 with more power became an option, and sales expanded with orders from such countries as Austria, Iran, Thailand and Argentina. Fairchild had rolled out its first PC-6B with a -20 in June 1966, but AiResearch had entered the picture. The president of Air America wanted Turbo Porters for CIA operations in Laos and specified the TPE331. This model became the PC-6C, flown in September 1966. The installation was 80 pounds heavier than a PT6, but even so, Air America soon had a PC-6C fleet, and Fairchild made this version its principal offering. This was not to discount the importance of the PC-6B in Southeast Asia. Continental Air Services in Vientiane, Laos, had a Porter fleet of 18 in 1970, 16 having PT6s. These averaged 194 hours of flying per month in 1969. Serviceability was outstanding and TBOs were being extended, which Continental attributed to P&WC training and growing experience in the field. Continental added Twin Otters in 1970. Elsewhere in the Orient P&WC products made steady inroads. In Thailand, two King Airs were used on aero surveys. The Thai Police used PC-6Bs and the Bell 212, and the S-58T found various roles with the government and on offshore oil exploration. In 1987 there were some 80 PT6-powered aircraft in Thailand alone.

Some Fairchild Turbo Porters were sold, including three to Intermountain Aviation to serve a US Forest Service smoke-

A Garrett TPE331-powered Turbo Porter. This was a COIN demonstrator marketed by Fairchild Hiller. Note the distinguishing rear location of the Garrett exhaust. This demo was at Wright Patterson AFB in May 1968. In the end it was the PT6 which dominated Turbo Porter sales. (Larry Milberry)

jumping contract. Five went to Colombia to support oil exploration in the jungles. Of 450-plus PC-6s built to date for customers in over 50 countries, some 250 have the PT6.

Another important Swiss project was launched in 1965—the re-engining with a PT6 of the Pilatus PC-3 trainer. The PC-3 was the standard Swiss Air Force trainer and had a 240-hp piston Lycoming. The PC-3 with the PT6, which first flew in April 1966, gave better performance but suffered from increased fuel consumption, which cut endurance to an hour. The prototype was displayed at the 1966 Hanover Air Show but did not win any sales. A few years later there was a better climate for the PC-3, brought about by the need in many countries to replace aging trainers. Redesignated PC-7, it was shown at the 1975 Paris Air Show, with increased fuel capacity, aerobatic capabilities and hard points for external stores. Pilatus started an initial batch of 35 aircraft in 1977.

A PC-6B with the -27 engine sits on a rugged strip in the outback of Ecuador. (Pilatus)

A Swissair PC-7 trainer lands at Stans Airport. It uses the PT6A-25A adapted for aerobatic flying, the same engine used in the Beech T-34C. (Anton Heumann)

The first production version flew with a 550-shp PT6A-25A in August 1978, and sales began coming in. By 1988 more than 400 PC-7s had been ordered and the aircraft is today one of the world's leading military trainers. It gave rise to the PC-9 with a PT6A-62 of 950 shp, first flown in 1984, and adopted by a number of countries.

The First PT6 Ag Conversion

In the 1950s New Zealand sheep farmers began using aircraft to spread fertilizer on their hilly grazing lands, a process known as top dressing. Beavers became popular; then California entrepreneur Wendell Fletcher designed an airplane especially for the job—the Fletcher FU-24. Over 100 FU-24 kits were shipped to New Zealand, where Air Parts (an agricultural consortium) assembled them and made spares. The FU-24 was manufactured in New Zealand beginning in 1965. One of the members of Air Parts, James Aviation, proposed a PT6 for the FU-24. P&WC provided a -20, while Air Parts raised the funds to build a prototype (the Fletcher 1060) with the engineering assistance of Air New Zealand. Many mods were incorporated, including replacement of the 27 cu.ft. hopper with one of 48 cu.ft. The aircraft first flew on July 19, 1967. It was the first use of a PT6 in an Ag plane. On one demonstration it took off simultaneously with a piston FU-24, dropped its load, made a quick turnaround and was back on the ground again while the piston aircraft was still on its first drop. Another day, the PT6 FU-24 operated in hills where the gradient was too steep for piston operations, spreading 10,000 lb of fertilizer on four flights. While the PT6 was successfully demonstrated in the Fletcher, only one example was ever operated. It was later fitted with an inertial particle separator to reduce engine wear from ingested dust,

Close-up of the business end of the turbine Fletcher. Its PT6 was the first small turbine in New Zealand. The engine was pushed noticeably forward to maintain c of g. Note the stubby prop blades and dual-wheel main gear. The exhaust stacks were provided by Pilatus from PC-6B stock. The second photo shows the Fletcher flown by Tom McClunie of James Aviation. He spent a year at P&WC learning the PT6. Here he has a passenger in his single-seat Fletcher. Nearest the camera is the much-admired Mike Saunders, P&WC's tech rep on the project. The Fletcher operated with the PT6 until August 7, 1979. Severely worn after 5,326 hours of flight, the aircraft was grounded and cannibalized. The PT6 was sent to P&WC for overhaul. The aircraft was restored to piston configuration and is on display at Hamilton Airport. (via Fred Cowley, Florence Saunders)

grass, etc. Meanwhile various other P&WC-powered aircraft turned up in New Zealand and Australia. The Twin Otter became popular in the sixties and since then such types as the King Air, Bandeirante, Cheyenne and Bell 212 have appeared. PT6 overhauls for the region were handled by Trans Australia Airlines in Melbourne and Hawker de Havilland in Sydney.

Brazil and the PT6

In the mid-1960s P&WC found a market in Brazil. Like Canada, this large, geographically diverse country had long appreciated the benefits of air transport. The first flight there was made in 1910, and the famous Brazilian aviator Alberto Santos-Dumont had made Europe's first powered, heavier-than-air flight in 1906. In the interwar years Brazil established a small aero industry, and in the late forties the Ministry of Aeronautics set its sights on building a modern industry. One of its first moves was formation of the Centro Téchnico de Aeronáutica (CTA) in 1946, which would train Brazil's first aeronautical engineers with the help of MIT in the US. CTA set up modest facilities at São José dos Campos, about 50 miles from São Paulo, and brought in the famous French designer Max Holste to head activities.

In 1965 Thor Stephenson visited Brazil with a Canadian trade delegation and learned of CTA's ambitions. In 1966 Dick McLachlan was sent to confer with CTA, which in July 1965 had launched a program for a light utility turboprop. He found CTA's facilities sparse, with no production capabilities, but was intrigued at the enthusiasm of everyone despite the problems involved in designing and building a state-of-the-art aircraft. P&WC threw its support behind CTA and the PT6 was chosen for its engine. The project was known as IDP-6504 but was later named "Bandeirante," the Portuguese word for pioneer. The first prototype flew on December 11, 1968, with -20 engines. Intitially there were funding problems, with few in Brazil's private sector anxious to take part. The government stepped in, and a flight test program with several prototypes was conducted, which led to major redesign and adoption of the more powerful PT6A-27.

In January 1970 Embraer (Empresa Brasileira de Aeronautica) was formed to carry the Bandeirante to production. The aircraft was designated EMB-110, and in April Embraer ordered its first 40 PT6A-27s. The redesigned aircraft weighed 11,000 vs 9,900 lb and had double the seating (18 passengers). It flew on August 19, 1972, and first deliveries were to the Brazilian Air Force six months later. Sales were soon made to commercial and military customers in various countries, and the Bandeirante was off on a winning course. With more than 100 aircraft completed by 1977, the 680-shp -27s were

Eager young CTA engineers get their first look at a PT6 during the 1966 visit to São José dos Campos of the Twin Otter demonstrator. The sight of the Twin Otter with its simple, lightweight engines spurred on the CTA people in the dream for their own small airliner. (via Harry Mochulsky)

(Above left and right) The prototype IPD/PAR 6504 nears completion at São José dos Campos. Its -20 engines are fitted. The second view shows early taxi trials. (P&WC Archives)

(Left) The "6504" prototype takes off on its first flight from the CTA airstrip at São José dos Campos, October 22, 1968. Later named Bandeirante, it was the forerunner of a family of turboprop transports used worldwide. This aircraft was presented to the Aeronautical Museum in Rio de Janeiro in 1975. (Embraer)

replaced by 750-shp -34s, and gross weight climbed to 13,000 lb. The thousandth PT6 to Embraer was delivered August 24, 1981, and the millionth Bandeirante hour was logged April 12, 1982. Nearly 500 Bandeirantes had been delivered to the end of 1988, by which time the once tiny CTA had grown to a huge complex at São José dos Campos employing 10,000 workers and building a wide range of aerospace products.

Agusta of Italy

Giovanni Agusta S.p.A. of Italy has long built helicopters, beginning with licence production of the Bell 47. Later, the

Dick McLachlan, Dusty Miller and Gordon Hardy discuss the PT6 with Embraer's top man, Colonel Ozires. (P&WC)

The 400th Bandeirante, an EMB-110P with -34s, was delivered to the US carrier Southeastern in April 1982. (Embraer)

An EMB-110B of the Brazilian air force. The Bandeirante serves in many roles, in this case as a photo survey aircraft. Note the ventral camera pod. This kind of adaptability is attributable to the solid design of the airframe, but also to that of the PT6 in its various models. (Embraer)

Bell 204 was added and 24 Sea Kings were built for the Italian navy and 26 Chinooks for the army. The company had been interested in a twin-engined version of the 204 and had tried its own mods with twinned T58s and Astazous, but no production resulted. Then, in 1968, P&WC proposed the PT6T-3/6 TwinPac Bell 212, and Agusta immediately ordered engines. Its first 212 was demonstrated at the 1970 Farnborough Air Show, and the Shah of Iran purchased the first two for his personal flight. Orders rapidly came in, including a batch of 212s for the Italian navy to use aboard small vessels and at coastal bases. Engines for the Agusta 212s were assembled by Alfa-Romeo, which later became a supplier of parts to P&WC. Today the 212 and 412 are in production at Agusta's Milan factory and are sold globally. In 1988 there were over 120 TwinPac-powered 212s and 412s in the Italian military alone.

East Bloc Connections

Czechoslovakia's postwar aircraft industry turned out a variety of successful products, but few were marketed in the West. Among projects in the early 1960s was a small turboprop engine, the M601, but progress was slow. In 1966 the Czech national aircraft works, LET, enquired about PT6 licence possibilities. It was developing a Twin Otter-class utility plane and thought that western engines would enhance sales beyond the East Bloc. It also approached Turbomeca, but the French would limit sales of licence-built Astazous to the Czech domestic market.

In January 1967 Dusty Miller visited LET, where he heard, to his amazement, that as many as 5000 engines could be needed. The M601, based generally on the PT6, was still being developed, but if it fell through LET wanted a P&WC licence to fall back on. LET ordered some PT6A-27s for its prototype L410 and asked P&WC to do what it could to promote the aircraft to the Soviets at air shows such as Hanover. Meanwhile Aero-Dienst trained Czech mechanics on the PT6, and Miller, Georges Ostiguy and Serge Page assisted at the LET factory. The prototype L410 made its first flight on April 16, 1969. Dusty Miller later flew the aircraft and was asked by the chief designer for his opinion. The Czech grinned broadly when Miller told him that the L410 flew just like a Twin Otter!

The Czech airline, Slovair, evaluated the L410 in 1971 but no big orders came in. That changed when the M601 finally got into production. The L410 with M601s first flew in 1973 and production picked up, reaching about 100 aircraft per year, most for the Soviets. Only 28 PT6-powered L410s were built. P&WC never did have a licence agreement with LET, but what is almost a PT6 clone (the refined M601) now powers more than 400 L410s.

In 1976 Dusty Miller was marketing PT6s to the Polish national aircraft organization, PEZETEL. He learned of the capabilities of Poland's aero engine industry, and that the WSK factory at Rzeszow was manufacturing Russian gas turbine engines under licence. Discussions revealed that WSK was eager to make parts for P&WC in order to advance its own technology and to earn US dollars. Their prices for a wide range of machined parts and complex weldments justified the assistance that would be needed to ensure that Polish-made parts would meet stringent P&WC standards. The low-cost PT6A-100 series was now introduced, with WSK to

One of the few L401s with the PT6A-27. LET took advantage of the -27 while awaiting development of the Walter M601. (via D.S. Miller)

P&WC personnel (with families) during a familiarization visit to Poland in the summer of 1977. From the left (in rows) starting from the front are Helen and Bob Wysocki, Stefan Klosowski, Dianne Vaillancourt, Tony Wysocki, Jean-Paul Vaillancourt, Henry Soloniewicz, Chantal and Romy Kowalczewski, Martial Todorovic, Bert Walser, Zoltan Batory, and Ed Zielinski. (Bob Sachs)

be an important parts source. This series was adapted for a number of new aircraft including the Cessna Caravan, Piper Cheyenne I, and some agricultural conversions.

A long-term agreement was negotiated and signed with PEZETEL just before Christmas 1976. The "Polish project" was organized with a program manager (Bob Sachs) based in Longueuil, and an assistant program manager (Stefan Klosowski) who moved to Rzeszow. Other P&WC personnel moved there with their families to assist and to monitor WSK's inspectors and production people during start-up. Those moving to Rzeszow were assisted with orientation briefings, Polish language training, relocation expenses, education allowances for their dependents, and the supply of certain commodities not obtainable in Poland. WSK provided furnished apartments and got the P&WC people priority for such things as cars and fresh meat. The first contingent from Canada included: Stefan Klosowski with his wife Halina and daughter Izabela; Bob Boyko (chief inspector) and his wife Dagmar; Bert Walser (purchasing); Jean-Paul Vaillancourt (inspector) with his wife Dianne and children, Luc, Marc and Joanne; Zoltan Batory (inspector), Romy Kowalczewski (quality assurance metallurgist) and his wife Chantal; and Tony Wysocki (inspector) with his wife Helen and son Bob.

In many cases the Canadian families benefited from the opportunity of living in Europe. The Vaillancourt children attended Polish public schools, learning their basic subjects in Polish. Izabela Klosowski attended the University of Krakow, graduated as a medical doctor, and married a Polish doctor. Besides the P&WC residents in Rzeszow, another two dozen quality, production and purchasing specialists spent months in Poland throughout 1977–79, assisting WSK with the many start-up problems. For WSK, this was the first experience with the Western gas turbine industry, and for P&WC its first exposure to an Iron Curtain supplier. Today WSK is an important supplier to P&WC, and the initial agreement has been renewed.

Sometime after the Canadians had settled in Poland turmoil erupted with the Solidarity movement. Ed Zielinski, P&WC's No.2 man in Poland (and fluent in Polish), drove to the Canadian Embassy in Warsaw to check out the situation, but found that the ambassador was stranded in Vienna. By now several telexes from Longueuil urging the Canadians to pull out had arrived. Slim Lawton, chief P&WC man in Poland, was held up in London. In case the trouble worsened, Zielinski hired a bus to take his charges to Czechoslovakia. Just then Lawton arrived in Rzeszow and the exodus, now deemed unnecessary, was cancelled. Although the Polish workers at WSK went on strike, the P&WC line continued operating. By not pulling out, the Canadians maintained their rapport with the Poles, and WSK soon became a significant supplier of parts.

The Chinese Market

In 1970 Canada established diplomatic relations with the People's Republic of China, and it was not long before P&WC

made its first excursion to Beijing. In August 1972 Thor Stephenson, Ken Sullivan, Dusty Miller and Bob Sachs attended the first Canadian Solo Trade Fair there, bringing along thousands of red-covered technical brochures printed in Mandarin, and cutaways of the JT15D and TwinPac. In a scientific community still reeling from the Cultural Revolution the display was a sensation. Ken Sullivan relates: "The trip was full of surprises. We saw a country as big as Canada but with almost a billion people. The people were not mobile—most stayed in their home villages. Regional travel was limited, and only government officials got to fly. We recognized the importance of getting our foot in the door to this large potential market. Our first experience with Chinese air travel was educational. We flew from Beijing to Canton on a turboprop. Once seated, everyone was given a free fan. That was our air conditioning. Midway through the flight we made a stop and were led off the plane to the dining hall in the terminal, where we were served a six-course dinner. That was our 'in-flight' catering!"

Following the 1973 Paris Air Show, Dusty Miller returned to China. In 12 days he made 11 presentations on the PT6, ST6, TwinPac and JT15D. The Chinese were reluctant to discuss applications they were planning but were high in their praise for P&WC products. On a follow-up visit, Miller, accompanied by Gerry Gillies, Colin Wrong and R.F. Thompson, sold the Chinese three ST6s, six PT6 TwinPacs and seven JT15D-4s. The ST6s were for trials in a 60-ton ore-carrying truck used on steep grades. The TwinPacs were to evaluate in the Mi-4 helicopter and the JT15Ds were for a small twin-jet transport that was still on the drawing board. In 1974 and 1975 a Chinese delegation visited Longueuil to witness the assembly and test of their engines; then the engines were shipped to China, where P&WC continued to offer training and support, but in an atmosphere which found the Chinese still reluctant to deal openly with Westerners. In the end, none of the installations attempted by the Chinese resulted in viable applications, but useful hands-on experience was gained. Around 1980 China purchased eight Bell 212s for utility work, then an IFR 212 for offshore oil exploration in the South China Sea. Meanwhile, Bell and Aérospatiale were vying for a helicopter/engine assembly plant in China. The French won with the Turbomeca-powered Dauphin.

Thor Stephenson (centre) confers with Bob Sachs and Ken Sullivan during a P&WC visit to China, as Mao looks on placidly. (D.S. Miller)

P&WC's big success in China came with the indigenous Y-12 light transport. It evolved from the underpowered piston Y-11. Dusty Miller discussed updating the Y-11 with the Chinese at Farnborough one year. They were interested but admitted their shortage of foreign currency to pay for engines. Miller suggested a swap—PT6s in return for the JT15Ds purchased earlier but by then in mothballs. A deal was struck, and the prototype Y-12 flew (with PT6A-11s) on July 14, 1982. The following year it flew with -27s. To make its product more marketable outside China, the Harbin Aircraft Corp. began having its Y-12s finished (e.g. with interiors and

Dusty Miller of P&WC visits Harbin Aircraft during TwinPac trials with the Mi-4. The installation resulted in such a long nose that pilot visibility was impaired and the scheme was dropped. Mr. Wong, chief engineer at Harbin, and Mr. Sun Zhao Qing are with Dusty. (P&WC Archives)

A Chinese delegation visits Longueuil in October 1979. Along with Ken Sullivan, Dusty Miller and Donald Lowe, they are viewing a PT6T-3B TwinPac. (P&WC Archives)

One of China's indigenous Y-12s powered by PT6s. (Harbin Aircraft)

(Right) An early Arava with -27s. (P&WC Archives)

avionics) in Hong Kong by HAECO. The first export sale of Y-12s was to the Sri Lankan government, and a market for several hundred Y-12s is foreseen. Meanwhile, China has produced the Y-7 based on the Soviet An-24. Discussions have been held about using the PW124 on this 50-seater.

Israel Aircraft Industries

The IAI Arava was the first new project of Israel's aircraft industry. Israel had close ties with France and had built the Magister trainer under licence. It also had an engine factory set up with the help of Turbomeca. Dusty Miller first got wind of Israel's plan to build the Arava while flipping through the pages of an aviation magazine. A report mentioned that the new light twin would use Turbomeca turbines. Miller immediately contacted Dov Sa'ar of IAI by mail, then filled him in with a series of letters on the latest PT6 technical and marketing data. Miller notes: "What finally caught IAI's attention were the PT6's ease of starting, quietness, good performance in icing, and our product support organization. Plans for the Turbomeca were dropped, and we received an initial order for three PT6s. Believe it or not, we closed this deal for the price of a few postage stamps!" The Arava was displayed at the 1971 Paris Air Show, and enough orders came in for production to go ahead. Besides Israel, the main Arava market became Latin America. As the aircraft grew from 12,000 pounds to about 16,000, it moved from the PT6A-27 to the -36, which is especially good for hot and high operations. Today there are some 60 Aravas in service.

TwinPac and Sea King

P&WC's early emphasis on a version of the PT6 for helicopters culminated on May 25, 1965, with certification of the PT6B-9 turboshaft. It was developed alongside the -6 turboprop, the main difference being the single stage reduction gearing in the turboshaft that produced a higher shaft speed of 6230 rpm. The Lockheed 286 became the only installation of the -9. Stanley Hiller explains why: "The helicopter industry of the 1960s was swayed by Army thinking. The Key West Agreement of 1947 had taken the Air Corps away from the Army and created the USAF. Thereafter, the Army only operated small airplanes and helicopters. The PT6 had intermediate power, but the Army thought it was too much. It preferred the Allison T63 at an announced unit price of $4000. When it created the specs for the LOH competition, they were tailored to a helicopter in the T63 range. All three LOH finalists had T63s. Later civil derivatives had similar Allison 250s."

P&WC had another idea for the PT6—using it to re-engine larger helicopters. One target was the S-58, of which more than 1700 had been built, most for the military, but some for airlines such as Sabena and Chicago Helicopters. Replacing an S-58's hefty R-1820 with three PT6s would yield a weight saving of 500 pounds, adding to payload and giving multi-engine reliability. Chicago Helicopters followed the idea, until Washington abandoned subsidies to helicopter airlines.

P&WC had participated in the LOH program but Allison won the competition. Its T63 became standard equipment in thousands of OH-58s, similar to this Canadian Army example. (Larry Milberry)

Another market was the US military, which was expanding its forces for Vietnam. Jock Graham pursued this market, but the military wasn't interested in revamping older machines. The idea of a PT6-powered S-58 was shelved.

Birth of the TwinPac

P&WC had been in touch with Bell Helicopters since earliest PT6 days and in February 1960 Bart Kelly of Bell visited Longueuil for serious discussions. Bell was gearing up to build a batch of T53-powered helicopters for the US Army

(H-40s, later called HU-1s, then UH-1 Iroquois, Model 204 for civil models). Kelly's visit led to a joint P&WC–Bell proposal to Ottawa to supply a UH-1 with twin PT6s for tri-service use. Ottawa was interested but didn't go for the idea as it was negotiating for a fleet of turbine Vertols.

In 1960 the 700-shp T53 in the UH-1 was replaced by an 1100-shp model giving rise to the UH-1D (Model 205), of which 7000 were built. In 1965 Bart Kelly again visited P&WC, this time to discuss a two- or three-engine UH-1. At first nothing resulted, and for its first twin-turbine helicopter Bell installed Continental engines developed from Turbomeca. Bell touted this version to the military, citing its twin-engine safety, but Continental, contrary to Bell's expectations, was unwilling to spend millions on certification without up-front orders. On the other hand, P&WC was—it offered Bell a

A standard Bell UH-1D "Huey" (Lycoming engine), thousands of which were built. Bell's desire for a twin-engine version of the UH-1, and interest from P&WC in collaborating on the project, led to the now-famous "Twin Huey". (Larry Milberry)

twinned PT6 having two power sections coupled to a single output gearbox and providing 1800 shp at maximum continuous rating. P&WC was encouraged when Ottawa agreed to support the R&D effort. This help was also extended to Bell. Both engine companies committed themselves to the project in November 1967 and Bell announced the Model 212 (military UH-1N) with twinned PT6s at the October 1968 NBAA convention in Houston. The US Navy brought the Continental and P&WC teams to Washington and negotiated two separate engine contracts. The Canadian one was placed through the Canadian Commercial Corporation in Ottawa. The Navy then locked up the contracts and made its recommendations. Dick Richmond made several trips to Washington trying to determine when the contract would be let. He

A version of the PT6T TwinPac of 1800 shp. (P&WC)

A Coast Guard Bell 212 lands at Vancouver, October 1986. With its twin-engine safety margin (PT6T-3B), the 212 has found wide appeal in the utility and search-and-rescue roles.
(Michael Macgowan)

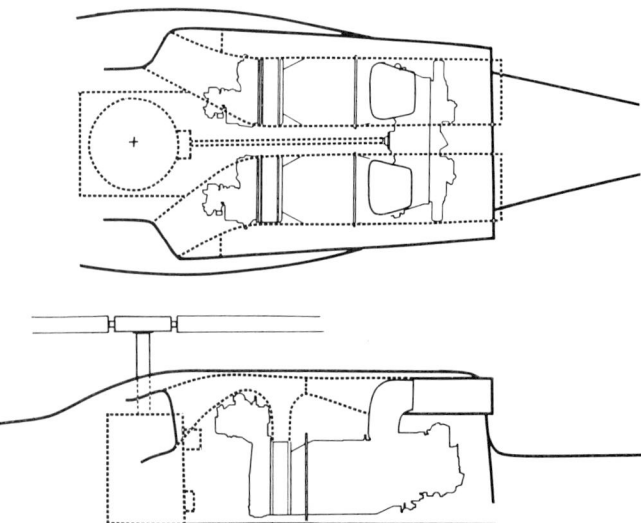

Overhead and side schematic views of TwinPac installation in Bell 212. (P&WC)

learned that each week the Defence Department published a list of new contracts, then received a tip to study a particular week's release. There he discovered a brief entry, "Helicopter engines to CCC". The Navy had been trying to keep its choice as quiet as possible to avoid controversy over letting a defence contract to a foreign supplier. Now work could go ahead at full speed on the new PT6.

Preliminary design of a gearbox had begun in April 1967. The new engine was designated PT6T-3 TwinPac (military designation, T400) and was first run in June 1968. It flew in a 205 testbed in May 1969 and by the summer of 1970 military and civil certification and first deliveries had all taken place. Development had cost $13 million and had been a gruelling experience for P&WC. Recalls Pete Peterson, TwinPac project engineer: "I was sceptical about the practicality of a coupling gearbox. We knew how to make turboshaft engines, but it was another matter to get the gearbox and controls working. This was our first serious helicopter installation and for a year I didn't even take weekends off." The TwinPac had the same design as the 715-shp PT6A-27 but had to achieve 900 shp per side. This required turbine blades withstanding higher temperatures. Normally the grain in a blade develops randomly, but a stronger blade results if the grain is oriented along its length. Effective new casting techniques were developed. Peterson adds: "Reducing weight was also vital. Our aim was to put weight as close as possible to the rotor shaft. We knew that for Bell's sake we had to work within the existing centre of gravity. By using a coupling gearbox, Bell could retain its original transmission." The gearbox would take the 33,000 rpm from each engine and reduce it to 6230 rpm on a single output shaft. The unit took three complete design efforts. Developing light yet high-strength gears was a major challenge. Separate oil systems were needed for each power unit and the gearbox so that failure of one unit would not contaminate the oil supply to the others. Developing engine controls was another difficult task. In a helicopter the main rotor maintains a constant rpm regardless of loads in the engine—for example, on takeoff, hovering or in transition flight. Peterson states: "We designed a governor to measure the rpm of the rotor shaft. If it sensed that the rotor was going too fast, it reduced air pressure in the pneumatic lines and this made the fuel control reduce the fuel flow to the engine." The two engines had to work together. A torque equalizing control would monitor output speeds and adjust output from one power section to compensate for any loss from the other. Since a helicopter cruises on half the power needed for takeoff, loss of one section could be made up by the other.

On the marketing side, there was Ottawa's interest, but the US Navy now wanted the engine for an uprated Cobra gunship. The USAF decided to pool its order for engines for UH-1N search-and-rescue helicopters with the Navy's, and Ottawa finally placed its own order for 50 UH-1Ns worth $50 million. The initial combined USAF–USN–USMC order was for 141 helicopters. In time, more than 100 TwinPac Cobras were delivered to the Marines and more than 200 to Iran. Selling the TwinPac in the US was complicated by the Senate Armed Services Committee, which objected to buying engines from a foreign supplier. A rider was attached to a supply bill stating that engines for US military helicopters had to be made domestically. To deal with this, P&WC formed a subsidiary, P&WA of West Virginia. Using parts from Longueuil and the components usually purchased from US suppliers, the new company supplied the hundreds of engines since ordered for the US military. The West Virginia plant was later expanded for overhaul of the PT6T, PT6A and PW100.

A USAF UH-IN Twin Huey during rescue qualifications with the 49th Fighter Interception Squadron in upstate New York, September 1986. (Larry Milberry)

The Bell AH-1J and AH-1T Seacobra attack helicopters use the P&WC TwinPac of 1800 and 1970 shp respectively. The "J" first flew in November 1969, the "T" in May 1976. This "J" visited Ottawa on August 15, 1970. (Larry Milberry)

The advanced Bell 412, still with the PT6T-3B, but with four-blade rotor. Licence production was taken up by Agusta of Italy and IPTN of Indonesia. (Gary Vincent)

A special order was placed for four VIP UH-1Ns for the presidential flight in Washington. These use a hybrid PT6T with especially high acceptance criteria. The "VH-1Ns" are used to carry dignitaries in the US and elsewhere.

The Sikorsky Sea King

In late 1962 Ottawa ordered the first of 41 Sikorsky CHSS-2 Sea King antisubmarine helicopters from P&WC to replace the HO4S-3s and bring to the RCN the new role of operating ASW helicopters from small naval vessels. The order launched Canada's first large-scale helicopter industry and prompted P&WC to form its Helicopter and Systems Division. It would market Sikorsky products and services to the Canadian government and build them under licence. With no more than its background as a helicopter sales/service/overhaul operation, P&WC was to become a manufacturer and work with the military on a major new weapons system.

The Sea King project was the end result of years of competition to win a lucrative contract. The favoured contender was the Kaman Seasprite, a medium-sized aircraft with several advanced features first flown in July 1959. Sikorsky had been moving more cautiously into turbines, introducing the S-62, first flown in May 1958, with a rotor system and other components already in use. Like the Seasprite, it had a single GE T58 turbine. To counter the Seasprite, Sikorsky offered several options including the turbine S-58 and the S-65 using S-58 dynamic components and a new amphibious fuselage. For the latter, three PT6s were proposed as one power option. Meanwhile, the RCN, strapped for dollars, was considering the sturdy old piston S-58! By 1962, Kaman and Ottawa had failed to come to terms over price and production sharing. Meanwhile P&WC learned that its position in the competition would improve if it would increase the Canadian content part of its proposal. The company could make a case for doing assembly on some structural components and the gearbox. Some jigs from this work could be used later for overhaul, so did not have to be charged against the contract. These and other details were added to the proposal, and Ottawa began looking harder at the Sea King. It had first flown in March 1959 powered by two T58s. Designated HSS-2 by the USN, it was developed into the S-61. Jock Graham recalls: "We didn't think our chances were too good with the HSS-2. What finally turned the competition in our favour was our willingness to place a large amount of work in Canada."

P&WC had some good people with helicopter experience from its overhaul shop but recognized that many others would be needed to support the program and meet RCN require-

Sea King production at Longueuil (Plant 2) at its peak in the mid-1960s. In the foreground is 4005, the first P&WC-assembled Sea King. Its GE T58 engines are uncowled. (P&WC Archives)

A Sea King component in the jig at P&WC. (P&WC Archives)

ments. To begin the Sea King era, the company sent a core of men to Sikorsky to train in helicopter engineering and manufacturing. To head the new operation, Dick Richmond hired Bob Raven from Canadair. A native of Windsor, Raven had graduated in 1939 from the University of Detroit as an aeronautical engineer. He worked at Malton with National Steel Car, then went on to Boeing and CPA on the West Coast. Canadair hired Raven to work on the North Star, sending him for a year's apprenticeship to Douglas in California. Following the North Star, he became project engineer on the Sabre program.

The Sea King was more than the assembly of stock HSS-2s. The project had to be adapted for small shop operations. The various RCN electronic systems had to be integrated into a complex package. System work totalled half the value of the contract. The manufacturing and production organization drew on experience gained in the P&WC helicopter overhaul shop. Charlie Seager was project engineer.

The immediate need at Longueuil was for an organization to assemble Sea King kits from Sikorsky, manufacture vital components and procure electronics. George McLean, longtime foreman of the helicopter overhaul shop, headed this effort, assisted by Eric Bentley, who had been in the helicopter shop since 1956. Bentley had grown up in Longueuil, served with Fairchild before the war, and later worked for Canadair and Wheeler Airlines. He relates: "Joining P&WC was more exciting than I had planned. My first helicopter ride was out of the yard at Plant 2. It really shook me up! Flying had always meant roaring down a lake in a bushplane on floats or skis. I can't describe what it was like to suddenly be hovering above the ground in a helicopter!"

McLean shuttled between Sikorsky and P&WC laying the groundwork for the organization while Bentley took charge of training: "When the program started, I took seven men to Sikorsky for several weeks of training. Included were Phil Anderson, Frank Johnson, Mike Gervais and Lionel LaLumière. We had to learn Sikorsky's techniques. I put one man in Sikorsky's gearbox and rotor department, another in subassembly and sheet metal, and others in assembly, electrical, hydraulic and flight test departments." The Canadians worked their way through the HSS-2 production line, working with many of the experienced Sikorsky people. Each afternoon they met to update the tooling list and re-draft procedures. The meeting place was the office of Jack Horner, a talented engineer whose father, H.M. Horner, had become chairman of UAC and a director of P&WC.

Bob Raven's engineers were also busy at Sikorsky. Most were men hired from Canadair who had been working in the missiles and systems division. Sid Masters headed the structural design office, Harold Bruce handled the systems (electronics, avionics, hydraulics), Bob Sparks and Pete Belanger were in mechanical design, and Stu McCorky and René Parent were in electrical design. Initially there were five men on Raven's team but that number soon swelled.

Sid Masters had begun with Imperial Airways in 1939, later moving to Bristol before coming to Canadair on the Argus. When he joined the Sea King program he found that P&WC had no airframe structural engineers: "We set up our own organization and went to work. The groundwork had been laid

Early in the Sea King program at P&WC, test flying was carried out from Plant 3, the old Fairchild airstrip near Longueuil. Seen in the hangar are a Sea King and the testbed Beech. The old flight test hangar was demolished (below) over the winter of 1987–1988. (P&WC Archives, Larry Milberry)

A Sea King during landing trials aboard helicopter frigate HMCS *Assiniboine*. On later DDH 280 Class frigates, two Sea Kings can be accommodated. (DND)

4001 (built by Sikorsky) formates with an H04S, the piston ASW helicopter replaced by the Sea King. (via R.L. Rogers)

by Charlie Seagar, the key man in all the early liaison work with Sikorsky. There would be many challenges along the way to making the Sea King a good working system for the Navy. Perhaps the most challenging was developing the rotor blade de-icing system. If there was ever a failure in a heating element on one blade, and ice built up on it, disaster could follow. Our job was to develop the automatic controls and sensors to make the system perform safely. We did a lot of research in Ottawa with the NRC icing rig, and ultimately solved all the problems." P&WC also had to work with the Navy to perfect the Sea King's electronic packages—sonar, navigation, avionics. The crew stations were different from the USN Sea Kings, and design of these helped P&WC build its capabilities as a "systems house."

A significant difference between the USN HSS-2 and the RCN CHSS-2 was the modification that enabled the Canadian version to operate from DDH-class destroyers. The main hauldown system devised to get the helicopter onto the DDH's small deck in any sea/weather conditions, day or night meant that the helicopter had to be strengthened, especially its undercarriage. The airframe had to sustain a sink rate of 12 feet per second and the severe side loading imposed when it was secured to a rolling and pitching deck. Fairey Aviation and the RCN also developed a system to keep the tail of the Sea King from pivoting, but this was not successful. Sid Warren, a designer with the Helicopter and Systems Division (as P&WC's helicopter organization was later known), came up with a new design in 1967 for a device that could do the job. When the helicopter landed, the pilot lowered a probe. As soon as the tail swung and the probe found a slot in the deck, it would engage automatically to firmly anchor the tail.

The first four CHSS-2 Sea Kings (Nos. 4001-4004) were made at Sikorsky in Stratford, and P&WC began manufacturing from 4005 onwards. Only the basic sheet metal hull, or "tub," arrived in assembled form from the US. All other parts came in kits or were manufactured at P&WC. As the program gathered steam, the organization grew and Canadian content increased. P&WC began fabricating sheet metal components, starting with tail cones, stabilizers and pylons, then adding sponsons and eventually almost every removable airframe item. Sikorsky masters were used for making larger pieces like the sponsons, which had to be interchangeable with parts made in Stratford. Rotor heads and transmissions arrived in boxes ready for assembly and test. A test rig was constructed at Plant 2 to run in gear boxes and transmissions.

At first the assembly line took up a modest space in one bay of Plant 2. As production built up, this grew to two bays, but even more space was needed. By 1965 there was a crisis over accommodating the Sea King, spare parts production and the PT6 program. Dick Richmond, who had overall responsibility for the plant, had his hands full making room for everything. To solve the problem, a 158,000 square foot plant was leased for the engine business, and construction began on a 130,000 square foot helicopter facility at St. Hubert. The latter was Plant 5, and Sea King production began moving there in May 1966.

Test Flying the Sea King

P&WC test pilot John MacNeil would need help once Sea Kings began coming off the production line at Longueuil. He was already busy with PT6 test flying in the Beech 18 and routine helicopter test flights from overhaul. To spread the workload, the company hired Ross Lennox, who had been with Okanagan and had some 4000 hours on the S-55. From a DEW Line site on Baffin Island one week, Lennox found himself at Stratford on February 17, 1963, where he took an intensive HSS-2 course that finished with 50 hours of instruction in the air and 150 with the production flight test department. During this time, Sikorsky completed the first CHSS-2 and John MacNeil and Bud Hartigan took it flying on May 9, 1964. Lennox and Hartigan flew the second one on May 31. About this time, the first RCN pilots arrived at Stratford to accept the Sea King. These included Buck Rogers, who was the first RCN pilot to train on the Sea King (at Key West in early 1963), and Seth Grossmith (later a pilot with P&WC).

The first three Sea Kings were ferried to NAS Patuxent River, Maryland, for training of further RCN pilots. Buck Rogers was in charge of this phase, and trainees included Colin Curleigh, Ted Fallen, Nils Florens, Gordon Grey and Robbie Watt. At the same time, Larry Zbitnew enrolled in the USN test pilots school at Patuxent and later flew on such trials as the Sea King haul-down system. Joe Sosnkowski of the RCN also did much early flying on the haul-down system. Training was complete in July 1963. After training of air and

The first RCN Sea King to arrive at HMCS *Shearwater* near Halifax on August 1, 1963. The crew, from the left, are Lt. R.L. Rogers, LT. R.A. Watt, CPO Sopko, L/S McQueen and (kneeling) unknown. (DND)

groundcrew, HS-50, the RCN's first turbine helicopter squadron, flew from Shearwater to join HMCS *Bonaventure* in January 1965 for the first Sea King cruise. P&WC tech rep Sig Hubenig was aboard to assist on the three-month operation. One of the young HS-50 pilots was John Searle, who later was P&WC chief pilot, succeeding Ross Lennox in 1982.

Ross Lennox's stay in Stratford gave him more than a taste of Sea King flying: "One of my first jobs was taking a helicopter over Long Island Sound for tests with HF radio, doppler and radar altimeter. By sending out four signals and measuring returns, the doppler could detect the helicopter's movement, providing over-water navigation data. On routine flights, we'd go over the Sound, letting down every so often in a 40-foot hover, the sonar dipping height. Then we'd try transmitting back to Stratford, to practice what the Sea Kings would later have to do routinely—transmit back to their home ship."

Ross Lennox recalls the Sea King test flying days: "John MacNeil and I would do the first flight on most new machines, but as he got busier on PT6 work I counted more on Seth Grossmith to help. He was with the Naval Air Technical Liaison Office, an intermediary between us and the RCN. Together we'd follow each Sea King through production to be sure that everything was coming together as planned. The flying was a steady mix of production (to make sure that each helicopter was basically sound in engine and airframe), development (we were constantly increasing Canadian content, and adding or changing equipment such as a FOD deflector) and experimental (for example, testing an external magnetic anomaly detector or MAD boom)."

The FOD deflector resulted from the Navy's worry about ingesting birds into the engines. This work began in late 1965 and proved more difficult than expected. The airflow against the deflector was enough to buckle the cabin roof, and that had to be strengthened. Then it was found that the airflow around the deflector created eddies that could allow ice build-up. The icing rig at NRC was used, and one result was that inlet screens were added. The doppler system from Marconi required extensive trials at Longueuil and Stratford. For the rotor blade deicing trials, a USN HSS-2 was borrowed and sent to Ottawa to work with the icing rig; while P&WC conducted trials at St. Hubert in late 1968. That year the

Sea King 4005 makes its first flight at Longueuil, piloted by John MacNeil. (P&WC Archives)

unified Canadian Forces came into being, and the CHSS-2 became the CH-124 Sea King.

On April 9, 1964, Sea King 4005, the first turned out by P&WC, made its first hover from the helipad behind Plant 2. From first hover on, a detailed sequence of tests (for example, hovers at various weights) marked the beginning for each new Sea King. After initial tests at the helipad, flying moved across to the old Fairchild hangar beside Plant 1. (In December 1965 Sea King flight testing moved to the old CF-100 alert hangers at St. Hubert.) Along with the P&WC pilots, the Navy had its own staff at Longueuil, acceptance pilots included, to assist throughout the production phase. That ended on April 1, 1969, when 4041, the last RCN Sea King, made its first flight. Following delivery of the 41 Sea Kings to the RCN, the aircraft were often seen back at P&WC. First came a "mini-mod" program, begun in September 1970, to update various systems and keep the Sea King fully capable in the ASW role.

The old P&WC airstrip at Longueuil. The square building near the taxistrip is the company hangar, part of the old Fairchild operation. Across from the hangar is a veneer factory which operates to this day. In the lower left is another phase of Fairchild, today used by Weston Bakeries. Everything between Weston's and the veneer factory has been demolished. The airstrip was abandoned some years earlier. (P&WC Archives)

In 1974 came the Sea King Modernization Program, which included such installations as crashworthy fuel cells and a dorsal radome. Gross weight was increased from 19,100 lb to 20,500 lb.

There had been hopes at P&WC of building Sea Kings for other customers, especially as Sikorsky's facilities were backlogged with orders for the US Navy. P&WC fought for a German order, but it went to another Sea King licencee, Westland in the UK. Jock Graham explains: "I don't think we got up the learning curve fast enough. The orders from Ottawa had come in bits and pieces, never at a pace where we could offer Germany a good price." A worthwhile small program that P&WC did enter was the manufacture of pylons, fairings and sponsons for the Sea King, S-61 and CH-53. It provided some tooling to the Germans when they set up for CH-53 production. There was also some work for Helicopters and Systems, subcontracting engine nacelles, for example, for Beech King Airs.

In the late 1970s business aviation was serving P&WC very well. While the Sea King had provided many good contracts, that business was not steady. After reviewing its operations, the company concluded that it could do better in the long run by focusing on engines and turning the floor space occupied by helicopters over to them. It terminated Sea King overhaul and modification work in 1979, and that work was taken over by IMP of Halifax. SPAR in Toronto took over gearbox maintenance. Helicopter and Systems personnel were transferred to other P&WC departments. Meanwhile, the Sea King will remain in service until 1995, then be replaced by the new EH 101 from Westland and Agusta. If Canada's EH 101s use the Rolls-Royce/Turbomeca engines being proposed, then P&WA, as North American agent, will have a stake in the new program.

All in a Day's Work

On October 29, 1965, two workmen were atop a 425-foot chimney at the Miron Frères site at St.-Michel in suburban Montreal. Without warning, their scaffolding collapsed and Claude Landry fell to his death. His companion, André Grenier, was left stranded and terrified. Soon a crowd gathered below, but there seemed no means of rescue—not, that is, until word spread to radio station CJAD, which operated a small

Pilots MacNeil and Lennox hover precariously close to the stack where André Grenier was stranded. (P&WC Archives)

helicopter for traffic reporting. CJAD was ready to launch its chopper to attempt a rescue, but weather made it impossible to get airborne. At the same time, word reached P&WC, and the front office was quickly on the phone to flight operations, asking John MacNeil if he had a helicopter available.

MacNeil reported that a Sea King was ready to fly, as soon as permission was received from the Navy. LCDR Seth Grossmith obtained the necessary clearance and at 17:30 MacNeil was airborne along with Ross Lennox and crewmen Jerry English and Rocky Marquis. Equipped with some makeshift rescue gear, they were hovering near Grenier within a few minutes. There were power lines near by, a stiff wind

The last Sea King worked on by P&WC's Helicopter and Systems Division came through Plant 5 in July 1983. Since then, overhaul has been done by IMP, Halifax. (P&WC)

As Sikorsky's Canadian agent, P&WC took part in occasional special events. Ross Lennox was captain of the first S-61 to fly the Atlantic. The route was St. Hubert–Baie Comeau–Knob Lake–Fort Chimo– Frobisher–Cape Dyer–Sondestrøm–Angmacassallik–Reykjavik–Hofn–Faroe Islands–Prestwick. Here Okanagan president Glenn W. McPherson and Thor Stephenson bid farewell to Lennox (at foot of stairs), Thomas Sheer (Okanagan pilot), Tom Harrison (P&WC tech rep) and Keith Rutledge (Okanagan engineer). (via R. Lennox)

In 1971 Ross Lennox flew this S-64 Skycrane on a demonstration tour along the east coast of Hudson Bay. The big machine offloaded supply ships and made proving flights with the ingenious "people pod." The S-64 is powered by a version of the P&WC-designed JT12 turbojet. In recent years S-64s have had various contracts in Canada, expecially in the BC logging industry. (via R. Lennox)

An S-58T of the Los Angeles County Sheriff's Bureau overflies rugged terrain. The Bureau uses its S-58T in mountain rescue, medevac and SWAT roles. This aircraft (N64CH) was converted to a TwinPac in 1972 and acquired by the LACSB in 1984, which plans to use it to the turn of the century. Photographer Henry Tenby reports, "The PT6T-3 is appreciated by the sheriff's department since most rescues are done at 5000-8000 feet, which on hot days can equal 8000-12,000 feet density altitude. The TwinPac gives a welcomed safety margin in these situations. Remember, the S-58Ts operate at close to full power, and, compared to doing rescues at sea level, there is nowhere near the power reserve." (Henry Tenby)

was blowing, rain was coming down and darkness was approaching. While the pilots focused on keeping the Sea King in a steady hover, the crewmen lowered the makeshift sling. Grenier pulled it in, then clung to it for dear life as English and Marquis hauled him into the cabin. A few minutes later MacNeil set down at St. Hubert and Grenier was taken to hospital. MacNeil and his crew were praised for their work, and Sikorsky awarded him its famous "Winged 'S'", the most coveted award in the helicopter world.

TwinPac S-58

Since the fifties there had been talk of a turbine S-58, but market conditions were never quite right for a go-ahead. Finally, the S-58T's day arrived. In the early 1970s, Bob Daniell of Sikorsky recommended that the company do something about declining commercial sales. Several factors intrigued him: the traditional popularity of the S-58, which had been flying since 1954, the availability at good prices of many surplus S-58s from military operators, new market openings for a medium helicopter, and the P&WC TwinPac.

Sgt. Davis of the LACSB inspects the PT6T-6 while pre-flighting N724SB, March 26, 1988. The smaller photo shows installation of the R-1820 in a standard S-58. The bulky piston engine provided 1525 hp compared to 1800 shp from the compact PT6T. (Henry Tenby, Gary Vincent)

After consultation with Longueuil, Sikorsky launched a program to convert S-58s to the TwinPac, and Derek Emmerson and Alex Stewart of P&WC took over installation design. The TwinPac had been built to fit behind the rotor mast of a Bell UH-1 and drive forward into the transmission. The S-58 had its engine in front, so the TwinPac had to be turned 180°, which involved several key alterations. The first S-58T flew at Sikorsky on August 19, 1970, using the PT6T-3.

An immediate S-58T market was in offshore oil exploration with early operators including Bristow Helicopters in the North Sea, Chevron in the Gulf of Mexico, Airfast in Indonesia and Okanagan in several regions around the world. Carson Helicopters of Philadelphia used a fleet of S-58Ts offloading cement from ships at Jedda. This job put the TwinPac to the test—flying at high power settings in hot and humid conditions, and being subject to corrosion from the salt air and erosion from the ever-present cement. In Canada the S-58T proved a tough general-duty machine. In Ontario the provincial power authority acquired two S-58Ts for powerline construction. One was bought from Sikorsky, the other converted

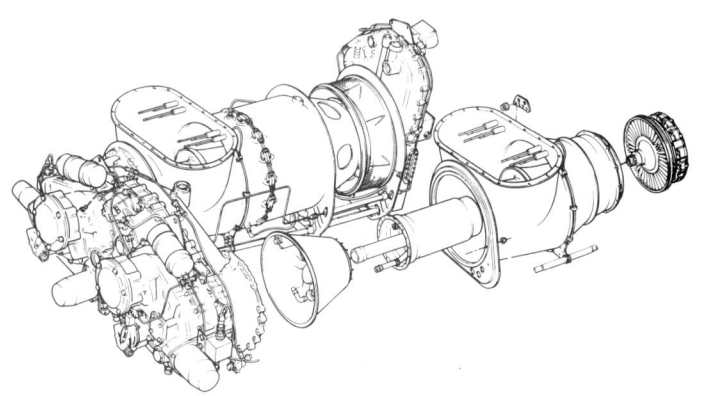

PT6T-6

A 1973 application of the PT6T-3 was in Sikorsky's ABC research helicopter. (Sikorsky Aircraft)

The S-76B with two PT6B-36 turboshafts. Derived from the popular S-76 II with two Allisons, it first flew in June 1984. It has double the useful load of the II. Its separate -36s represent a more powerful option to the P&WC TwinPac. The first EMS S-76Bs (two aircraft) were sold to the state of New Jersey for 1989 delivery. (Sikorsky Aircraft)

at P&WC. Okanagan bought its first from Sikorsky, then did several of its own conversions using S-58s purchased in Germany. By 1981 there were over 100 TwinPac S-58Ts all over the world. Sikorsky sold its licence to California Helicopter International, which did further conversions. Sikorsky turned to development of a new commercial design, the S-76, which first flew in March 1976 with twin Allison 250s. The S-76B followed in 1984 with two 960-shp PT6B-36s.

In 1988, California Helicopters was promoting the S-58T for use in the growing EMS (emergency medical services) field. EMS operator Faith Flight of Florida has used an S-58T since 1984 and praises such features as its high rotor blades (out of harm's way) and the separate cockpit which keeps the pilots away from the often-busy hospital environment in the cabin. The spacious cabin can accommodate a full range of medical gear, yet operating costs are only half of those of an EMS S-76. The TwinPac-powered "DC-3 of the helicopter world" may have yet another life extension of its own!

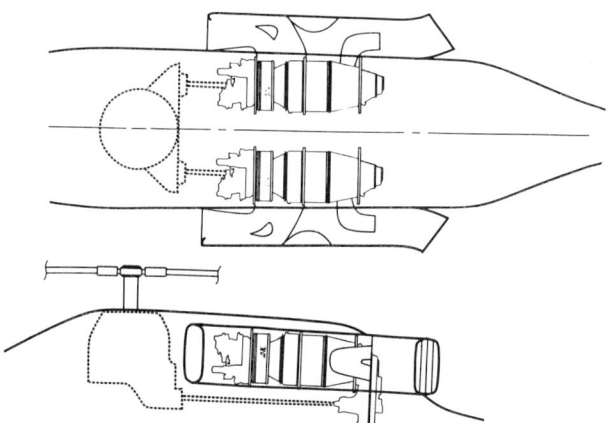

Engine installation in the S-76B.

From JT15D to PW300

Emergence of the PT6 set Pratt & Whitney Canada on a busy new course. So successful was the engine that it can be fairly argued that it changed the face of the aviation industry and engendered new aircraft designs. The PT6 was the engine that aircraft manufacturers had been waiting for, but P&WC knew that a single product line was insufficient in a competitive and ever-changing industry, especially with the piston spares business breathing its last. It also knew that the growth of a strong design and development organization required new and challenging projects.

A look at the mid-1960s market showed a gap at the lower end of the horsepower scale (85–340 hp). General aviation was producing 10,000 light aircraft yearly and required 15,000 engines. There was certainly room for a small turbine for the high end of this range but P&WC chose not to enter this market. Just beyond was the world of the PT6, where the fastest aircraft were the King Airs at about 270 mph. Then came the big jump to 500-mph executive jets. As yet, the market for corporate or commuter aircraft in the *400-mph* range had not opened up. P&WC had discovered this and explored engine design for such aircraft with Design Study 30 in 1958. This resulted in a turbofan layout. The "fan's" propulsive efficiency fit the speed between turboprop and turbo jet, but DS-30 was set aside as P&WC placed all its R&D energy on the PT6. Just then there seemed no market for a small turbofan, for even contemporary turbocharged piston engines were offering adequate power at 20,000 feet.

With the success of the King Air, Cessna was in a dilemma. There were thousands of customers flying its piston line for business. The Cessna 310 was a winner with over 2000 in use by 1965. The "push-pull" Models 336 and 337 had caught on, but Cessna had neither a larger model nor turbine power to offer its clients. The King Air was monopolizing the market, and Cessna was losing out. Cessna began studying the market between the turboprop and turbojet. The concept was for an aircraft powered by fan jets that would cruise at 400 mph. Its handling characteristics would let the pilot of a light piston twin readily step into the cockpit. A simple design with unswept wings would permit operations into airports forbidden to the "hotter" jets. Cessna had been kept well-briefed by P&WC about all its developments, including the 1800-pound thrust DS-32. Only a slightly more powerful engine would be needed. Cessna responded positively to the P&WC initiatives then contacted P&WA to gain its backing for the project. P&WC immediately launched development of a 2000-pound thrust turbofan, the JT15D. This designation came from the P&WA nomenclature handbook. "D" signified fan. The "15" simply denoted the P&W's fifteenth jet turbine design. Detailed design of the JT15D-1 began in June 1966 and an early change was from 2000 to 2200 pounds thrust as Cessna

Jim Crompton makes adjustments to an early JT15D. A pioneer design in the category of small fan jets, the JT15D produced 2200 lb thrust in its initial version. (P&WC)

refined airframe design. Turbine inlet temperature was increased, which in turn required cooling in the turbine inlet guide vanes. P&WC's research into centrifugal compressors and John Vrana's invention of the pipe diffuser had placed the company in the forefront of centrifugal compressor design. A single centrifugal compressor replaced a previously planned four-stage design in the new engine.

An on-going worry was the engine's large diameter, dictated by the single-stage turbine driving the fan. The engine ran for the first time with this arrangement on September 23, 1967, but at the insistence of Elvie Smith, who now headed engineering, work had already begun on re-design. Two turbines replaced the single unit to give the desired narrower engine nacelles; the two smaller stages would also facilitate future engine growth. The engine exceeded rated thrust on the second build in November.

The JT15D was operating in the test cells, but there were still no orders. Cessna was advanced in its design (now known as the FanJet 500), but in late 1967 delayed final go-ahead pending market studies. P&WC was preparing to flight-test the new engine. As its Beech 18 would not be suitable, a new test bed was found in the Avro CF-100. Many of these fighters were still in use as electronic warfare trainers with the Canadian Forces. The CF-100 could cruise at Mach 0.8, could reach 50,000 feet and had sufficient room for instrumentation. Aircraft 100760 was borrowed from the air force and arrived at P&WC's flight test centre at St. Hubert on November 22, 1967. Engineers designed a low-slung installation for the JT15D on the port underside of the fuselage. As this allowed for little clearance, engine intake doors were devised to prevent ingestion of foreign matter while the aircraft was running on the ground. P&WC's CF-100 "trimotor" made its first flight on August 14, 1968, and the JT15D was started for the first time in the air on August 22.

On September 16 P&WC received the initial order for 50 JT15Ds from Cessna, and the FanJet 500 mock-up was unveiled at the NBAA convention in Houston in early October. Del Roskam, president of Cessna, predicted that 1000 of the new jets would be delivered within 10 years. On September 15 the following year, the airplane made its first flight and was christened Citation after the famed race horse. The maiden flight was noteworthy for the quietness of the

Some of the key names responsible for the success of the JT15D/Citation combination look over technical drawings: Bruce Torell (P&WA), Malcolm Harned (Cessna), Elvie Smith, Ken Sullivan, Jim Taylor (Cessna), Thor Stephenson and Russ Meyer (Cessna). (Cessna Aircraft)

A JT15D mock-up used for fitment tests during modification work on the CF-100 test bed. (P&WC)

FROM JT15D TO PW300   251

CF-CPW, Citation No. 2, and the first of its type in Canada. It served P&WC for many years before being sold to a charter company. (P&WC)

engines. The engine layout had been skillfully tailored with this in mind; the fan had no stators ahead of it, and the stators aft were arranged in number and spacing to minimize noise. The FAA later specifically exempted aircraft "as quiet as the Citation" from noise curfews at certain airports. The Citation was indeed a neighbourly aircraft! The JT15D-1 received DOT and FAA certification in May 1971.

The Citation was certified on September 9, 1971, and soon began entering service. It was followed by the Citation I, and in 1976 the Citation IA was introduced with the JT15D-1A and increased wingspan. In 1977 the type was certified for single-pilot operation. The Citation II was certified in March 1978, having the JT15D-4, lengthened fuselage and wingspan, and increased fuel and baggage capacity. The Citation III flew in 1979, but, with gross weight over 19,000 pounds (compared with 12,500 for the Citation II), a bigger engine was required, and the aircraft was powered by AiResearch TFE731s of 3650 pounds thrust.

By mid-1988 there were more than 1350 Citations and Citations Is and IIs in service. Del Roskam's prediction 10 years earlier was not as overly optimistic as some may have thought. The first growth version of the JT15D-1 was the -4.

For the first decade of production, the -4 (shown here with Donald C. Lowe) was the primary version of the JT15D. (P&WC)

Airline operator Carl Millard poses beside his Citation (JT15D-1) and one of his ancient but still useful Beech 18s (R-985 Wasp Junior). (Larry Milberry)

The addition of an axial boost stage to the low rotor shaft in the -4 increased the primary air flow by 20% and the overall pressure ratio by 15%. The -4 is an engine of slightly lower bypass ratio than the -1, but with a rated takeoff thrust of 2500 pounds. The -4B, -4C and -4D versions followed. The -4 first ran in the test cell on January 28, 1972 and within two hours had covered the full range of power. The maturity of P&WC's engineering organization was paying off.

The next application of the JT15D was in a French design, the Aérospatiale Corvette. Like the Citation, it was a small "bizjet" with two engines. The first Corvette flew in July 1970 with JT15D-1s but crashed as a result of a control problem. The second example flew with -4 engines in December 1972. The Corvette never really "got off the ground" in sales. The first customer was Air Alpes (one of the earliest Twin Otter operators) but few additional sales followed. Some claim that this was because of the head start made by the Dassault

The Beechjet 400 with JT15D-5s originated with Mitsubishi. (Beech Aircraft)

Falcon, which first flew in 1963 (with P&WA JT12s, but only in the prototype). While only 40 Corvettes were built, the figure for the Falcon exceeds 500.

In 1963 Mitsubishi of Japan entered the corporate aircraft field with the MU-2 light turboprop. With experience from this aircraft (over 500 built), Mitsubishi began work on a turbofan bizjet. After a seven-year engine selection process, it chose the JT15D-4 to power its aircraft, the Diamond I. The first prototype flew on August 29, 1978, and customer deliveries began in January 1982. Major components were made in Japan, then shipped to Mitsubishi's San Angelo, Texas, facility for assembly and completion. This proved costly in time, especially with the language barrier and long communications line between Japan and Texas. Although Diamond I "booked" orders had reached 200 at one point, this rosy picture faded. The aircraft grew heavier in design, so its announced performance could not be met. Delivery dates slipped. Then the economy took a downturn and orders were cancelled for corporate jets of all kinds; deliveries worldwide of 1450 aircraft in 1981 fell to 380 for 1985. Piper closed two of its three plants. Beech was taken over by Raytheon, Cessna by General Dynamics. The character of the business was changing as conglomerates swallowed up family-run companies. Mitsubishi solved its performance problem with the uprated JT15D-5. Its modified aircraft was the Diamond II (first flight, June 1984). The company, however, disenchanted with corporate aviation, sold out to Beech in 1985 having marketed just six Diamond IIs. The aircraft is today's Beechjet 400.

Work on the JT15D-5 had begun in late 1977 to meet customer requirements for more power and improved fuel economy. The -5 grew progressively from 2700 to 2900 to 3200 pounds thrust before DOT certification was granted in December 1983. A demonstrator engine had flown in April 1978 and prototypes were delivered to Cessna in December 1982. The -5 demonstrated a 25% increase in altitude cruise thrust with a 3% improvement in specific fuel consumption compared to the -4. In 1987 Cessna announced the -5 for use in its new Citation V.

JT15D development has involved over 20 experimental and 17 customer prototype engines and is still under way using four test cells for performance analysis and endurance testing.

The bird cannon (the long barrel at the left) aims directly at the intake of a JT15D. The battery of lights was needed for filming. (P&WC)

Specialized facilities have been built for icing, foreign object ingestion, sound survey and inlet distortion tests of complete engines. Each production engine built is test run for four hours. During performance testing of the -4 in 1973 an engine was shipped to a UK facility for tests at simulated altitudes as high as 68,000 feet. The major structural cases of the engine have been analysed and tested using strain gauges, temperature survey and various destructive test techniques. Drop tests of complete engines and test cell running at more than 120% engine rating have demonstrated the high safety margin of the intermediate, gas generator and exhaust cases, as well as the whole engine. The JT15D can operate with all commercial jet fuels, on kerosene, JP4 and JP5 jet fuels, and for up to 50 hours in any TBO period on all grades of aviation gasoline. No change to engine combustion or fuel control components is necessary when switching from one fuel to another. Foreign object ingestion tests have included subjecting the JT15D-1 to

The testing of engines at P&WC has to be merciless at times. The results are sometimes catastrophic. Such testing gives the durable, efficient engines that have made the company famous all over the world. (P&WC)

Technicians work on JT15D assembly in Longueuil. This work has since been transferred to Mississauga. (P&WC)

Rotor assembly for the prototype JT15D-5. (P&WC)

sand, gravel, water, cleaning cloth, sheet ice and birds.

Protection from foreign object ingestion was a new worry for P&WC. The PT6 was not subject to this as its compressor inlet was at the rear and had a screen protecting it. To test the JT15D's survivability, an air gun was used to shoot dead chickens weighing as much as four pounds. On the first shot, the bird struck the nose bullet, breaking the bearing support structure of the fan stage. A redesigned support survived the next shot. The next tests involved showering the fan with a quantity of small birds, simulating flying through a flock of starlings. Considerable damage was done to the outer diameter of the fan where the blades have thick leading edges. The appropriate components were strengthened. In the next engine model, the fan blades were strong enough from the outset and required no stiffening. In the JT15D-5, small birds posed little hazard, but larger birds were still a threat, and their impact tore pipe diffusers loose and penetrated the engine external casing. Thicker casings had to be used. The JT15D became the first engine in its class to successfully complete such bird tests and comply with the FAA foreign object requirements.

Citation Adventures

Paired with the JT15D, the Cessna Citation has long since proven to be a winner. In 1988 it received the prestigious Collier Trophy for Aeronautical Excellence based on its consistent safety record. Many pilots have come to know this Cessna-P&WC thoroughbred, one of the most experienced being Lloyd Hubbard. He started flying early in the war, doing a tour on Blenheim IV bombers. After the war, he flew Sabres in the RCAF and was one of the famous Golden Hawk air demonstration pilots. Later he flew CF-100s, and in 1968 he left the RCAF to join P&WC, initially on the CF-100/JT15D program. He remained with P&WC for nearly 10 years, part of the time as chief pilot.

Since he left P&WC, Lloyd Hubbard has spent another decade flying the Citation. He and the Citation just seemed to get along like the best of friends. One flight was from Manila to Seattle via Okinawa, Sapporo (Japan) and Adak in the Aleutians. En route from Sapporo, an equipment failure forced Hubbard and fellow pilot Mike Rizzuto to navigate by dead reckoning, and they arrived at Adak only a few minutes later than flight-planned. On a flight that originated in Florida,

Hubbard had a double engine failure at 39,000 feet while doing a demonstration flight with a prospective customer. The engines would not re-light. Hubbard called air traffic control and got a vector to the airport at Grenada, Mississippi. The Citation emerged between cloud layers at 4000 feet. Back into cloud, the aircraft entered a furious rainstorm, but ATC had steered it within visual range of the end of the runway. Hubbard touched down safely, but he and his two passengers had to wait for 30 minutes for the rain to let up. When he stepped onto the tarmac, Hubbard admits that his knees were still a bit shaky. An investigation uncovered the cause of the incident—the fuel supply had been contaminated by water when he had refuelled earlier in the day.

In another adventure Hubbard was flying high over Washington state when he noticed a US Navy A-6 Intruder loitering about a mile away. The A-6 turned in to do a run on the Citation. Hubbard thought, "I'm going to show you something, ol' buddy!" He pulled right into the A-6, head on. It turned away but the Citation got on its tail. For the next few minutes, the Navy pilot tried his best to escape but until Hubbard broke off to land was fighting a losing battle. The Navy pilot just wasn't prepared to tangle with a nimble Citation flown by a veteran Sabre jockey! In 1988 Lloyd Hubbard was flying a Citation II for Siete Oil and Gas Corp of New Mexico.

The JT25D

By the early 1970s, small pioneer jet engines such as the JT12 were obsolete. They were noisy gas-guzzlers but popular because of their dramatic performance improvements. The door was open for replacements, and Garrett moved in with the TFE731, certified in August 1972, with the first examples going to Dassault for the Falcon 10. At about 3200 pounds thrust, the TFE731 was more powerful than the JT15D.

In 1971 P&WC had formed its advance design engineering department to specialize in conceptual or preliminary design. To date, there had been little need for such a department—so long as the PT6 remained the company's only product. Introduction of JT15D, TwinPac and PT6 growth versions changed this. Under Ron Morris, 15–20 people from various specialties came together with the goal of generating dozens of design studies each year. These would test the feasibility of

Ken Sullivan discusses the JT15D in 1984 with retired company secretary James Ross and director of materials Maurice Ménard. (P&WC)

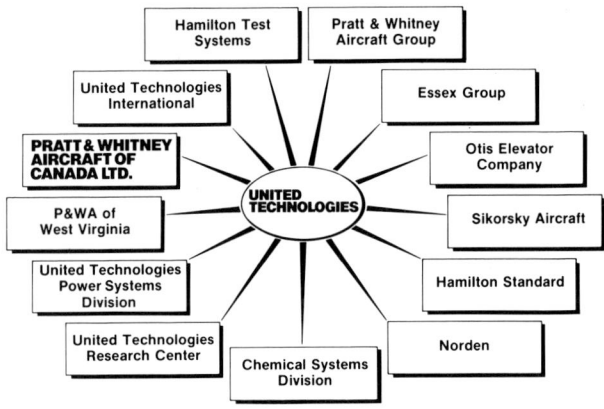

The United Technologies companies as of 1977.

new engines in terms of cost, weight, performance and customer interest.

In 1973 Cessna, concerned about Lear's plan for a transcontinental jet, had talked to P&WC about an engine in the TFE731 range but with greater fuel economy. A study was made of a low-cost project based on a PT6A-41 gas generator and a JT15 fan, but uncertainty over the looming energy crisis brought this project to a halt. P&WC next considered an engine beyond the growth potential of the TFE731. This was study ADS509 (later designated the JT25D) of 5000–7000 pounds thrust, but as design and market studies progressed, it was clear that the JT25D would be costly. P&WA suggested that Longueuil pool its resources with Rolls-Royce (with whom P&WA was collaborating on a large fan engine). The Bristol Engine Division of Rolls-Royce was testing its own RB401 in the power range of the JT25D. The RB401 became the lead project because of its advanced development status, lighter weight and smaller diameter. P&WC contributed in such areas as fan blade design.

A budget of over $185 million was foreseen to bring the RB401 to certification, the sum to be split about equally between the partners. The Rolls-Royce/P&WC engine was not to be. It was determined that Washington would rule that there would be disadvantages to US users (the US was envisioned as the prime market) in an engine developed jointly by two foreign nations. This made a joint venture almost impossible and the partnership was dissolved in 1976. A few years, later joint-venture engines became common as development costs put most programs beyond the reach of single manufacturers.

Turbofan Trainer

In the mid-1970s P&WC and the Agusta subsidiary SIAI-Marchetti began discussing a new turbofan military trainer. After eight years of study, the S.211 was announced and the JT15D was the engine selected. SIAI-Marchetti was aiming at the market between turboprops and larger jets. As rival Aermacchi had long since captured the Italian domestic market for jet trainers, the S.211 would have to be an export model. The first S.211 flew in April 1981, the first production model in October 1984. Singapore ordered 30, of which 20 were supplied as kits to be assembled overseas. An inverted oil system was added to its JT15D-4C to permit full aerobat-

The SIAI-Marchetti S.211 with the JT15D-4 is one of the lightest jets (3560 lb empty) and is intended to accelerate *ab initio* pilot training for smaller air forces. (via P&WC)

ics. The result of mating the S.211 and the P&WC turbofan has been an economic yet high-performance trainer/light attack aircraft. In 1987 an aerobatic version of the -5 was offered to SIAI-Marchetti in response to their study for an improved S.211.

PW100: A New Turboprop

In the mid-1970s the focus of advance design and marketing moved to a larger turboprop (1500–2500 shp). With no competitor in this area, there was the luxury of time for studying requirements. A new turboprop business plane that would outperform existing turboprops and compete directly with corporate jets seemed the first likely application. The energy crisis would, P&WC thought, spur on such a project. Meanwhile, the commuter market seemed to be responding slowly to the Civil Aeronautics Board's rule changes allowing use of larger aircraft. The first drawings of a potential new P&WC turboprop (three-shaft, 15:1 pressure ratio) were made by Mike Stoten in August 1974, but the concept was dismissed as impractical. A few months later, however, the idea was revived.

To find a market, the new engine would have to be low in cost and have a high pressure ratio of 15:1. On big fan engines such as the JT9D it was now common to use axial compressors to achieve high pressure ratios. Vane angles changed depending on engine speed, with different settings for starting, idling or running at maximum rpm. GE had used variable geometry in its T700 turboshaft helicopter engines—a complex and costly approach that had been funded by the US military. P&WC viewed variable geometry as expensive and failure-prone and preferred a pair of centrifugal compressors in series, running on separate shafts, with a third shaft providing output power. It had never carried a three-shaft engine into development. With three concentric shafts running at high speed in different directions, there would be many technical questions. To find answers, P&WC built its first technology demonstrator engine, the TDE-1 of 1665 shp. It didn't have to be the exact size, so long as test results could be scaled to customer demands. Design of the TDE began in October 1976. The first gas generator ran in December 1977 and the first complete engine in August 1978.

P&WC's original gas turbine team had been technically talented but green in the art of engine development. For the PT6 they had used test rigs to certify design assumptions. This was time-consuming and was displaced in 1961 by P&WA's build-them-and-break-them approach. Now new engines moved quickly from preliminary design to prototype with a minimum of rig testing. The use of a "proof of concept" TDE would speed development further and save money. There were only minor problems with the TDE. Shafts were sometimes cut to pieces at high whirl speeds. Had this happened in full-scale development it would have cost millions. Total TDE costs were a modest $5.5 million. Many design adjustments were incorporated before full-scale development was given the green light in mid-1979. By then, the results of airline deregulation in the US had made the commuter airlines the prime market for the new engine; a profusion of new bizjets cut short the hope of a large turboprop competing on the corporate market.

Philippe Nadeau makes adjustments to the TDE-1 technology demonstrator engine for the PT7A-1. In this early photo for publicity purposes, the pipe diffusers, so much a trademark of the PW100 today, were air-brushed out. (P&WC)

US airlines had long been closely regulated by the Civil Aeronautics Board along the lines of public utilities, but by the 1970s the time had come for air transport to take its place with competitive segments of the economy. The Airline Deregulation Act of 1978 ended 40 years of government control and opened the door for commuter airlines to take over service to small and medium communities. The larger carriers were allowed to concentrate on higher-density ones. Raising the limit on commuter airliner capacity from 30 to 60 seats stimulated design of aircraft larger than the stalwart Twin Otter and Beech 99. The bigger commuters had been mainly older types such as the F-27 and HS748, but new ones began appearing, including the Short 330 and the Dash 7. In this revolution, P&WC saw markets for two new engine families: a more powerful PT6 and a new concept based on the TDE.

P&WC felt that more power than could be produced by the TDE (obtainable by increasing air flow) would be necessary. The market it sought had long been dominated by the Rolls-Royce Dart in the 1500-2250-shp range. Designed in 1945, the Dart had a fine operational record but was aging. GE was interested in the same market with the CT7, a version of its T700. P&WC knew it would be advantageous to offer an engine with a range of shaft horsepowers, for aircraft manufacturers would be turning out a variety of commuters in the

An HS748 at work in northern Ontario. The 748 was one of a series of 1950s transports using the Rolls-Royce Dart turbine. By the 1970s they were all needing replacement. The appearance of the PW100 was just what the airframe manufacturers had been waiting for, and spurred them on to new designs such as the ATP, itself based on the 748. (Larry Milberry)

 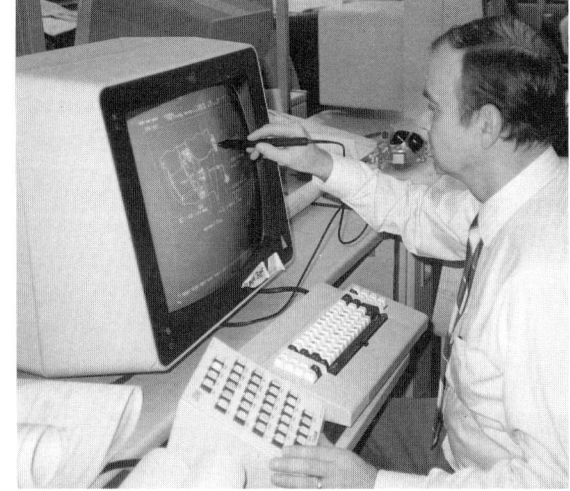

The computer began making inroads at P&WC in the 1970s and is today the company's main tool for design and analysis. In the second photo, Wolfgang Doehl works with CAD equipment. Both photos are in P&WC's R&D building at Plant 1. (Larry Milberry)

30-60-seat class. Thus its starting point rose to 1800 shp with growth potential to 2600. As work progressed, P&WC released sufficient information to the aircraft manufacturers to encourage them in their own design studies.

The engineering team for the new engine (designated PT7A-1) was formed in February 1979 and full-scale efforts began in June. Manpower requirements on the project quickly mounted: 34,000 hours on design for the third quarter of 1979; 105,000 for the last quarter of 1980. The gas generator made its first run that December, the first complete engine the following March.

Dave Cook was the first project engineer on the engine. He had started his career at Armstrong Siddeley, moved to Orenda, then to P&WC in October 1957. He had been project engineer on the original PT6A-40 and took over the TDE from Pete Peterson in the late seventies: "We did several complete re-works on the PT7, including redesign of each component at least once. We did this in response to failures and non-performance of parts that one finds in any development program. It was serious work, but looking back, some of the problems we faced seem rather amusing. In one period the oil system was a major irritation, and we didn't enter the test cell without an umbrella, there was so much oil dripping from the ceiling. We gave ourselves four years to take the engine to certification. Despite many difficulties, we met that goal."

To keep the PT7 on schedule and in step with changes wrought by deregulation, P&WC had to add space and hire new engineers. Longueuil was stretched to capacity, forcing the company to rent space in a shopping plaza near St. Hubert, four miles from Plant 1. Later, a vacant furniture store became temporary quarters for the product support group. There was a temporary problem as the labour pool in Montreal dried up. The company opened an office in Toronto, Canada's second major aerospace centre, which provided much of the engineering talent needed for the new engine program. In 1980 P&WC employment hit a new high of 8300.

Besides space and labour shortages, P&WC also faced heavy costs. Engineering expenses were absorbing 22 cents of every sales dollar. The project would require new machines and tools. Changes in engineering and production were required—computer-aided design (CAD) and computer-aided manufacturing (CAM) were becoming a way of life, offering improved flexibility and repeatability. "Robotics" was a new buzz word and could not be overlooked. Budgets had to cope with demands for the computers needed by every department. United Technologies insisted that senior managers accept such

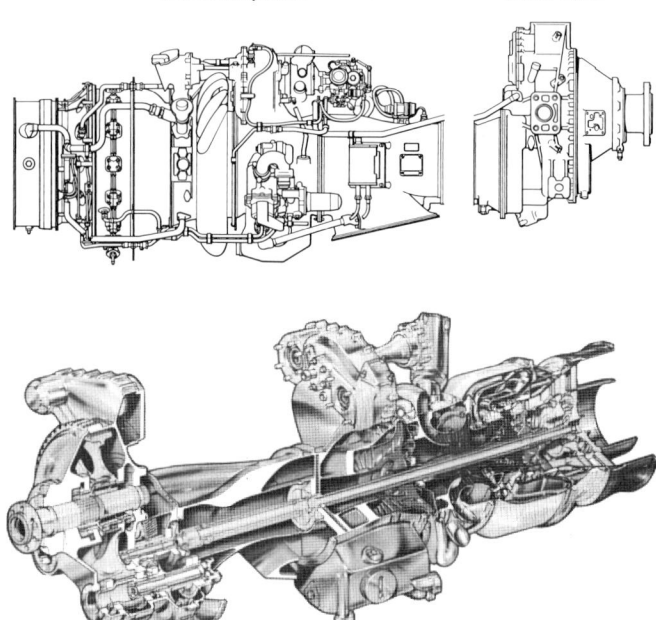

The top diagram shows the PT7 in its two basic modules. Below is a cutaway showing the two modules joined by a cast structure.

new technology. Recent graduates had been trained on computers, but to the older employees computers were new, even threatening. A program was established that sent senior personnel to Hartford on a rotational basis for three-day computer immersion courses. This lasted two years. As each manager returned to his desk he found a complete PC set-up in his office. The equipment could be taken home for after-hours practice.

The resulting PT7 had the basic qualities of high performance, growth capability, rugged simplicity and high reliability with long "critical parts" life, all at a price manufacturers and airlines could live with. It had turbomachinery (compressor, turbine and accessories section) in a module and a reduction gearbox module coupled by a cast structure to eliminate alignment difficulties, simplify mounting and form part of the air inlet duct. The turbomachinery (with a design life of 30,000 cycles) had only six major rotating components to form a three-shaft/two-spool core, with compressed air produced by two centrifugal compressor stages driven independently by single-stage compressor turbines. The advanced compressor reflected years of research and experience with the PT6 and JT15D, and provided high pressure ratio and efficiency over a wide operating range. The two-spool allowed both compressors to run optimally without complex variable geometry or overboard bleed. Pipe diffusers and a reverse annular combustor assured clean, efficient and quiet combustion. High pressure stator and turbine blade cooling were used.

A third concentric shaft coupled the two-stage power turbine to the reduction gearbox. The gearbox was slightly elevated to provide the best installation compromise for high- and low-wing aircraft. The free turbine concept as used in the PT6 permitted operation at peak component efficiency while reducing external noise. The PT7 was designed to grow by the mid-1990s from 2000 to 4000 shp. Simultaneous development of new turbomachinery and gearbox sections was the heart of company strategy to create this engine family. By mixing and matching the various modules, a broad range of power could be offered.

Work on the PT7 (including work on the initial growth versions) was well advanced before the first customer was signed. Once again, P&WC was in new sales territory, as Ken

Sullivan recalls: "Everyone was in the guessing stage in early PT7 days. We heard arguments advanced for various size aircraft. We were doing our best to anticipate which directions the market would take. We could talk about the success with our technology demonstrator, but were still talking from paper about the PT7. The customers had many concerns—they wanted to be assured that our engine would be available when they needed it, and at the power, weight and price that suited them."

Careful market research enabled P&WC to present detailed data to potential clients. It was able to demonstrate, for example, PT7 operating cost advantages. Much had been said by competitors about fuel burn, specific fuel consumption and operating costs, but always of engines derived from military projects, then adapted to civil use. The PT7 was *purpose-designed* for the airlines, a valuable advantage. By the time P&WC had signed its first sales contract for a PT7, it had determined that average stage lengths for the commuter airlines would be increasing 2% annually. Deregulation was opening up longer stage length "city pairs" to commuters. Larger commuter aircraft would fly at higher altitudes where they would achieve their best speeds and where the turboprop ran more efficiently. The combined power and prop advantages of the PT7 would provide a 30-40% higher rate of climb (e.g. to most economical altitude) than any competitor.

In September 1979, Embraer chose the PT7 for its new EMB-120 commuter, and in 1980 de Havilland Canada chose it for the Dash 8. This was just as general aviation began a slump brought on by a downward turn in the global economy. By mid-1981 corporations and commuter carriers were deferring orders for new aircraft, and manufacturers and distributors were staring at unsold inventories. Several airlines which had sprung up in the US, with fleets of Dash 7s and other new types, foundered and went under. Corporations began closing their private flight departments. By September 1981 P&WC was projecting a 25% drop in sales for the next year, but with 14-18-month lead times on most engine parts, it was already well into procurement for 1982–1983 production. The company had to do some quick belt-tightening. Some hourly workers were laid off and other staff put on a four-day week. There was a 10% pay cut for management and technical staff (realized by placing them on a 4.5 day week). Negotiations with the union brought a work-sharing plan which stabilized the number of hourly workers while reducing total hours. There was cost-cutting across the board. Such measures were all in place within three weeks of P&WC detecting the economic slide. For such a large and complex company, its ability to react quickly (as it had during the strike) proved vital. So did its big gamble to *accelerate*, not retard, work on the PT7. This decision would help P&WC capture a large slice of the market for new-technology commuters that was beginning to emerge. Employment in the company had fallen to 6300 by 1982.

The PT7 first flew in the Viscount test bed on February 27, 1982. By the time the first production engines became available, P&WC had eight in-house experimental engines in the test cells at Longueuil, more than 15 prototype engines for the first flights of customers' prototype aircraft, and several engines per customer for the completion of development certification work. As if engineering didn't have enough work to do, it also had to contend with interference from competitors. One day P&WC received a list of some 30 technical questions from a prospective customer. It was clear that these had really originated with one of the competing engine makers, but P&WC still had to provide the answers!

Market requirements had helped shape the P&WC design concept, but a team effort by marketing, product support and engineering was needed to win each sale. All the aircraft involved were subject to US FAR-25 requirements, the same standards that governed certification of the biggest airliners. Companies such as GE, Fairchild and Fokker had worked to these rules for years, but much of this was new territory for P&WC (the PT6 usually powered aircraft in a lower gross-weight category that were not subject to FAR-25).

PT7 (PW100) Development

Design team formed	February 1979
Design start-up	June 1979
First run gas generator	December 1980
First run complete engine	March 1981
First flight in test bed	February 1982
Prototypes available	January 1983
Engine certification	December 1983
First production delivery	January 1984

The PW115 awaits its first flight in the Viscount test bed. It is in a Brasilia nacelle. The second photo shows a DHC power plant system test stand with a PW120 and Hamilton Standard 14SF propeller. (P&WC, DHC/Charles Bryant)

264 POWER: THE PRATT & WHITNEY CANADA STORY

Even while the TDE was in the test cell, P&WC knew of the large commuter designs of Embraer and DHC and was sizing the PT7 accordingly. Embraer had been studying a Bandeirante successor since 1974, and gained further experience with the EMB-121 Xingu, introduced in 1976 with PT6A-28s. Often when Embraer and P&WC sat down to talk business, the matter of a larger aircraft would come up. Deregulation in the US was the catalyst that finally moved Embraer to begin work on the EMB-120, planned as a simple yet elegant passenger commuter with an impressive cruise speed of 300 knots.

Embraer foresaw a 1200-1300-shp turboprop for the 120, but P&WC convinced it of the merits of 1500 shp. A special feature of the engine would be an advanced hydromechanical unit/electronic control system developed by Hamilton Standard to eliminate the need to compute power settings. The power setting curves would be stored in an electronic control unit memory that would sense and control engine acceleration. With it and the HMU providing all essential fuel control functions, pilot workload would be greatly eased. Not all was smooth sailing for P&WC, however. GE, eager to place the CT7, offered Embraer the use of its large airline marketing organization to help sell the 120. This was tempting, but the Brazilians maintained their loyalty to P&WC based on the successful Bandeirante/PT6 combination—especially the PT6 record in TBOs. P&WC suggested that the PT7 would have a 6000-hour TBO.

In 1980 P&WA introduced a new system of engine designation. The old one was becoming too cumbersome, as shown by one model, the JT9D7R4D1. United Technologies president Bob Carlson introduced a simplified system adapted from his background in the heavy equipment business. Henceforth, P&W engines would have the "PW" prefix. For P&WA civil engines a four-digit number starting with an even digit would be used (odd digit for military engines). P&WC engine designations would be three-digit, the first representing the engine family, the other two shaft horsepower. For example, PW115 means 100 series family and 1500 shp. The PT7 became the PW100. Hartford also decided to replace the traditional P&W eagle with a modern, stylized one. To many of the company's customers, the old eagle was a prized symbol found not just on engine nacelles but on tool boxes, tie

The famed symbol of P&WA in one of its pre-1981 versions.

The short-lived stylized Pratt & Whitney emblem.

The traditional P&W emblem, now restored after a brief period in retirement.

The first complete PW115 in the test cell, and a PW120 on a mobile test stand during acoustic trials at Mirabel airport. Microphones on the mast pick up the sound. (P&WC)

clips and belt buckles all over the world. The new eagle took a lot of abuse and the old one, grounded in 1981, made a comeback in 1987.

Embraer had been counting on continuity of engine designation in its marketing plans for the 120—the reputation of the PT6 would be carried over to the PT7. Embraer was a bit surprised at the change to PW100 and P&WC made a trip to Brazil to explain. While there, Ken Sullivan also discussed Embraer's plan to name the 120 "Itaipiu" after a large power project in southern Brazil. Sullivan explained that "Itaipiu" was a bit of a tongue-twister outside Brazil and offered his own suggestion, "Brasilia," a name that would clearly identify the origin of the new commuter. Embraer quickly agreed— both engine and airplane now had new designations. The Brasilia flew in 1983 with the PW115 (Series 100, 1500 shp) and won FAA certification in July 1985. Before long Embraer

A side view of a production PW100. (P&WC)

moved up to the PW118 and by late 1988 more than 150 Brasilias had been ordered.

Market analysis had shown de Havilland Canada a gap for commuter aircraft between Twin Otter and Dash 7. In September 1980 it had the go-ahead for a design to fill that gap with the Dash 8, for which it selected the PW120. Prototype engines were needed by late 1982 for initial flight testing of both the Brasilia and Dash 8, with certification and production engines to follow within a year. SAAB of Sweden was also working on a new commuter, the 30-seat SF340. SAAB had joined forces on the project with Fairchild, and it was at Fairchild's Long Island facility that the SF340 engine competition focused. Derek Emmerson was involved: "The SF340 competition was where we really started to see how GE did business. GE offered guarantees covering reliability and operating costs and excellent terms to secure a launch cus-

FACING PAGE (From top left) Fitting the first PW 115 to the Brasilia prototype, May 12, 1983. The massive Embraer complex in Brazil; it expanded quickly with the success of the company's P&WC-powered aircraft. The Brasilia roll-out, June 29, 1983—it had first flown two days earlier. (Embraer)

(Top) The aesthetic Brasilia in flight with a Tucano (PT6A-25C) at the 1988 Farnborough air show (Richard Beaudet)

(Centre right) A Brasilia of SABENA affiliate Delta Air Transport. More than 150 Brasilias had been sold by late 1988 when the monthly production rate had reached five aircraft. (Embraer)

(Right) A West German Brasilia at Glasgow, July 1987. (Wilf White)

A Dash 8-300 awaits its PW123s at Boeing Canada in June 1988. (Larry Milberry)

The left PW120 fitted to Dash 8-200 No. 107 on the line at Boeing Canada in Toronto. In October 1986 P&WC won top honours in the Canada Awards for Excellence competition for the PW100. The decision to push ahead with the PW100 during the recession of the early 1980s was a bold gamble that worked. Dave Caplan has noted, "Our company would today be just a shadow of itself had we not invested heavily in R&D during the recession." (Larry Milberry)

Air Atlantic and Air Nova operate Dash 8 fleets in Canada's eastern provinces. The new breed of PW100-powered commuters has won the praise of travellers around the world. (Larry Milberry)

In the form of this Talair Dash 8, DHC and P&WC have combined to send the latest in Canadian technology to Papua New Guinea. (DHC)

tomer. This forced us to look at our own methods, and change our approach to marketing." GE won the SF340 program with its 1735-shp CT7. It also won the 45-passenger CASA (Spain)/IPTN (Indonesia) CN-235.

The next competition was for the Avions de Transport Régional ATR42, which evolved from two earlier ventures, the Aeritalia AIT-230 of Italy and the Aérospatiale AS-35. The two companies pooled their resources and agreed to split the work about 50-50. The engine competition was intense, and at the Paris Air Show in June 1981, the PW120 was announced the winner. The ATR42, planned as the first in a series of commuters, flew in August 1984. While ATR began with the PW120, P&WC agreed to develop versions as required. The stretched ATR72 was announced in 1985 with the PW124. A novel feature for the ATR42 would be "Hotel Mode" wherein a propeller brake would be fitted to the right engine. On the ground, the brake could be engaged and the engine run to provide the services associated with an APU. ATR estimated that this would save $17,000 for an operator running its airplane 2000 hours a year.

In 1981 British Aerospace, already vastly experienced with its BAe748, requested proposals for its new commuter. Fokker

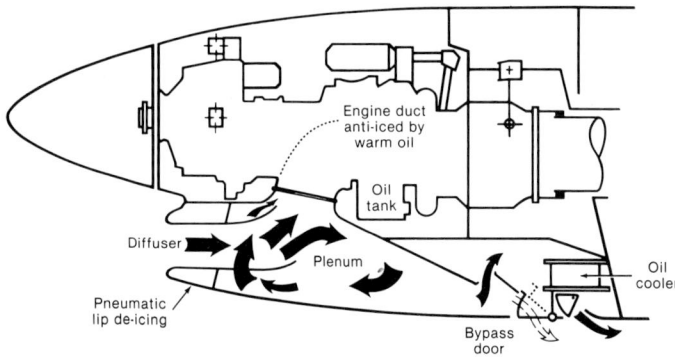

PW120 cowl and engine intake on the Dash 8 prevents ice and other foreign matter from entering the engine.

The ATR42-300 production line at Toulouse, with the nearest aircraft fitted with their 1800-shp PW120s. (Aérospatiale-Aeritalia)

An ATR42 belonging to a Taiwan airline at Glasgow. It carries pre-delivery French registration letters. Orders for the ATR42 exceeded 250 by the end of 1988, and there were already orders for the ATR72 (PW124s). To meet demand, P&WC was building about 30 PW100s monthly in 1988. (Wilf White)

272 POWER: THE PRATT & WHITNEY CANADA STORY

of Holland was also anxious to enter the market, and announced its contender based on the tried and true F-27 Friendship. P&WC had anticipated these projects—neither manufacturer was likely to sit and watch the market built up by the 748 and F-27 be taken over by newcomers. In each case, P&WC would be competing against the Dart, which Rolls-Royce was offering in a modernized version. Rolls also considered a new engine based on the RB401 study. For P&WC, the BAe and Fokker entrants would mean a 2400-shp engine, but the basic PW120 was still being developed and was not yet certified. The PW124 would have to be started up immediately even though not all the important lessons from PW120 development had yet been learned. The PW124 was launched in July 1982, its added power to be achieved by "boring out" the low-pressure compressor to increase air flow through the engine by 15% and the pressure ratio by 23%. A new gearbox was also needed to drive a six-blade propeller.

At the 1982 Farnborough Air Show, British Aerospace announced the PW124 as the engine for its forthcoming ATP (Advanced Turbo Prop) commuter but it was not able to confirm an engine order until government approval. (Program launch came in March 1984.) Fokker was yet to sign, and Ken Sullivan went to Amsterdam to win over the Dutch. He had their agreement in December 1982 and a contract the following May. The new aircraft was named the Fokker F-50. P&WC executives remember this episode as the high point of Ken Sullivan's sales career, for it assured plenty of new business into the 1990s. Sullivan saw it another way: "When I got back from Europe I disappeared into the country for a week. This was the last program I could see for the PW120 for some time, and the next few years would be mainly ones of consolidation. I had always been more interested in winning new installations, rather than in supervising day-to-day, after-sales activity. When I couldn't see a new challenge, I realized it was time for me to retire."

Another departure in early PW100 times was that of Don Lowe. He left to join UTC in September 1980. Lowe had felt a need for P&WC to have a more technically minded leader for the new era it was entering. He backed Elvie Smith as his replacement as president and CEO. To Lowe, Smith was the ideal choice, with his long association with the company and natural leadership qualities. Smith went on to become chairman and CEO in 1984. Lowe spent two and a half years at UTC then returned to Canada, holding key positions at Allied Chemicals, Kidd Creek Mines and Bombardier. In 1986 he became chairman of Canadair.

A Fokker/Fairchild Friendship (left above) with a version of the 2050-2300 shp Dart 511. Nearly 800 Friendships were built. Their popularity with airlines led Fokker to re-engineer the basic airframe, add PW124s, and come out with the Fokker F-50. (Larry Milberry)

Ansett of Australia was launch customer for the Fokker F-50 (left), though DLT of West Germany was the first to operate the type in August 1987. (via P&WC)

The first installation of a PW123AF in the Canadair CL-215T was made in late October 1988. The CL-215T will replace the tried and true version with R-2800s. (Canadair Ltd.)

(Above) The first 64-seat BAe ATP was flown August 6, 1986. Design is based on the earlier HS748, and uses the PW124A (or 126) with six blade BAe/Hamilton Standard propellers. This British Airways ATP was at Farnborough in September 1988 (Richard Beaudet)

(Right) Testing a Hamilton Standard/British Aerospace 6/5500/F propeller for the ATP. The engine is a PW124. (P&WC)

When Lowe left P&WC, he also supported L. David Caplan's advancement. Caplan, a McGill graduate and chartered accountant, had joined the company in 1964. He became vice-president of finance and administration in 1976 and executive vice-president in 1980. He became president in 1984, and president and CEO in April 1985.

Airline Support

To prepare for the PW100, P&WC's product support organization was reorganized. Its focus had been corporate aviation, but now it would be dealing with regional airlines. When Gene Schweitzer retired in 1980, the company hired him back to lay the groundwork for PW100 operations: "A lot of people felt that this was P&WC's major entry into the airline business. I took a different view. Going out and calling on big US commuters like Henson and Ransome was much like visiting TCA and CPA 35 years earlier. They were larger, but at a similar point in their development."

Under Schweitzer's guidance P&WC formed the basis for its airline support group, set up to assist airlines with every aspect of beginning PW100 service, from selecting tools to organizing maintenance records to deciding what procedures best suited its aircraft and routes. The group included such experts as former chief engineers from the airlines, marketing people and engineers. P&WC also laid the groundwork for an international network of service centres for the PW100. The first US independent base was Neosho Teledyne. Aérospatiale set up a similar base at Le Bourget.

Elvie Smith (left) and Dave Caplan had become the senior executives at P&WC by the mid-1980s. (P&WC)

Gene Schweitzer travelled Canada and the world for P&WC for 40 years. His final assignment was to lay the groundwork for the PW100 airline support group. In this snap he is seen on the Amazon in the 1960s where he was assisting a Twin Otter operator. (P&WC)

PW100 Installations 1982–1989

Country	Aircraft	Engine	First Flight
Canada	Viscount	PW115	February 1982
Canada	Boeing/DHC Dash 8	PW120 (2000 shp)	June 1983
Brazil	Embraer EMB-120	PW118 (1800 shp)	July 1983
France/Italy	ATR42	PW120 (1800 shp)	August 1984
Netherlands	Fokker F-50	PW124 (2240 shp)	December 1985
UK	BAe ATP	PW124A (2520 shp)	August 1986
Canada	Dash 8-300	PW123 (2380 shp)	May 1987
France/Italy	ATR72	PW124 (2400 shp)	October 1988
Canada	CL-215T	PW123AF (2380 shp)	1989

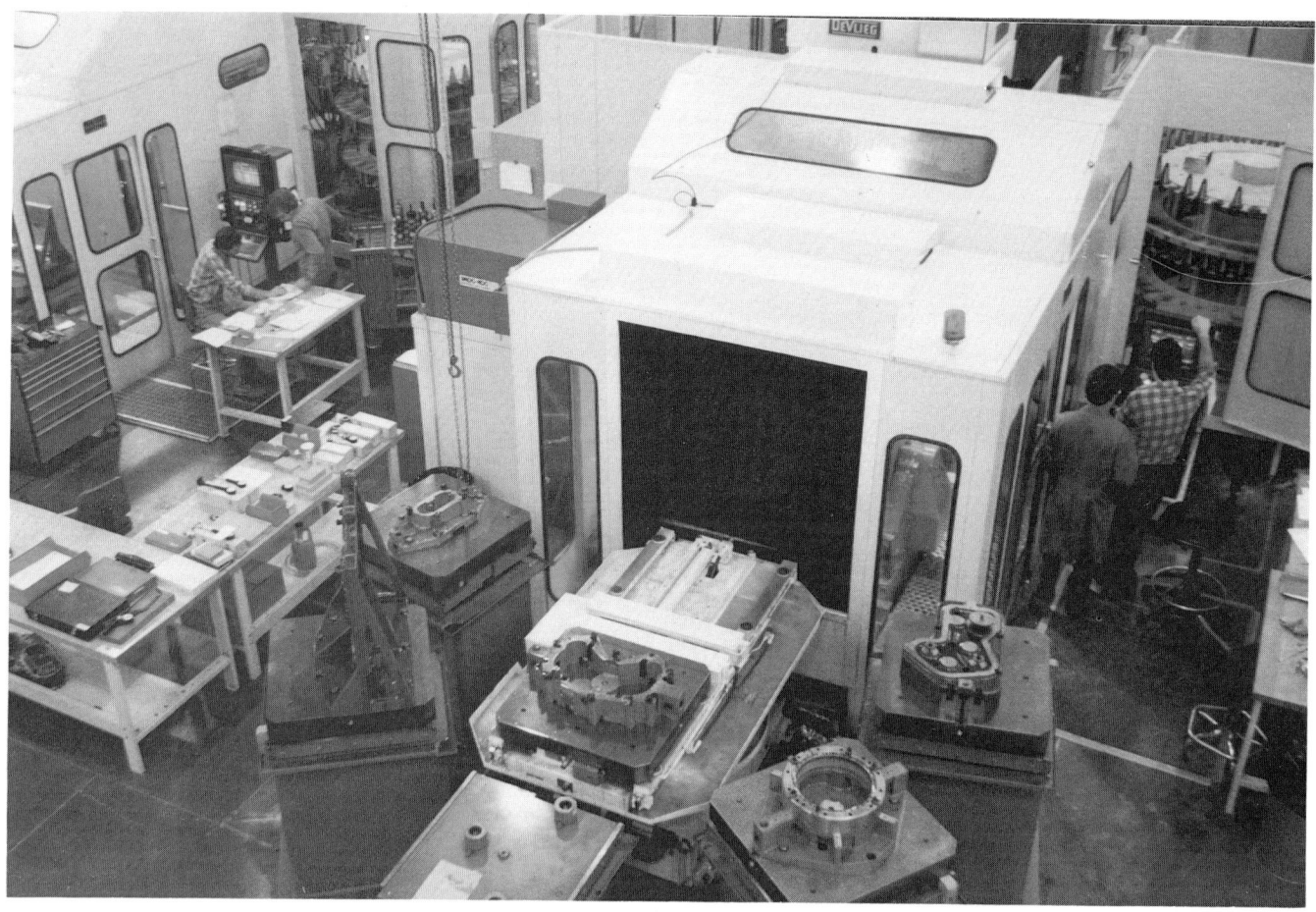

A work area in Plant 41, located in the Aerotech Business Park adjacent to Halifax International Airport. The large modules at top left and centre are machining centres. In front are box-like pedestals. Parts to be machined are bolted to the pallets atop the pedestals and are carried to the centres via AGV, where they enter the machining centres on the pallets. The average tooling operation in Plant 41 takes 45 minutes and includes some 70 different tools. A new tooling set-up takes about 45 minutes in Halifax compared to 4-8 hours in the plants at Longueuil. (Larry Milberry)

Expansion to Halifax

The P&WC plant in Halifax had its beginnings in 1981 when expansion was still foremost in management's mind. The plant would be based on the principles of CIM (computer-integrated manufacturing) and be a key in the company's goal of improving quality while remaining price-competitive. Compnay growth having thus far been concentrated in Longueuil, management was mindful of a responsibility to support other parts of Canada, as Ottawa was providing considerable funding for R&D through repayable loans. Roy Abrahams, head of manufacturing, and Dave Caplan, vice-president of operations, began inspecting sites. Then came the recession of 1982, and P&WC was forced to trim back. By this time sites in Halifax and Winnipeg looked best for future growth and Halifax was selected. There was time to prepare,

(Left) A light-alloy casing manufactured at Plant 41 during proving operations in 1988. (Larry Milberry)

(Above) Retired P&WC vice-president Bob Raven discusses developments at Halifax with Plant 41 manager Doug Renton. (Larry Milberry)

Elvie Smith, Nova Scotia Premier John Buchanan, and Dave Caplan attend the opening of Plant 41 at Halifax International Airport in November 1987. (P&WC)

and the next two years were spent in planning, including visits throughout North America, Europe and Japan to study applications of computer processing to manufacturing. People from manufacturing, quality assurance, human resources, materials, finance and other departments worked as a team in the planning and definition phase of the new project. Team members were chosen from among self-starters. They were pulled from their regular jobs to make their contributions towards a new vision for P&WC and the technical aspects for what Halifax would be. The team functioned from 1984 to 1986.

Halifax would incorporate many features seen in Japanese factories, where CIM was already a factor. There would be AGVs (automated guided vehicles) to transport materials throughout the plant, a computerized tool management system supplying 4000 tools to a variety of machining centres, and a 65-foot-high ASRS (automated stock recovery system). The factory would be designed as a pleasant working environment and have such features as a gymnasium. The plant would function virtually every day of the year, 24 hours a day.

In mid-1985 hiring began of those who would manage and direct Halifax. They defined such concepts for the new plant as "multi-skilling" wherein individual workers would be skilled at several tasks. P&WC anticipated young workers entering the labour force with hands-on computer experience. They would not have to "unlearn" older attitudes from pre-computer times. They would be bolder, appreciate the chance to participate, and adapt to multi-skilling. Since computers would be doing most of the work, personnel would appreciate multi-skilling to relieve tedium. Multi-skilling would also provide individuals with enhanced job protection, as they would have diverse talents. Workers would be self-disciplining. There would be no supervisors on the floor.

From the outset, there was a goal of staffing the plant with Nova Scotians. Assistance was provided by Ottawa through the Departments of Regional Industrial Expansion, and Industry, Trade and Commerce. The provincial ministers of education, labour and industry were also key participants. The Nova Scotia Institute of Technology worked with P&WC to develop courses to train students in a flexible manufacturing system. NSIT added a new wing to accommodate those entering its programs. The first course graduated from the 44-week program in 1986.

P&WC actively supports the creative arts. (Top) In 1987 it co-sponsored the spectacular Leonardo da Vinci show in Montreal. (Above) Families and friends enjoy works of art at the annual "Les Femmeuses" in March 1988. (Below) Pierre Henry of P&WC's public relations department presents long-time employee Max Vejins with a prize in the company photo contest, September 1987. (P&WC)

(Above) The toddlers' race gets under way at a P&WC Recreation Club picnic. (Centre left) Test cell engineer Pierre Hamel explains the mysteries of his job to young visitors at Family Day. (Centre right) Skiing at Mont Ste-Anne (Below left) Oldtimers hockey. (Below right) Virginia McGrath, Margaret Gibbons, Kathleen Ettrich, Belle Burke and Caroline Demers in the "Marathon des Étoiles," June 1986. (P&WC)

Family day in May 1988 included an air show at St. Hubert. The historic P&WC Beech 18 is in the foreground, now re-converted to standard configuration. Part of the fun included rides in the Dash 7. (P&WC)

Dave Caplan escorts Boeing Canada, DH Division, president Ron Woodard at St. Hubert on "de Havilland Day" sponsored by P&WC on April 7, 1987. (P&WC)

(Far right) Elvie Smith and Quebec education minister Claude Ryan during the opening of the test cells at l'École nationale d'aérotechnique in April 1988. (P&WC)

James Ross, the last of the original P&WC employees, reviews the manuscript of the company history in August 1988. Looking on are senior P&WC employees Betty MacDonald and Jean Demers of public relations. Betty joined P&WC on May 26, 1942, at $8 per week. She was Mr. Ross's secretary for some years and in 1989 retired as secretary to the Service Centre manager. (P&WC)

In 1988 Longueuil saluted P&WC on its 60th anniversary. Longueuil mayor Roger Ferland and National Assembly member for Laporte and Minister of Municipal Affairs André Bourbeau look on as Dave Caplan signs the city's guest book. (P&WC)

Ground was broken for the 125,000-square-foot Halifax plant in July 1985, with the opening taking place late the following year. Through 1987 finishing touches were added and the plant was equipped. Through 1988 the main goal at the plant was to prove tools and introduce the first few parts to production. The official opening was on November 5, 1987. Some 25 parts will eventually be made in Halifax, including light-alloy casings for reduction gearboxes. Full production is slated for 1989. Total cost to bring the new plant on line was $90 million.

Further PT6 Growth

As technology and analytical methods advanced, the timing was right in early 1984 to move to the PT6A-67. Its redesigned compressor, broadened and lengthened airfoils and new impeller provided 10% more airflow. The power-to-weight ratio of the -67 (4:1) is today the best of all turboprops in the western world. The first new aircraft to fly with -67s were the radical twin pushers: the Piaggio Avanti, and the Beech Starship. Piaggio later turned to the -66, similar to the -67 but with counter-rotating gearboxes. The first production sales were for the updated Short 360-300. Until the -67 the PT6 wasn't a serious contender to replace big radials such as the R-1820, R-1830 or R-2000. Now the PT6 had the power to do so. Conair in BC and IMP in Nova Scotia each launched a program to convert the Grumman Tracker from R-1820s to -67s for firefighting and military applications. Both aircraft flew in 1988.

The PT6 originally grew in 10% increments in days when development was something new at P&WC. It was the P&WC connection with Beech which resulted in the first growth versions for the King Air series. In the seventies, P&WC moved to 20% increments. This cut in half the development dollars needed to reach a particular rating, and engineering resources were freed to use elsewhere. By the mid-1980s there were several fast commuter aircraft on the drawing boards and conditions again looked ripe for PT6 growth. A jump in power of 35% over the 1100-shp -67 was thought prudent. The "super" PT6 would be a bit longer and wider and was designated PW400. It was targeted at the forthcoming Embraer/FAMA CMA-123, Short/DHC NA90 and Dornier Do328.

The CMA-123 was decided first. A fast 19-seater with twin

(Above) The Beech 2000 Starship makes its public debut, February 24, 1986. The Starship, with its 1100-shp PT6A-67 "pushers," will cruise at over 300 knots. The second view shows the unorthodox design of the futuristic Starship. (P&WC, Beech Aircraft)

Piaggio of Italy also adopted pusher configuration (with PT6A-66s of 800 shp) for its own exotic new business plane, the Avanti. (Piaggio)

The original PT6 pusher design was the beautiful 400-knot Learavia Lear Fan 2100. Largely of composite materials and powered by two PT6B-35F turboshafts, it flew in January 1981. Financial problems ended the project. Seen here is the prototype N626BL ("BL" for Bill Lear). Twelve engines for the Lear were delivered. (Chris Day)

282 POWER: THE PRATT & WHITNEY CANADA STORY

Conair of Abbotsford, BC flew its turboprop Firecat, the first PT6-powered Grumman Tracker, on August 7, 1988. The crew included contract test pilot Al Baker of Calgary, co-pilot Gordon Darnbrough, and flight test engineer Bill Schinstock. With 1424-shp PT6A-67AFs replacing its R-1820s, the aircraft flew a series of proving flights before leaving for the *Securité Civile* in France on August 24. Conair's turboprop Tracker is aimed at the fire fighting market for which it envisages about 60 conversions. Conair collaborated with IMP Aerospace Ltd. of Halifax which in 1987–1988 was developing its own turboprop Tracker, but for the military market. In this photo, Conair's Tracker conversion team has turned out to pose with the latest addition to the company's family of fire bombers. (Conair)

(Below left) IMP flew its own turboprop Tracker on September 15, 1988. Like the Turbo Firecat it has the PT6A-67AF. IMP is proposing its conversion to several countries using Trackers in civil government and military roles, e.g. on fisheries and environmental patrols, and in anti-submarine work. (Larry Milberry)

(Below right) IMP turboprop Tracker project engineer Warren Munroe (second from right) poses with the crew of the prototype following its second flight September 16, 1988: contract test pilot Al Baker, IMP test pilot Al Stoddart and FTE Michel Greffard. Stoddard reported of the aircraft, "It feels very much like a Tracker. Pilots will just love it." (Larry Milberry)

The latest King Air is the "300," powered by the PT6A-60 of 1050 shp. Meanwhile older King Airs have never been busier, having found new life as commuters. In the second photo, an Air Ontario Beech 200 Super King Air and a Voyageur Airways King Air A100 wait at Toronto for their early morning flights. (Beech Aircraft, Larry Milberry)

A far cry from the sleek Starship and Avanti is the boxy Gafhawk 125 utility plane. Developed in California for the oil industry, it flew with a single PT6A-45R. As with a number of such private ventures, the project was finally abandoned. (Chris Day)

Pilatus remains a steady user of P&WC engines in its PC-6B Turbo Porter (below) and its PC-7 and PC-9 (shown above). (Pilatus)

A Bandeirante nears completion (above) at Embraer in 1988. Its -34s have just been installed. Embraer also built the EMB-121 Xingu light transport (below), first flown in 1976 and powered by -28 or -135 engines of 750 shp. (David W.H. Godfrey, Embraer)

Brazil's spectacular Esquadrilha da Fumaça fly the EMB-312 Tucano. Here the team pulls up past the old airship sheds at Santa Cruz airbase near Rio de Janeiro. The high-performance Tucano, seen at right in a close view, has the PT6A-25C, especially modified for aerobatics. (Embraer)

(Top to bottom) The ever-popular Short 330 (-45 series) and 360 (-65 series) commuters owe much of their success to their Canadian power plants. In these views a Thai 330 gets airborne, a -45 on a 330 is shown with cowls open and a pair of 360s fly formation. (Shorts)

(Below) Another important user of the larger PT6A is the Beech 1900C with two 1000-shp -65Bs. The 1900 comes as the "C" model for cargo and the King Air Exec Liner for commuter use. The 1900 first flew in September 1982 and has since found a steady market, including the USAF and the Egyptian Air Force. It can carry 19 passengers faster than 200 mph for over 600 miles. (Beech Aircraft)

The DC-3 lives! The first of several PT6-powered DC-3s was produced by Specialized Aircraft Co. of California. The project was headed by the late Jack Conroy, famous for his aeronautical innovations. His "Tri Turbo-Three" with -45s and Hartzell propellers offered improved speed, range and payload, but the program ended when the FAA demanded costly certification trials, as if the aircraft was a totally new design. The Conroy DC-3 flew in November 1977. It's seen here visiting Toronto. The sole aircraft was still busy on charter work in Alaska in 1988. Carl Millard recalls a flight with his friend Jack Conroy. They took off in the turbo DC-3 and flew non-stop Miami to Los Angeles in 14 hours, 2 minutes, cruising on two engines. (Gary Vincent)

Aero Modifications Inc. of Waco, Texas, produced this turbo DC-3 with -65Rs. It flew in August 1986. The payload is 11,200 lb vs. 6000 lb in a standard DC-3. Much of the weight saving in such conversions comes from the light PT6. For example, a -65R weighs 515 lb, about one quarter the weight of an R-1830. (via Pat Boyle)

Warren L. Basler of Oshkosh, Wisconsin, has also produced DC-3 conversions, using the -45 (shown here with its over-the-top exhausts) and -65R. A -67 conversion is imminent. (via P&WC)

Brakes on (and tail lifting!) an AMI DC-3 does an engine run at Waco. The shaded area on the aircraft is a fuselage extension. (via H. Reichel)

By far the commonest DC-3 is still the old piston-pounder with R-1830s. Hundreds operate worldwide. Overhaul shops still meet operator needs and spare parts are plentiful (even new). This DC-3 is roaring down a gravel strip in northern Quebec, having delivered three tons of freight. (Larry Milberry)

(Left) The products of P&WC have made aviation history and can be found in many museums. On September 5, 1988, a PW100 was donated to the Science Museum in London, England, and on October 26, 1988, the prototype Dash 7 (PT6A-50s) joined Canada's National Aviation Museum. Here are the two Lockheed XH-51s in the US Army Aviation Museum. They were among the earliest PT6 installations. The one in the background also has the J60 (see photo on page 144). (Larry Milberry)

A famous threesome—test pilots Ross Lennox, the late John MacNeil, and John Searle. Each held the chief pilot job at P&WC. (P&WC)

Viscount crew for the 100th test flight of the PT6A-50, October 1977. Standing are Arthur Trudel, George Greff, Ross Lennox, Alan Tring and Peter Kent. In front are Claude Rivest and Daniel Ketelson. (P&WC Archives)

pusher engines, it was planned to offer it with a choice of engines, much as large airliners are marketed. P&WC realized it was unlikely to gain more than half the business. This made the CMA-123 an uncertain commercial venture from P&WC's vantage point, and it withdrew from the competition. When other potential applications were put on hold, work on the PW400 was curtailed. The company continues studying the power range between the PT6A-67 and PW400. At the first encouraging sign from an aircraft maker, a new project can be launched.

Flight Test at P&WC

Since early PT6 years, P&WC has operated an active flight department for flight test of engines and for corporate transportation. The first aircraft operated was the RCAF Beech 18 used to test all the prototype and early production PT6s. It flew for nearly 20 years and more than 1000 hours before being retired in June 1980. Ross Lennox with flight test engineer Dan Ketelson made the last flight. The aircraft was turned over to the aviation trade school at St. Hubert as a ground training aid and a Beech King Air was acquired to replace it.

The King Air was supplemented by an ex-Air Canada Viscount acquired in November 1973. It already had 27,875 flying hours. It was needed to test the PT6A-50, for which the Beech was too small. CF-TID remained in service until the fall of 1988, by which time it had tested many of the PW100 series. In P&WC service it made more than 535 flights for some 1250 hours. In August 1976 P&WC purchased another Viscount (CF-THX). Ross Lennox and Lloyd Hubbard ferried it down from Winnipeg, where it was stripped of useful parts to help keep 'TID flying, then scrapped.

The first Viscount flight with a -50 was on May 10, 1974. In November 1977 the nose nacelle was modified for a regular PT6, and again in February 1981 to accept the PW115 for Embraer. First flight with the PW115 was on February 27, 1982. Later that year, further mods were done to test the PW120 and PW124. First flight with the latter was on December 30, 1984. Besides engines, the Viscount also tested propeller systems, including those for the Brasilia, Dash 7, ATP and Fokker 50. John Searle recalls that it was a propeller flight that brought some added interest to the Viscount story:

The P&WC Avro CF-100 (No. 100760) with a JT15D underslung. '760 made its final flight (1:25 hours) on June 28, 1982. Today it is a historic monument at CFB St. Hubert. (P&WC)

Pilot George Greff and flight test engineer Dan Ketelson just after completing the last flight of the CF-100. (P&WC)

The Plant 5 complex at St. Hubert about 1970 with the old RCAF alert hangars and small flight operations building in front. The Sea King production building is beyond. It later housed the Viscount and is today a busy engine overhaul centre. The small building at centre left is the guard house. (P&WC)

Viscount 'TID has served P&WC for 15 years, testing a range of PT6 and PW100 engines and various propellers. Here it flies with a PW120. (P&WC)

The Lear with a standard Garrett TFE731-2-2B of 3500 pounds thrust on its right side and a JT15D on the left. (P&WC)

"It was a warm summer's day. George Greff and Steve Wade were flying, with Dan Ketelson as FTE. There was considerable interest, as this was one of the early PW100 flights. Five project engineers were along with the crew to see their new engine in action. Engineering requested that the test engine be running (but not at high power) during takeoff. A power setting was selected that would ensure minimum drag from the prop. The takeoff was a bit long and the climb was slow. In fact, the Viscount was seen weaving around obstacles on the departure path. The pilots quickly realized that prop drag was the culprit and made the appropriate adjustments. For months after this, there was no rush by people from other departments to fly with us as observers!"

Searle adds. "All things considered, the Viscount was a real workhorse. In common with the Beech 18 and CF-100 it had

The test bed King Air B200 with a PT6A-42/3-blade prop on the right side and a -67/4-blade prop on the left. Personnel are Paul Peden (pilot), André Dominici (FTE) and Marc Patino (FTE). The second photo shows the B200 with stacks fitted for noise trials. The aim was to separate exhaust from prop noise. Microphones were inserted down the fuselage. Another was on the boom on the wing. Pilot Steve Wade had few compliments to bestow on its flying characteristics. (P&WC, via J. Searle)

Plant 5 today (compare with page 291). The large new hangar in the foreground houses the Boeing 720 and other P&WC Flight Command aircraft. The company's Citation and Lear are on the ramp. (Larry Milberry)

the advantage of an independent test engine. Inevitably, all three aircraft came to the limit of their usefulness. The Beech 18 was extremely limited at 30,000 feet (it was damn cold, too!) The CF-100 could not provide a suitable environment for data recording equipment, and once the Air Force retired its last few 'Clunks,' it was not practical for us to continue flying the beast. The Viscount ran out of engine hours, and its airframe calendar life severely limited the ability to adequately pressurize. It was also too slow and couldn't reach the speeds and altitudes of our customers' new designs."

P&WC's test bed King Air 200 is serial number BB2 (C-GARO). The oldest surviving 200 (BB1 was tested to destruction on the ground by Beech), BB2 was modified from a Beech 100. Ross Lennox and Steve Wade delivered the aircraft to St. Hubert in October 1979. Like the Lear, it was a prototype, so came to P&WC with some peculiar mods. John Searle notes: "One mod is a pair of extendable landing lights in the wing underside near the tips. They shine downward when stowed and are useful for stirring up flying saucer rumours." The left nacelle of BB2 carries test engines. It has had the -41, -42, -60, and -65 installed over the years, and more recently flew with the -66 for the Avanti and the -67 for the Starship. By 1988 BB2 had logged about 450 test flights with P&WC.

Like any test bed, the King Air has had its "moments." On one flight, just at lift off, the test engine went beyond normal power and did not respond to throttle movements. The crew gained some speed and height, shut down the engine and made a routine single-engine landing. The trouble was a failure in the fuel control unit. John Searle recalls: "On another occasion we were to examine a propeller low-pitch stop setting of about 5° below normal. This worked OK except on final approach at about 1500 feet as we retarded the power to idle. Wow! The left wing simply stopped flying and we snapped into a spiral dive. Once we got the engine spooled up again things returned to normal."

Flight test acquired a Learjet in November 1981. This is C-GBRW (ex-N26GL, serial number 36-001). Pilots Ross Lennox and John Searle ferried it from Wichita, where Gates Learjet had been using it as a prototype for the Lear 26, 35 and 36. P&WC wanted a Lear with a Longhorn wing, which offered improved high-altitude performance and had no tip tanks (thus avoiding the aggravation of fuel sloshing around and adding to the yaw problem during engine handling). One such wing was at Dee Howard's in San Antonio being tested with a wing-root "glove," and P&WC acquired it. John Searle remarks: "The Lear is a lot of fun to fly. When we were at Wichita to train on it we worked with Lear's experimental chief pilot, Ed Chaplin. The first time I got the controls, the 'brand X' engine failed twice, the second time on short final! Back at St. Hubert we removed one of the standard engines and replaced it with a JT15D. We continued having trouble above 40,000 feet with the standard engine. Since it provided our cabin pressurization, we became very proficient at quick descents! We did our own study of the engine problem and found that the engine-driven fuel pump was cavitating. We replaced the pump and that solved our problem. We then had a very reliable test bed." By mid-1988 the P&WC Lear had flown more than 560 hours/275 flights on engine test work. This included cold weather trials with the JT15D at La Grande, Quebec, and at Thompson and Churchill, Manitoba. As the Lear was always experimental, it had an unfinished interior and was kept in that configuration.

Both the King Air and the Lear were limited by the need to put the test engine in an existing nacelle. For one thing, the requirement to shut down the test engine in flight for cold soaking and relights causes loss of altitude. Another problem is too much instrumentation for the Beech's small cabin. Nonetheless, both aircraft have served well, especially the Lear, which has produced remarkable results. The reliability of the data, coupled with the Lear's sparkling performance, make it an ideal aircraft and it will continue as the JT15D test bed for years to come.

With development of the PW100 and other new engines, there was a lengthy study, inspired by Doug Fage, manager of flight operations, to find a new flying test bed for P&WC. International Aero Engines, a consortium of P&WA, Rolls-Royce, Japan Aero Engines, MTU and Fiat Aviazione, was developing the V2500 of 25,000 pounds thrust, and P&WC was contracted to do the flight test phase. This was the company's first flight test program with another maker's engine. As it would require a large test bed, an ex-Middle East Airlines Boeing 720B was purchased. Registered C-FETB ("enormous test bed"), the aircraft was overhauled and

The Boeing 720 on its first V2500 flight in May 1988. Under the starboard wing is the V2500, and the nose has been modified in preparation for PW100 installation. (Richard Beaudet)

Close-up of the V2500 being installed at St. Hubert. (P&WC)

One of the engineering stations aboard 'ETB. 15-20 engineers and observers may be carried aboard the aircraft. (Richard Beaudet)

The Viscount is towed from the hangar under the nose of 'ETB. Two very strange birds! (Richard Beaudet)

modified by Aviation Traders at Stansted, near London, over a two-year period and test flown on October 9, 1986. The three-hour flight was made by pilots Jim Gannett (captain, retired from Boeing) and Capt. Webster (copilot) of the UK airline Tradewinds. John Searle and flight engineer Phil Handy of P&WC were along to observe. The P&WC personnel flew the 720 for 8.7 hours on the following two days, and on October 12 Jim Gannett, John Searle and P&WC pilots Gerry Thorneycroft and Lorne Scott ferried 'ETB non-stop to St. Hubert. Searle notes: "Thorneycroft did the takeoff and circled once to wave farewell to the maintenance crew. The ceiling was low and it was noon on a Sunday. There was a small church nearby, and it appears that we underlined the parson's fire and brimstone sermon with a suitable roar."

On October 14, the 720 was demonstrated to P&WC employees at St. Hubert: "It seems we made an impression on the local populace with several orbits at low altitude. Apparently roaring red beasties belching smoke are classed an unusual events!" That day, the 720 was pulled nose-first into a

Crew members on 'ETB during a typical V2500 flight in 1988: Mike Solomides (P&WA controls engineer), Brian Beabout (P&WA test engineer), Paul Finklestein (P&WA performance engineer), Ken Kennedy (P&WA instrumentation engineer), Scott Carr (P&WA operability engineer), Mike Miller (P&WA test director), Dan Ketelson (P&WC flight test engineer), John Searle (P&WC chief pilot), Jim Gannett (contract test pilot), Lorne Scott (P&WC pilot), Phil Handy (P&WC flight engineer), Steve Calabrese (P&WA test engineer) and John Bewick (Rolls-Royce senior flight test engineer). (Richard Beaudet)

A model of 'ETB showing its three test engine stations. (Larry Milberry)

small hangar. The vertical tail was removed and a large tent was set up to cover the rest of the aircraft. In the following months a crew installed a series of airframe modifications to make the aircraft a full-fledged test bed. The interior was completely reworked, and a new hangar was later constructed for the aircraft. On January 27, 1988, 'ETB made its first flight with all its airframe mods. Gannett and Searle (pilots), Handy (FE), Scott (recorder) and Kant and Fage (observers) comprised the crew. An initial series of flights lasted until February 19, by which time 28.1 hours had been logged from St. Hubert. The aircraft was then grounded for further work.

On May 7 'ETB flew initially with the V2500 mounted on the inside starboard engine mount. The test engine ran for 2:17 hours. A flying program was scheduled at altitudes as high as 42,000 feet. The last flight was on June 16, by which time the V2500 had been aloft for 49.4 hours. The Airbus A320 was the first aircraft to operate with the V2500 and first flew with them on July 28, 1988. The first major A320/V2500 order was placed by Air India. Besides flight testing the V2500, 'ETB will be used on other programs. It will replace the Viscount for testing PW100 and PW300TP engines on its nose mount, while small turbines such as the JT15D and PW300 will be mounted in a pod on the starboard fuselage near the nose.

P&WC in Mississauga

In 1979 P&WC opened Plant 21, a 42,000-sq.ft. operation in the Confederation Building in Mississauga, west of Toronto. This became the centre for JT15D developments. Engineers and technicians were recruited locally and overseas to staff the facility. Today, Plant 21 is involved working on the JT15D, PW200 and PW300.

In 1981 Plant 22 was opened near the western edge of Pearson International Airport in Mississauga. It combined a workshop area to support engine development programs, an engine assembly area, two JT15D test cells and offices. There were 73,000 sq.ft. The new facility's first certification program was the JT15D-4C for Italy's S.211. Next came the more complex JT15D-5, for which preliminary work was done at Longueuil before being concentrated in Mississauga. By late 1983 two turboprop test cells were added and the -5 had been tested and certified.

The turboprop cells were first used to handle overflow work from Longueuil, and in 1987 one was converted for JT15D production test runs. JT15D assembly/test had been transferred to Plant 22 in December 1986, engines being assembled from kits of parts shipped in from Longueuil. Each engine is then test run for four or five hours before being shipped.

A fifth cell was commissioned at Plant 22 in 1988 to support the PW300. That program will use nine test engines in 5000 hours of running. Meanwhile, MTU added a further 1600 hours of test cell running and conduct altitude tests as well. Flight testing on the Boeing 720B will also be done as the PW300 heads for certification.

Computerization has brought rapid advances in test technology. Computers record 100-200 parameters at once (before, only a dozen or so could be recorded). Test equipment is also much more reliable than in earlier years. Modern

A unique wood carving of 'ETB done by Léon Oligny, a sheet metal specialist in flight operations (Larry Milberry)

298 POWER: THE PRATT & WHITNEY CANADA STORY

communications means that engineers needn't be present to monitor a test—a mainframe computer sends data to performance engineers at Plant 21.

Plant 22 has 11 test rigs used to test combustors, gas generators, structures, oil systems, gears and controls. There are also five spin pits. In 1986–1987 a further 112,000 sq.ft. were added to Plant 22.

Company ties with the National Research Council date back to the 1930s when the NRC type-tested the "H" Wasp. The PT6 benefited from the use of NRC facilities, and such links continue. The NRC and P&WC each have an altitude control chamber in Ottawa. These share a common control room and can simulate flight conditions up to 50,000 feet and -65°C. Controlled conditions and precision instrumentation provide data that would be difficult to obtain with a flying test bed. In 1987–1988 engines at the NRC/P&WC facility included the P&WA/Textron Lycoming T800, the PT6A-66, the PW901A and an APU demonstrator based on the PW205B.

The PW200 Series

The commercial helicopter industry was booming in the late seventies and early eighties, stimulated largely by gas and oil exploration. With the market hot, Ottawa felt the time had come to establish an indigenous helicopter industry. After all, Canada was the world's second largest market for civil helicopters, with as many as 150 machines imported annually. Bell Helicopter Textron of Fort Worth had a new concept based on the 206L Long Ranger, but using the four-blade rotor system developed for the US Army's AHIP program and a growth version of the widely used Allison 250. The design was named "Twin Ranger." In September 1982, plans switched to a new composite-fuselage design, the Bell 400. As Bell and Ottawa held discussions about bringing the Twin Ranger to Canada, it only seemed natural that P&WC, famous in the helicopter world for its TwinPac, should be included.

In 1983 the Canadian and Quebec governments announced that they would support a new Bell plant at Mirabel, north of Montreal, to manufacture the Twin Ranger, for which Ottawa agreed to back P&WC in developing a new power plant, the PW200. In a separate deal, Ottawa and Ontario agreed to assist MBB of Germany in establishing a factory in Fort Erie

Techs work on an experimental PW209T. This project was set aside with postponement of the Bell Twin Ranger. (P&WC)

to build BO-105 helicopters. While the first 105s would be Allison-powered, later models would use the PW200.

As usual, the challenge for P&WC would be to produce the engine while keeping costs under control. It adopted a new approach to design: to enhance quality and minimize costs, a formal pre-detail-design phase was introduced between advance design work and submission of the project to the design office. Principal architect of this approach was Allan Newland. His plan was to establish an effective configuration for the engine by synthesizing aerodynamics, structures, production, cost, reliability, maintainability and durability considerations. This was far from the informal discussions that shaped design of the early PT6, and a step beyond the make-them-and-break-them approach.

Work on the PW209T (TwinPac, 937 shp) began in October 1983. A six-month preliminary design phase (with detailed input from Bell and MBB) was used to study alternatives and substantially define general layout. Design reviews were held to debate major achievements, controversial solutions and important details. The aerodynamic definition was completed at the end of preliminary design, and aerodynamic design at the end of May 1984. The final product would be just 13 inches in diameter and have 45% fewer parts than a typical PT6. In its TwinPac form it would weigh 470 pounds complete. The single centrifugal compressor was made of titanium and had an 8:1 pressure ratio. Thin-wall castings replaced sheet metal parts in many areas of the engine. The gearbox had all the accessory drives at the same end as the output shaft, requiring very detailed engineering. An aerodynamic demonstrator engine ran in December 1984 and detailed design was completed in April 1985.

The PW209T ran on schedule on November 26, 1985. When it was torn down after 30 hours it was in excellent condition. It was reassembled and put on a successful 150-hour block test. In all P&WC's history the company had never produced a new engine that had shown such favourable results. Pre-detail design had allowed the company to produce an excellent product *on paper* long before it was run for the first time. The engine was soon being expanded from its original 937 shp. Outside diameter of the power section was increased to provide a 10% increase in power to meet a request for "hot and high" operations. Then it grew a further 25% within the same dimension for the higher power (590 shp) needed in the single-engine PW205B version (first run February 2, 1987). Enthusiasm over the 209T was dampened when, in mid-1987, Bell put the Twin Ranger on the back burner. The high times of a few years earlier had faded and civil helicopter sales plummeted as energy exploration fell. At P&WC interest moved to the 205B, for which the BO-105LS became the first installation.

The PW300

The recession of the early 1980s cut demand for corporate aircraft and left manufacturers in several years of uncertainty. As conditions stabilized, an analysis of industry requirements gave P&WC the confidence to begin development of a larger turbofan—the PW300. With the recession, aircraft at the lower end of the price/performance scale owned by smaller companies were vulnerable to the failing economy. Least affected were medium-size jets such as the BAe 125 and Citation III owned by corporations with the resources to weather the downturn.

The recession also altered the rules governing new aircraft development. Existing models had to be updated or replaced or would lose market share, but few companies had the profits

The PW205B made its first official flight on October 13, 1988, in a twin-engine BO105 LS-X 01. This aircraft is a test model used by MBB Helicopter Canada Ltd. at Fort Erie, Ontario, to evaluate the PW200 series for use in upcoming MBB designs. Gross weight of

the LS-X is 5232 lb. In the secod photo John Blackie of P&WC explains features of the PW205B to artist Tom Bjarnason. Tom was visiting Plant 22 doing research for the painting for the jacket of this book. (MBB, Larry Milberry)

to invest in entirely new designs. It made more sense to develop growth versions of existing models. The principal limitation was the engines available. Various medium class aircraft had the TFE 731, some 10,000 of which had been delivered in models from 3,500 to 4,500 pounds of take-off thrust. If there was room to compete in this market, it would be at powers above those of the 731. Since the growth potential of the 731 was unknown, this had to be calculated. Computer modelling at P&WC confirmed that the 731 was near the top of its growth cycle, but that didn't mean there was an automatic market for a larger turbofan. Airframe makers had to be convinced that improvements in range, speed and cabin volume would justify spending development dollars. Only then would they consider switching engines. Sizing the new turbofan was a major gamble. If too big or too small, P&WC could fail. Each of the candidate aircraft was modelled by advanced design, and alterations were made to their lengths

(Top left) A test cell employee at Plant 22 monitors instruments. Behind the glass a PW300 gas generator runs at full power. (Bottom left) Experimental test engineer Doug Romans and quality control inspector Daniel Littman work on a PW300 impeller rotor and high turbine shaft. (Above) Joe Fernandes and Tony Matthias work on a PW300 mounted on its special carrier. (Larry Milberry, P&WC)

and weights to help determine the most desirable characteristics of a replacement engine. High power, low weight and superior fuel economy were essential ingredients. Meanwhile P&WC was aware of interest by its competitors in a similar new engine. It would be paramount to get an engine into the test cell well ahead of other manufacturers.

Through 1985 the basic concepts governing design were settled, nearly all the work being done at Plant 21. Sizing the engine evolved from market studies and close discussions with aircraft manufacturers. A design point was set of 1,113 pounds thrust at 40,000 feet at Mach 0.8. Fuel consumption would be 0.675 lb/hr/lb thrust in cruise to better the 731. Engine weight could be no more than 950 lb, otherwise aircraft with rear-mounted engines could not be stretched without costly rebalancing. Cost had to be compatible with a corporate jet selling for $8–12 million.

P&WC had earlier changed its approach to engine development with the PW200. It used to take 18 months from the time the sketches emerged from advanced design to first engine run. Now a preliminary design phase had been inserted between advanced design and detailed design. The number of specialists critiquing the layout design was broadened. For example, the manufacturing organization and suppliers were shown specifications for various parts and asked how they would make them. Designs changed to accommodate manufacturing technology. Aircraft manufacturers continued their input, as did analytical specialists. "The excitement level was shown by the attention people paid to detail," notes Robert Pottinger, manager of turbofan programs. "It was the first time that individuals could put their thumb print on a design."

Experimental testing of high-risk technology was increased *during* preliminary design to accelerate learning. Development dollars were spent closer to the start of the new program so that long-term risk and cost could be reduced. Fifteen ideas were experimentally tested. Three failed and new approaches were taken. For example, the original bearings could not withstand the instability caused by the loss of a fan blade. On the other hand, concept testing on a high-speed dynamic rig showed that the number of bearings could be reduced from three to two. "Concept testing was successful, especially when we had failures," recalls Pottinger. "We discovered problem areas before we had entered detailed design. If we had waited until hardware was running, design changes would have become too expensive. On the PW300 we came closer to weight and cost targets than ever before."

The PW300 emerged from preliminary design as a high-technology turbofan of 4750 pounds of take-off thrust capable of on-going derivatives. It would have immediate 30% growth potential, either in one leap or in 10% increments. A turbofan version for commuter aircraft was planned, as was a turboprop core engine. Estimates of PW300 development costs were astronomical—about $500 million, or equal to P&WC's net worth. Longueuil couldn't carry the burden alone, even with generous support from the government of Canada, and sought a risk-sharing partner. Motoren und Turbinen-Union GmbH (MTU) had the financial and technological resources to take a 25% stake. A subsidiary of Daimler-Benz A.G., the German firm brought a European dimension to the PW300 which would enhance international marketing. MTU had also worked with P&WA, had become a partner in Turbo Union (with Rolls-Royce and Fiat Aviazione) on the RB199 fighter engine, and is one of five members of the IAE consortium developing the V2500 commercial turbofan. MTU assumed responsibility for design, development and manufacturing of the PW300's low turbine and shares the test program. As to financial support from Ottawa, it was agreed that any federal money provided for the PW300 (26% of development costs) would be repaid by P&WC based on royalties from engine sales.

The PW300 has a four-stage axial, single-stage centrifugal compressor and pipe diffuser. The first two axial stages incorporate variable guide vanes developed with P&WA's assistance. The compressor is driven by a two-stage high turbine with cooled first stage vanes and blades. The straight-through combustor, the first employed by P&WC, helped reduce engine size. It features air blast fuel nozzles derived from PW100 technology. The high-technology unshrouded fan of 30.7 inches diameter is driven by a three-stage low turbine giving a bypass ratio of 4.5 Accessories usually left for the aircraft manufacturer to install were incorporated in PW300 design, saving weight and making the engine more attractive to potential customers. A dual channel "full authority digital electronic control" (FADEC) was developed for optimum high-altitude handling. This is a fully electronic unit with no hydromechanical backup. Advantages include better performance, lower weight and cost, and better reliability and durability.

The PW300 is the highest-technology engine P&WC has ever developed. In terms of performance, cost and thrust-to-weight ratio, it compares favorably with the large commercial fan engines. The greatest area of risk is achieving the prescribed weight at specified power. State-of-the-art materials have been used as well as advanced manufacturing methods such as super plastic forming of titanium sheet metal parts.

Design of the PW300 was well advanced in 1987 when General Electric and Garrett announced their collaboration to develop the CFE 738, a 5,600-pounds-thrust turbofan aimed at the slot between the Garrett's ATF-3 (5,400 lb) and GE's CF34 (9,000 lb). Market surveys were reviewed to understand what motivated GE and Garrett to develop a turbofan 20% larger than the PW300. The conclusion was that their sizing had been determined not so much by market studies as by the use of available, US-military-funded engine technology. CFE 738 is based on the core of the GE38 which is itself derived from the Army-funded GE27 technology demonstrator. Garrett will make the fan and the low-pressure turbine, drawing on its 731 experience.

The test core of the PW300 made its first run on September 28, 1987. A hybrid fuel control was run and advances in computerized test cell instrumentation allowed two thirds more data to be gathered in the same time. First complete run of the PW300 was in mid-March 1988 at P&WC's Plant 22 in Mississauga, Ontario. Full power was pulled on the first two builds of the first engine, the engine easily achieving rated takeoff thrust of 4700 pounds, and came within 1.9% of target specific fuel consumption. Eight engines were scheduled to be operational by November, some in test cells in Mississauga, others at MTU in Germany. Within a few months of its first run, the PW300 was attaining 5000 pounds of thrust. Challenges for 1988 included four major test programs. The four-pound bird test is a certification requirement designed to prove the structural integrity of the PW300 should it ingest a bird on takeoff. For the "fan blade off" test, P&WC will have to prove that the engine will contain the missing blade and not rip off its mounts. The blade must be knocked off while spinning at 11,000 rpm. Special explosive charges will be used which do not add any additional energy to the blade. The fan shaft shear test will use a specially designed cutter to cut the shaft. Altitude performance tests will be done on P&WC's new Boeing 720 test bed, the PW300 mounted on the starboard side of the forward fuselage in an instrumented nacelle. Prototype engines for customers will be available in mid-1989, production engines in 1990, the same year the PW300 will be certified. Although a launch customer was secured in the spring of 1988, this was still closely guarded information by year's end. The first PW300 is due for delivery in March 1990.

New Engine for the US Army

In 1984 the US Army announced a requirement for some 5000 multi-mission helicopters to replace its aging fleet. The design would evolve from a "light helicopter experimental" (LHX) program and have a new turboshaft engine, the T800. P&WC initiated its own T800 program in mid-1985. P&WA quickly entered the competition, and that July competitive contracts were awarded to two consortia: P&WA/Textron Lycoming and Garrett/Allison. P&WC joined P&WA/Textron Lycoming, taking responsibility for the centrifugal compressor and two-stage power turbine. The Army advanced $200 million to each consortium, but the teams were expected to put down $100 million each towards the project. The P&WA/Textron Lycoming team's demonstration engine ran on July 7, 1986, and the complete engine in October. Never had P&WC seen a new engine designed and built so quickly, even though companies with distinct engineering cultures comprised the team. The three-year T800 development phase ended in late 1988 with Army selection of the Garrett/Allison engine.

The T800-APW-800. P&WC was part of the consortium offering this new helicopter engine to the US Army. (P&WC)

A New APU

The Boeing 747-400, the world's largest and longest-range airliner, became the launch customer for the PW901A auxiliary power unit. When the new airliner was announced in October 1985, Boeing made it clear that it wanted some competition for its APU—Garrett APUs had been used on previous 747s. It wanted an APU based on a proven engine, and one that would fit the existing installation. P&WC sent Boeing a proposal in early 1986 based on the JT15D, with a 40% less fuel burn than the competition's APU. A contract was signed in June. Boeing gave P&WC a complex set of specs with requirements for design, performance and maintainability, yet still wanted prototype engines by December 1987.

P&WC took the JT15D-5 core compressor and turbine and added a new power turbine to drive a load-gearbox and load-compressor on the front of the engine. The gearbox powers two 90KVA generators and a blower to cool the APU compartment. The compressor provides air for main engine start and the environmental control system (cabin pressurization). No separate APU controls are found in the cockpit. Operation is slaved to main engine start or ECS controls. Fuel management is automatic, using a new fuel tank in the tailplane.

Design of the PW901A was nearly complete before a contract was signed. The first demonstrator engine ran in June 1987 and the first complete engine on December 13 that year. Other engines were added for development and certification, the seventh setting the production standard. The most demanding conditions for the 901 are cold days when the ECS must heat the cabin to a comfortable level in a short time. Engine starts require about 30 seconds at full APU power. They are usually done in sequence, but in field trials the 901 proved able to start all four 747 engines at once. International Civil Aviation Organization standards are that noise not exceed 90db within 20 metres of an aircraft. P&WC carried out noise testing at Mirabel airport using an APU in a 747 rear fuselage mock-up. Other requirements that had to be met were the FAA's for containment, fireproofing and cold starts. The first flight-certified PW901As were delivered to Boeing in January 1988. The first 747-400 was rolled out on January 26. Transport Canada qualified the 901 on July 28, and 28 production examples were at Boeing by year's end. PW901A project engineer Bill Dumbreck notes: "The 901 was useful in

An early version of the PW901A (top), then a typical production example. This engine represents the first time P&WC took full responsibility for an entire APU. The earlier TriStar APU program was run by Hamilton Standard. (Larry Milberry, P&WC)

P&WC Engine Production to October 1988

PT6 turboprop	21,149
PT6 turboshaft	5822
JT15D	3499
PW100	979
PW901A	17
Total	31,466
Total aircraft in operation	13,704
Total operating hours	158,711,189

This wooden scale mock-up of a Boeing 747-400 tail section was constructed for outside noise tests of the PW901A at Mirabel airport. (P&WC)

A PW901A mock-up installed in the 747-400 in July 1987. The 901 has been a lead project at P&WC and employed 200 people from its start. The first engines were shipped to Boeing in January 1988. (P&WC)

The prototype 747-400 (N401PW) on its maiden flight, escorted by Boeing's Canadair Sabre chase plane. Sales of the -400 exceeded 150 by late 1988. (Boeing Commercial Airplanes)

building P&WC's experience in project management. We had a schedule-driven program that led us to take a broad team approach to development. We expanded the project group to include coordinators from the analytical specialities, design, performance, test engineering and manufacturing. They gave their co-workers a clear picture of what was required, and replaced the old 'mail box' approach to getting information out. With the 901 we didn't have the luxury of completing each step in design and development in series. We did things in parallel, and learned how to compress time. For example, we would procure parts with preliminary drawings, then follow up with details of how they should be machined. Normally we would have waited until all the details were known. We are now applying this experience to the PW300." Retrofit markets for the PW901A are being sought, and P&WC is studying an APU derivative of the PW205B for smaller jet aircraft.

Pratt & Whitney Canada Today

With its proven line of engines and sales verging on $1 billion annually, P&WC is the leader in Canada's aerospace industry. The company's original foundation was well laid, and those who followed James Young, Ron Riley and Thor Stephenson have made sure to strengthen it. More than 30,000 P&WC turbine engines are today in service in every part of the world. The PT6, once a little project with a doubtful future, is, 30 years after its appearance, still in production. It has been built in many versions, often to a customer's special needs. It has not grown much in size, but it has in power—from 450 to 1400-plus shp. The group of young engineers who (perhaps brashly) invited a throng of VIPs to watch their first test cell run in 1959 did their work well. The PT6 will remain in production for years to come, and power both Starship and DC-3 into the 21st century!

The JT15D, mini-powerhouse for more than 1500 Citations and other types, continues in production and will soon be certified in the -5 version, 1000 pounds of thrust larger than the original -1. A range of customer-tailored PW100s is currently available. These are used on most of the free world's modern regional turboprop airliners. Two new engine families, PW200 and PW300, are in the wings, and the PW901A APU is flying in the first Boeing 747-400s. The company has made plenty of history since James Young formed it 60 years ago and placed the first order with East Hartford for Wasp engines to be shipped to Longueuil.

There are more than 9000 employees at P&WC, working at several locations, mainly on the South Shore (Longueuil and St. Hubert, in Mississauga and in Halifax. There are other sites such as P&WC Services with offices in Ottawa.

The market for P&WC's products has changed since earlier days when most PT6s were for corporate aircraft. Even in 1981, 65% of the 3000 engines sold were for this market; but following the recession of the early 1980s, the commuter carriers, boosted into prominence by deregulation, began taking over the small-turbine market. Three-quarters of P&WC engines certified since 1981 have been sold to the airlines. Viewing this changing scene, Dave Caplan has noted five strategies for continued success at P&WC: on-going development of new products for a diverse market, close relations with customers to understand and meet their needs, ensurance of a profitable and responsive aftermarket, improved product quality at competitive prices, and full use of the employees' potential by developing an environment of participation, involvement and teamwork. "Our challenge is not only to provide a reliable product with promised technical performance," stated Caplan, "but also to provide a level of product support fulfilling the expectations of the world's major airlines." To achieve this, P&WC has restructured its market-

A new test cell being completed at Plant 22, Mississauga in 1980. (P&WC)

JT15D assembly at Plant 22. About 15 engines were being shipped monthly in 1988. (Larry Milberry)

The PT6 will see many more years of production. The Polish PZL-130T Turbo Orlik trainer is one of the more recent applications. In cooperation with Airtech Canada, a Turbo Orlik was fitted with the aerobatic PT6A-25A. It first flew at Peterborough, Ontario, in June 1986. Test pilot Paul Hartman likened its performance to the Mustang and Sabre he had flown years earlier. (via P&WC)

ing and product support organizations with a careful focus on the needs of different market segments. Spare parts supply and overhaul have been similarly reorganized, with closer ties with the airlines. Spare parts remains a profitable activity with sales in 1987 of $278 million, most for the PT6 and JT15D, but with PW100 parts becoming steadily more significant. In overhaul, P&WC has continued to expand. Overhaul used to be centred in Longueuil in Plant 2, then in the old Navy building (Plant 4), then at the new facilities at Plant 5. Gradually the Bridgeport, West Virginia, operation took on more overhaul work. On June 25, 1988 it opened its expanded facilities for overhaul of the PT6, PT6T, JT15D and PW100 and is now a "full service" overhaul base. In July 1988 the full service overhaul base in Portsmouth, UK, was opened by H+S Aviation to handle PT6s and PW100s from Europe, the Middle East and Africa. Besides these three centres, a host of other P&WC parts and overhaul centres is spread around the world.

In research and development, P&WC continues to be major leader. In the 1980s R&D soared, and one year accounted for 27% of the company's total budget. This has since settled back to about 20% per annum (the average in Canadian industry is 3%). Its R&D investments in the PT7/PW100 in the late 1970s and through the lean early 1980s have made P&WC a firm believer that success in the future rests mainly on the groundwork laid through R&D. With the PW300, P&WC embarked upon its most exciting venture, with the R&D phase being critical and the most expensive to date. If success with the PW100 is any guideline, the PW300 will confirm the wisdom of P&WC's R&D philosophy. Meanwhile, P&WC still invests about 33% of its budget in improving current engine models, and in developing additional versions of them to meet customer requirements.

P&WC is much changed from earlier days. For decades it was run on the traditional production line model. Changes began to occur following the last strike, and any old timer touring P&WC today would know at a glance that a whole new operations philosophy has been adopted. He would see plants equipped with the latest in tools and other equipment, but also sense a new workplace atmosphere. In his day, a worker did his job with little emotional involvement. Inspectors kept a close watch on the finished products, and foremen were eagle-eyed, ready to "crack the whip" if someone was slacking off. A worker could spend his career putting in the years at one mundane job. Management was aloof from those on the shop floor; the production line ground away year after year. Change was slow to come but a number of factors combined in the mid-1970s to overturn the old ways of doing things.

The computer age came knocking on industry's door. Markets were beginning to take a new shape, and competitors were appearing with their own small turbine engines. Customers were becoming more demanding. They wanted higher technology engines that would be more efficient, yet still remain affordable. All these factors would have to be addressed should any engine maker hope to remain in business. At P&WC, management decided to deal with this situation in an evolutionary way. Changes began gradually. In one case, the company's personnel department began taking greater interest in the needs of employees. Whereas "hiring and firing" had been its role in simpler times, it now became a human resources organization with specialized staff to handle such areas as training, benefits, employee relations, wage and salary administration, security, safety and health. It moved from being regulatory to being service minded. Counsellors

P&WCers on the job in Longueuil. In the photo above a PW100 bull gear is being readied for some fine finishing on the work bench. In the second photo an employee punches instructions into a computerized laser cutting and drilling machine which saves 50-80% in time over earlier equipment. (Larry Milberry)

Gilles Ouimet with Mississauga employees Shirley Jubinville and Pritpal Girn. Ouimet has been active in establishing P&WC's new operational guidelines. As a young man he served in the RCAF (1962-1970), eventually as an engineering officer at Cold Lake maintaining a fleet of CF-104s. He left the air force to take an MBA at Queen's University, then worked in finance for Quebec Cartier Mining and Northern Telecom. He joined P&WC in 1977 as comptroller, became vice-president of finance in 1980, vice-president of operations in 1984 and senior vice-president of operations in 1985. In 1988 he was elected senior vice-president of marketing and customer support. Ouimet has directed the first major company restructuring in 20 years. (P&WC)

were hired to assist employees with job-related problems or more personal ones such as drug or alcohol use. A system called Dialog was introduced by which any employee could carry a complaint or question to senior management and be assured of a satisfactory answer within 10 working days.

By late 1987, P&WC had completed most of the engineering work on four big projects—the PW100, PW200, PW300 and PW901A. That year, engineering had peaked at 2450 employees. Now there was a surplus of 168 skilled engineers to whom P&WC was loath to give "the pink slip." The decision was to hold on to them, and human resources undertook to interview the engineers, determine their job preferences in other areas of the company and find them a position until the day when they could be re-absorbed into engineering. In earlier times, a large employer would have thought nothing of firing surplus help, but companies such as P&WC now

The engineering group on the PW300, photographed in late 1988. Left to right, Archie Tannock, Dick Hart, John Wilson, Roy Blinco (program manager), Brian Uffen, Dave Ellison, Bob Pottinger (program manager, turbofan/shaft engines), Peter Boyd, Dave Millington, Abdul Rashad. Not shown are the following members of the original team: Mike Pottinger, Dave Simpson, Geoff Best, Tony Hayward. (P&WC)

The 844th and last Twin Otter, ready for delivery with its PT6A-27s in November 1988. (Boeing Canada, de Havilland Division/ Charles Bryant)

understand that their future success depends not just on turning out products, but on building and maintaining a positive atmosphere among their employees.

In 1985 a new outlook on production was introduced which would eliminate much of the need for inspectors. Quality (achieving the standards demanded by the customer) would become the responsibility of each worker doing his or her job. There would be many small teams to look for areas needing improvement, then tackle them and come up with solutions. Red tape would be cut through to find and implement solutions. The corporate hierarchy would be flattened, and there would be greater interplay among departments. This approach was adapted from one devised by the Ohio steelmaker Armco, and was described as "a total company approach towards quality and productivity improvements through teamwork." Now called Quality Plus, it is known throughout industry as "Q+." At P&WC it has already proven itself in many areas of daily operations. In one case there was a materials handling problem in Plant 5—parts were suffering damage. A Q+ team was formed to study the problem. This resulted in a number of special containers being designed, along with a procedural manual for loading trucks and an instructive video, "Handle With Care." It was the workers themselves who had identified the problem and rendered solutions. In another example, smokers at Plant 41 were asked to form a committee to review a pending company-wide smoking policy. After looking at the situation from many angles, their own well-being included, they came to a straightforward solution… they would "kick the habit" and join the non-smokers!

Looking Ahead

In 1983, with sectors of the aerospace industry in deep trouble, Pratt & Whitney Canada had boldly announced its intention to double its market share to 30% and triple engine sales within a decade. Halfway into this new era, the company continues its progressive outlook. Its sights are set on becoming the industry's premier low-cost manufacturer of quality products. To reach its goal, P&WC has begun placing more emphasis than ever on the belief that the customer is "number one." This means focusing on *all* the customer's particular needs. These are more complex than ever before, as P&WC now serves a range of markets—aircraft manufacturers, airlines, utility

Some of the earliest P&WC facilities continue in use at Plant 2, Longueuil. The "red brick building" is still used for offices. In the view with the large shipping door open, we are looking at the side of the Walmsley plant first used by P&WC in 1928, with the original test cells across the lane. The reader may orient himself with the photos on pages 111-112. (Larry Milberry, Jean Demers)

operators, corporate aviation and the military. To please all these important customers, it is up to P&WC to ensure that engines are delivered on schedule to the manufacturers, that spare parts are readily available wherever they are needed in the world, and that overhauls are completed to the customer's satisfaction. In each area, the work and/or products must be of the best quality at a competitive price.

To attain its long-term goals P&WC has adopted the "team approach" to operations and has begun by integrating such areas as design, engineering and production, so that these function together and each person is aware of what is happening in the various other departments. The teamwork philosophy was proven at P&WC with the successful PW901A project.

Besides providing for its basic responsibilities as a modern-day industrial employer, P&WC now strives to be much more. It includes the employees in decision-making at many levels and is sensitive to their special concerns. The communication lines among employer, employees and union are always open, keeping P&WC responsive to change.

The road to establishing an overall new "company culture" has been a major challenge. While P&WC has been investing heavily in new technology, the anticipated turnaround in general aviation has been slow in coming. The rising Canadian dollar has made the company's products and services less competitive internationally. Competition from other engine makers has been tightening. Nonetheless, Pratt & Whitney Canada has not relented in its drive to steadily improve its products and continue investing heavily in the R&D, expansion and modernization necessary to guarantee the company continued prominence in the field of small and medium-size turbine engines.

A view downriver of P&WC in October 1988. Plant 2 in the foreground still includes many features from 1928 and World War II. The propeller shop is in the foreground. Towards the river can be seen Plant 1. (Larry Milberry)

Appendices

APPENDIX 1
P&WC Directors

	Elected	Resigned
Harold Earle Walker	Jan. 15, 1929	Jan. 17, 1929 (Provisional)
Jean Martineau	Jan. 15, 1929	Jan. 17, 1929 (Provisional)
Earnest Howard Cliff	Jan. 15, 1929	Jan. 17, 1929 (Provisional)
Frank Bernard Chauvin	Jan. 15, 1929	Jan. 17, 1929 (Provisional)
Hugh Wylie	Jan. 15, 1929	Jan. 17, 1929 (Provisional)
Thomas S. Stewart	Jan. 15, 1929	Jan. 17, 1929 (Provisional)
Christine Imrie	Jan. 15, 1929	Jan. 17, 1929 (Provisional)
James Young	Jan. 17, 1929	Apr. 30, 1964
Ross H. McMaster	Jan. 17, 1929	Oct. 31, 1960
Frederick B. Rentschler	Jan. 17, 1929	Nov. 8, 1933
S.A. McClellan	Jan. 17, 1929	Dec. 2, 1930
Charles W. Deeds	Jan. 17, 1929	Oct. 25, 1938
Montegu Black	Jan. 17, 1929	(Deceased Jul. 5, 1959)
G. Herrick Duggan	Jan. 17, 1929	(Deceased Oct. 8, 1946)
Hubert G. Welsford	Jan. 17, 1929	(Deceased Aug. 11, 1969)
Joseph F. McCarthy	Jan. 17, 1929	May 20, 1941
Donald L. Brown	Dec. 2, 1930	(Deceased Apr. 29, 1940)
Philip S. Johnson	Nov. 8, 1933	Oct. 17, 1934
Leonard S. Hobbs	Nov. 21, 1934	Apr. 26, 1939
CMDR Eugene E. Wilson	Oct. 25, 1938	Mar. 20, 1941
J. Carlton Ward, Jr.	Apr. 27, 1939	Apr. 18, 1940
Raycroft Walsh	Apr. 18, 1940	Mar. 20, 1941
H.M. Horner	Apr. 18, 1940	Mar. 20, 1941
Ronald T. Riley	Mar. 14, 1941	(Deceased Jun. 24, 1959)
George M. Black, Jr.	Mar. 14, 1941	Apr. 29, 1958
Paul P. Hutchinson	Apr. 16, 1943	Apr. 26, 1946
H. Mansfield Horner	Apr. 26, 1946	Oct. 23, 1968
W.R. Robbins	Oct. 16, 1946	Oct. 18, 1972
William P. Gwinn	Jun. 22, 1950	Oct. 1, 1972
John W. R. Drummond	Apr. 27, 1951	Dec. 31, 1965
Wright A. Parkins	Oct. 25, 1956	Apr. 10, 1958
H.H. Lank	Jul. 26, 1957	Apr. 21, 1983
L.C. Mallet	Apr. 10, 1958	Apr. 14, 1970
Thor E. Stephenson	Apr. 29, 1958	Jun. 23, 1977
George M. Black, Jr.	Nov. 3, 1959	Apr. 9, 1968
Victor W. Tryon, Jr.	Apr. 29, 1960	Jun. 17, 1976
H. Roy Crabtree	Apr. 26, 1961	(Deceased Jun. 2, 1986)
R.D. Richmond	Dec. 23, 1963	Mar. 31, 1970
G.A. Hart	Apr. 28, 1965	Apr. 21, 1983
F.R. Miller	Dec. 29, 1966	Jun. 17, 1976
A.E. Smith	Apr. 9, 1968	Apr. 10, 1980
W.P. Wilder	Oct. 23, 1968	Apr. 11, 1972
Marcel Bélanger	Apr. 14, 1970	
Elvie L. Smith	Apr. 14, 1970	
B.A. Shmickrath	Apr. 14, 1970	Apr. 1, 1971
H.J. Gray	Sept. 30, 1971	
Bruce N. Torell	Sept. 30, 1970	Aug. 16, 1979
Kenneth H. Sullivan	Apr. 11, 1972	Feb. 1, 1984
Edward L. Hennessy, Jr.	Oct. 18, 1972	Apr, 19, 1979
Michael L. Koerner	Dec. 14, 1972	
Donald C. Lowe	Nov. 1, 1975	Sep. 30, 1982
L. David Caplan	Jun. 17, 1976	
Jean-Paul Gignac	Dec. 16, 1976	Sep. 30, 1980
S.B. Brown	Apr. 19, 1979	Dec. 11, 1986
R.J. Coar	Jun. 21, 1979	Jun. 19, 1986
Robert J. Carlson	Aug. 16, 1979	Feb. 11, 1985
Alexander M. Haig, Jr.	Apr. 10, 1980	Jan. 21, 1981
David A. Golden	Feb. 19, 1981	
William C. Missimer, Jr.	Oct. 21, 1982	Aug. 18, 1983
Robert Gratton	Apr. 21, 1983	
Jack H. Warren	Apr. 21, 1983	
Selwyn D. Berson	Aug. 18, 1983	
J. Edward Newall	Jul. 1, 1984	
Robert F. Daniell	Oct. 18, 1984	
Arthur E. Wegner	Jun. 19, 1986	

P&WC Presidents and Chairmen of the Board

	President	Chairman
James Young	Jan. 17, 1929–Dec. 31, 1947	Jan. 1, 1948–Jun. 16, 1960
Ronald T. Riley	Jan. 1, 1948–Jun. 24, 1959	
Thor E. Stephenson	Jul. 1, 1959–Oct. 31, 1975	Nov. 1, 1975–Apr. 21, 1977
H. Mansfield Horner		Jun. 16, 1960–Oct. 23, 1968
Donald C. Lowe	Nov. 1, 1975–Feb. 28, 1980	Feb. 29, 1980–Sep. 30, 1982
Elvie L. Smith	Feb. 29, 1980–Feb. 29, 1984 (CEO Mar. 1, 1984–Apr. 17, 1985)	Apr. 18, 1985–
L. David Caplan	Pres./COO Mar. 1, 1984–Apr. 17, 1985, Pres./CEO Apr. 18, 1985	

P&WC Company Secretaries

Charles W. Deeds	Jan. 17, 1929–Oct. 6, 1938	Ronald T. Riley	Apr. 26, 1946–Dec. 31, 1947
Joseph F. McCarthy	Oct. 25, 1938–Apr. 18, 1940	James Ross	Jan. 1, 1948–Dec. 31, 1969
Hubert G. Welsford	Apr. 18, 1940–Mar. 20, 1941	Harry A. W. McGee	Jan. 1, 1970–May 1, 1988
Ronald T. Riley	Mar. 20, 1941–Apr. 26, 1943	Christopher Pascoe	May 1, 1988
Paul P. Hutchinson	Apr. 26, 1943–Apr. 26, 1946		

APPENDIX 2

Pratt & Whitney Canada Engines and Variants

Model	Certification	Take-off rating SHP & RPM	Main applications
PT6A-6	December 1963	550 shp, 2200 rpm	King Air, Turbo Beaver, Turbo Porter, Beech 18, misc. conversions
PT6A-6A	May 1965	550 shp, 2200 rpm	Turbo Beaver, Potez 841, Turbo Porter
PT6A-6B	December 1967	550 shp, 2200 rpm	Turbo Beaver, Turbo Porter
PT6A-6/C20	March 1973	550 shp, 2200 rpm	Beech C90 (17 engines)
PT6A-10	January 1980*	475 shp, 2200 rpm	N/A
PT6A-11	December 1977	500 shp, 2200 rpm	Piper Cheyenne/Cheyenne IA, Piper T1040, Piper Malibu
PT6A-11AG	March 1979	500 shp, 2200 rpm	Ayres Turbo Thrush, Weatherly 620TP, Schweizer G-164B Ag-Cat.
PT6A-15AG	September 1977	680 shp, 2200 rpm	Grumman Ag-Cat, Air Tractor AT400, IAR827, Turbo Thrush, Gulfstream Turbo Ag-Cat, Schweizer G-164B, Frakes Turbo Cat.
PT6A-20	October 1965	550 shp, 2200 rpm	King Air 90 Series, Beech 99, Turbo Beaver, Twin Otter, Bandeirante, Turbo Porter, Merlin 2, Beech 18, Commanchero 500, misc. conversions
PT6A-20A	March 1973	550 shp, 2200 rpm	Beech C90
PT6A-20B	October 1973	550 shp, 2200 rpm	Turbo Porter
PT6A-21	December 1974	550 shp, 2200 rpm	Beech C90
PT6A-25	May 1976	550 shp, 2200 rpm	Beech T-34C
PT6A-25A	November 1976	550 shp, 2200 rpm	Pilatus PC-7, T-34C-1, Turbo Firecracker, PZL-130T Orlik
PT6A-25C	April 1982*	750 shp, 2200 rpm	Embraer EMB-312, Tucano
PT6A-27	December 1967	680 shp, 2200 rpm	Beech 99A, Twin Otter, EMB-110, Frakes Turbo Mallard, Helio Stallion, LET410, Turbo Porter, Harbin Y-12-2
PT6A-28	March 1969	680 shp, 2200 rpm	Beech E90, 99A, A100, EMB-121, Cheyenne II
PT6A-34	November 1971	750 shp, 2200 rpm	EMB-110, EMB-111, Frakes Turbo Mallard, IAI Arava, Saunders ST-27 and ST-28, Turbo Titan, Avalon 680, Spectrum One, Omni Turbo Titan
PT6A-34B	August 1976	750 shp, 2200 rpm	Beech T-44A (King Air G90)
PT6A-34AG	February 1977	750 shp, 2200 rpm	Ayres Turbo Thrush, Turbo Cat, PZL Turbo Kruk, Fieldmaster, G-1645B, G-164D
PT6A-36	December 1973	750 shp, 2200 rpm	Arava, Beech C99
PT6A-38	May 1975	750 shp, 2000 rpm	Beech C-12A (Super King Air 200)
PT6A-40	January 1984*	700 shp, 2000 rpm	SOCATA/Mooney TBM 700
PT6A-41	October 1973	850 shp, 2000 rpm	Super King Air 200, Cheyenne III
PT6A-41AG		850 shp, 2000 rpm	G-164D
PT6A-42	September 1979*	850 shp, 2000 rpm	Super King Air B200, Beech 1300, Beech C-12F
PT6A-45A	April 1976	1173 shp, 1700 rpm	Frakes Mohawk, Short 330
PT6A-45B	July 1978	1173 shp, 1700 rpm	Short 330
PT6A-45R	May 1980	1198 shp, 1700 rpm	Short 330, Short C-23A Sherpa, Metro 3, Gafhawk 125, USAC DC-3
PT6A-50	September 1976	1120 shp, 1313 rpm	Dash 7
PT6A-60	April 1983*	1050 shp, 1700 rpm	Beech King Air 300
PT6A-60A	May 1983	1050 shp, 1700 rpm	Beech King Air 300
PT6A-61	March 1983	850 shp, 2000 rpm	Cheyenne IIIA, TP-600 Malibu
PT6A-61A	January 1984	850 shp, 2000 rpm	Cheyenne IIIB
PT6A-62	April 1985*	950 shp, 2000 rpm	Pilatus PC-9
PT6A-65AG			Ayres Turbo Thrush, Fieldmaster
PT6A-65B	December 1982*	1173 shp, 1700 rpm	Beech 1900
PT6A-65R	August 1982*	1376 shp, 1700 rpm	Short 360, Hawk Industries
PT6A-65AR	August 1982	1424 shp, 1700 rpm	Short 360, Short Super Sherpa, AMI DC-3
PT6A-66	July 1986*	850 shp, 2000 rpm	Piaggio P.180 Avanti

*First production date

Model	Certification	Take-off rating SHP & RPM	Main applications
PT6A-67	November 1986*	1200 shp, 1700 rpm	Beech 1900, King Air 300, Starship I
PT6A-67AF		1424 shp, 1700 rpm	Conair Turbo Firecat, IMP Turboprop Tracker
PTA-67R		1424 shp, 1700 rpm	Short 360-300, Basler DC-3
PT6A-110	January 1980	475 shp, 1900 rpm	Dornier 128-6 Skyservant
PT6A-112	August 1978	500 shp, 1900 rpm	Cessna Conquest I, Cessna Caravan II
PT6A-114	April 1984*	600 shp, 1900 rpm	Cessna Caravan I
PT6A-116	October 1984	700 shp, 1900 rpm	Cessna 435
PT6A-121	May 1982	615 shp, 1900 rpm	
PT6A-135	May 1977	750 shp, 1900 rpm	King Air F90, Taurus, Cheyenne IIXL, Xingu II, Commanchero, Riley Turbine 210 and 421, OMAC Laser 300
PT6A-135A	August 1983*	750 shp, 1900 rpm	Beech F90, Avtek 400, Dornier Seastar, OMAC Laser 300
T74-CP-702	June 1967*	550 shp, 2200 rpm	U-21A (Beech B90)
T74-CP-702	February 1968*	680 shp, 2200 rpm	RU-21A to E (Beech 90-100)
PT6T-3	April 1970	1800 shp, 6600 rpm	Agusta/Bell 212, S-58T
T400-CP-400	March 1970	1800 shp, 6600 rpm	Bell AH-1J, UH-1N, CH-135
T400-CP-401	June 1972	1800 shp, 6600 rpm	UH-1N, VH-1N
PT6T-3B	August 1979	1800 shp, 6600 rpm	Bell 212 and 412
PT6T-6	December 1974	1875 shp, 6600 rpm	Agusta/Bell 212, S-58T
T400-WV-402	February 1975	1970 shp, 6600 rpm	AH-1J, AH-1T
JT15D-1	May 1971	2200 lb	Cessna Citation
JT15D-1A	July 1976	2200 lb	Citation I
JT15D-1B	July 1982	2200 lb	Citation I
JT15D-4	September 1973	2500 lb	Citation II, Corvette, Diamond I
JT15D-4B	May 1983	2500 lb	Citation S II
JT15D-4C	July 1982	2500 lb	SIAI-Marchetti 211
JT15D-4D	December 1983	2500 lb	Diamond 1A
JT15D-5	December 1983	2900 lb	Cessna T-47A Citation, Beechjet
JT15D-5A		2900 lb	Citation V
PT6B-34		750 shp, 6000 rpm	
PT6B-35F	January 1982*	650 shp, 6188 rpm	Learavia Learfan
PT6B-36	June 1984*	960 shp, 6050 rpm	S-76B Mk II
PW115	October 1984*	1500 shp + 202 lb, 1300 rpm	EMB-120 Brasilia
PW118		1800 shp	EMB-120
PW119		1900 shp + 250 lb, 1300 rpm	Dornier 328
PW120	January 1984*	2000 shp + 250 lb, 1200 rpm	ATR-42
PW120A	January 1984*	2000 shp + 250 lb, 1200 rpm	Dash 8
PW124	November 1985	2400 shp + 250 lb, 1200 rpm	Fokker 50
PW124A	November 1985	2400 shp + 250 lb, 1200 rpm	BAe ATP
PW125		2400 shp + 250 lb, 1200 rpm	
PW125A	November 1985	2400 shp + 250 lb, 1200 rpm	BAe ATP
PW205B		590 shp, 6016 rpm	MBB BO105 LS-X
PW209T		937 shp, 6290 rpm	
PW300		5225 lb ATO (after takeoff)	
PW901A		377 lb/min (main engine start)	Boeing 747-400
ST6J-70	July 1968	620 shp, 2200 rpm	I&M engine
ST6J-77	September 1974	750 shp, 2200 rpm	I&M engine
ST6K-70	October 1968	620 shp, 6230 rpm	I&M engine
ST6K-77	May 1972	690 shp, 6230 rpm	I&M engine
ST6T-75	June 1973	1700 shp, 6600 rpm	Bell Voyageur ACV
ST6T-76	December 1974*	1850 shp, 6600 rpm	Textron Marine LACV-30 ACV
ST6L-73	November 1970*	820 shp, 33,000 rpm	APU for L1011 Tristar
ST6L-792	April 1975*	575 shp, 33,000 rpm	Industrial
ST6L-794	September 1979*	725 shp, 33,000 rpm	Industrial
ST6L-795	September 1979*	825 shp, 33,000 rpm	Industrial
ST6L-812	December 1976*	840 shp, 27,000 rpm	Industrial
ST6L-813	September 1980*	960 shp, 27,000 rpm	Industrial

*First production date

Index

Abrahams, Roy 276
ACLS 193, 194
ACV 199-201
Aerial Experiment Association 11
Aero Commander 166
Aero-Dienst 215, 225
Aero Engines of Canada 28
Aero Modifications Inc. 288, 289
Aérospatiale Corvette 253, 254
Aga Khan 152, 196-8
Agar, Carl 75, 77, 78
agricultural aircraft 177-82, 221, 222
Agusta 234, 258
Agusta/Bell 223, 225
Air America 219
Air Board 12, 13
Airbus A320 298
Airco 207
Airline Deregulation Act 260
airline support 275
Air Parts 221
Aircraft Repair 66
Air Rice 181
Airtech Canada 308
Alaska Coastal Eillis 170
Alberta Gas Trunk Line 209
Alcan 75
Alcoa 88
Allegheny Airlines 109
Allison engines 120, 199, 204, 205
Allison
 T56 109
 T63 142, 146, 148, 230
Amtrak 212
Andersen, Bill 144, 156
Anderson, Phil 236
Anglin-Norcross Quebec Ltd. 87
APU 192, 193, 207, 304-6
Aquarius 198
Armstrong, Bill 135
Armstrong Whitworth 20
ASH program 141, 142
ASP-10 193, 194
Atlantic Canada Aviation Museum 119
ATP 273, 274
ATR42 271, 272
Austin Airways 54, 60, 61, 66
Austin, Jack 54
Avro Canada 69
Avro
 Anson 35, 41, 42, 45, 46, 60
 Arrow 104-6, 129
 CF-100 104, 129, 250, 251, 291, 294
 Jetliner 103, 104
 Lancaster 69
Ayres
 Thrush Commander 179

NEDS 179
Ayres Corp. 182
Ayres, Fred 178, 181, 182,
Azores 109

Babb Co. 73
Baddeck 11
Baker, Al 283
Bannock, Russ 160, 190, 191
Barfoot, Lois 55
Barkley-Grow 54
Barlow, Lynn 212
Barnes, J. 130
Basler, Warren L. 288
Bates, Fred 127
Batory, Zoltan 226
Beabout, Brian 197
Beatty, Sir Edward 45
Beauregard, J.P. 115, 116, 122, 124, 129, 132, 140
Beech Aircraft 120, 176
Beech, Miles 87, 99
Beech, Olive Ann 148, 151, 155
Beech, Walter 148
Beech
 Diamond 254
 King Air 151-7, 186, 187, 190, 249, 284, 293, 294
 L-23F 148, 149, 156, 201
 NU-8F 149-51
 Queen Air 148, 151, 152, 155, 157, 158, 169, 176
 Starship 281, 282
 T-34C 158, 159, 220
 18 54, 163-5, 252
 18 (test bed) 133, 138-40, 281, 290
 99 176, 177
 200FJ 187
 1900 189, 190, 287
Bee Hive 27
Bélanger, Marcel 211
Belanger, Pete 237
Bell
 47 75
 212/412/UH-IN 230-4
 AH-1 Cobra 233
 OH-58 Kiowa 146, 230
Bell, Alexander Graham 12
Bell Helicopter Textron 193, 194, 199-201, 299
Bell, Ralph 47, 69
Bellanca 28, 54, 66
Bennett, Capt. D.C.T. 45
Bennett, John 127
Bennett, R.B. 27
Bentley, Eric 236
Bernier Museum 203
Best, Geoff 310
Bethlehem Steel 88

Bewick, John 297
Bird, Jim 88, 121
Bjarnason, Tom 300
Black, Conrad 56
Black, G. Montegu 26, 38, 101
Black, G. Montegu, Jr. 55-7, 101
Black Margaret 38
Black, Wayne 164
Blackburn Aircraft 116
Blake, D.O. 117, 131
Blanchette, Anita 55
Blinco, Roy 310
Blizzard, Cyril 127
BOAC 45, 63
Boehm, Rolf 216, 217
Boeing Airplane Co.
 40B-4 27
 707 104
 720B 294-8, 303
 727 104
 737 104
 747 208, 304-6
 B-17 48
 B-47 106
 T60 120, 148
Boeing, Bill 15
Boeing Canada 270, 280
Boland, Arden 68
Bolton, Ed 141
Bombardier 214
Boone, Doug 71
Booth, Frank 46
Borusigwich, G. 121
Boucherville 45, 86
Bourbeau, André 281
Bouthillier, Monique 55, 137
Bowery, Leroy 108, 158
box storage shed 50, 51, 70, 111
Boyd, Peter 310
Boyko, Bob and Dagmar 226
Brazil 222, 223
Briese, J.G. 121
Brintnell, Leigh 66
Bristol
 Bolingbroke 43, 44
 Mercury 43
 Jupiter 28, 68
 Orpheus 116
British Aerospace 271, 273
British Aerospace ATP 273, 274
British Commonwealth Air Training Plan 36, 37, 41, 42
Brooks, George G. 87, 88, 99
Brown, Johnny 168
Brownridge, Ross 191
Bruce, Harold 237
Buchanan, John 277
Bujia, Jim 137
Burke, Belle 279
Brunelli CBY-3 63

Burns, T. 121
Buy American Act 84

Calabrese, Steve 297
Calhoun, John 150
Canada Coast Guard 200, 232
Canada-US Pact 84
Canadair Ltd. 61-4, 123
Canadair
 C-4/North Star 61-4, 101
 C-5 63, 175
 CC-109/CL-66 108, 109
 CL-41 116-9
 CL-44 108-9
 F-86 85, 129
Canadian Aeroplanes Ltd. 12
Canadian Airways 27, 31, 100
Canadian Arsenals 84
Canadian Associated Aircraft 35
Canadian Car and Foundry 35, 36, 42, 94, 96
Canadian Commercial Corp. 231, 232
Canadian General Electric 38
Canadian National Railroad 212-4
Canadian Pacific Air Lines 63, 98, 100, 101, 104
Canadian Pacific Railroad 49
Canadian Pratt & Whitney Co. Ltd. 11, 17, 20-7, 48-50, 54-6
Canadian Propellers Ltd. 28, 37, 54-6
Canadian Transcontinental Airways 19
Canadian Vickers 14, 15, 17, 21, 28, 30, 35, 55, 57
Canadian Vickers
 Vedette 14, 17, 20, 22, 31
 Velos 14
Canadian Wright 28, 30
capital cost allowance 85
Caplan, David L. 270, 275-7, 280, 281
Cardiff, Bill 181
Carlson, Bob 265
Carr, Scott 297
Carrier, Thérèse 55
Cartierville 37, 61, 63
CASA CN-235 271
Caswell, Ed 68
Causey, Tom 66
CC-7 Cushioncraft 199
C.D. Howe 65
Central Technical School 73
Centro Técnico Aeronáutica 222
Cessna
 Caravan 190, 191
 Citation 249-57, 306
 T-37 120, 158
Chalk's International Airlines 172

Chaplin, Ed 294
Charest, Charles 137
Charles Walmsley and Co. 21-5
Charleson, J.C. 65, 151, 157, 177
Cheetah engine 41
Chetagne, Rosaire 71
Chinook engine 68, 69
China 226-9
Christensen, Nils 171
Chudors, Jim 77
City Class frigates 205
CJAD 243
Clark, John N. 104, 146, 164
Clay, Leroy 187
Cliche Commission 184
CMA 123, 281, 290
COIN aircraft 138, 139, 146, 158-60
Cole, W/C W.R. 130
Comeau, Laurier 212
Commuter Airlines 176, 177
computers 261, 262
Conair 283
Connor, Warren 140
Conroy, Jack 288
Consolidated
  PBY 41, 56, 62
  B-24 45
Convair 159, 160
Cook, Dave 261
Cordeau, Pauline 55
Cornwall, Pete 78
corporate aviation 103
Corporate Systems Centre 211
Coughlan, Joe 191
Cowley, Fred 142-4, 146, 163, 164, 167, 172, 207
Cox Turbo Otter 188, 190
Crabtree, Roy 211
Crompton, Jim 250
Crown Forest Products 170
CTA 161
Cummings, F/L Lloyd 81
Curleigh, Colin 239
Curtis, A/M Wilf 64, 92
Curtiss, Glen 12
Curtiss
  C-46 98-101
  JN-4 12
  OX-5 12
  HS-2L 13, 14
Cutbill, Tom 130
Czechoslovakia 225

Danahee, Ray 132
Dando, Frank 20, 23, 25, 39, 47, 51, 52, 66, 88, 94
Daniell, Bob 246
da Silva, Ozilio 161, 223
Darnbrough, Gordon 283
Dawson, Ken 37, 41, 43, 52, 60, 64, 137
Davidson, Grant 161
Davis, Sgt 247
Day, Alan G. 55, 57
DDH-280 202-5, 239
Dean, Robert 185
Deeds, Andy 165
Deeds, Charles 24, 26

de Havilland
  Comet 103, 104
  Gipsy Queen 68
de Havilland Canada
  DHC-1 68, 86
  DHC-2 2, 67-9, 73, 77, 86, 98
  DHC-2T 146-8
  DHC-3 73-7, 146, 147
  DHC-4 109-10
  DHC-5 193, 194
  DHC-6 146, 160-3, 176, 222, 310
  DHC-7 188, 189, 191
  DHC-8 263, 267, 270, 271
  Fox Moth 68
  Giant Moth 28
Defence Research Board 146
de Maurivez, Maurice 135
Demers, Caroline 279
Demers, Jean 281
Department of Defence Production 84, 104, 123, 125, 141
Department of Regional Industrial Expansion 278
Depression 27, 28
Desrochers, Fernand C. 116, 124
Desormeaux, Marthe and Raymonde 55
DEW Line 98-100
Dialogue 309
Dickie, Harold 86, 121
Dickinson, W.H. 121
Dickins, "Punch" 45, 73
Diefenbaker, John 129
"dirty dozen" 116
Doehl, Wolfgang 261
Dome Petroleum 209
Dominici, André 293
Dominion Engineering 25, 26
Dominion Skyways 30
Domtar 208, 209
Dornier Do328 281
Dorval airport 46, 52
Douglas
  A-26 128
  C-47/DC-3 61, 98, 100, 102, 289, 290
  C-54/DC-4 57, 61, 100, 101
  C-124 172
  DC-6 64, 99, 101, 102
  DC-8 104
  DC-9 104
  O-2B/MO-2BS 15-17
Downsview 74
Drummond, John 40, 52, 55, 57, 59-61, 65, 66, 68, 83, 84, 86, 87, 90, 92, 95, 99, 104, 106, 110, 133, 137, 167
DS-3J 117-9
DS-4J 116-7
Dubreuil, Joseph 58
Duggan, G. Herrick 26
Dumbreck, Bill 306
Dyment, Jack 33, 43

École aéronautique 139
Edinburgh University 115
Edwards Air Force Base 104
EH 101 343

Elliot, George 183
Ellison, Dave 310
Elsworth, Ken 116, 124, 129
Embraer 161, 222, 223, 263, 265, 281
Embraer
  Bandeirante 222-4, 265, 285
  Brasilia 264-9, 290
  Tucano 287
  Xingu 265, 285
Emery, Vic 158
Emmerson, Derek 178, 180, 189, 247, 267, 271
English, Jerry 243, 246
Ettrich, Kathleen 279
Eveleth, Skip 64

Fage, Doug 294, 298
Fairbanks, Dave 109
Fairchild airstrip 237, 242
Fairchild (Canada) 17, 27, 35, 54, 66, 111, 237
Fairchild
  71 27, 29, 30, 54
  82 29, 30
  FC-2 17-19, 27
  Husky 67, 123
  Sekani 29
  Super 71 29
Fairchild (U.S.) J83 engine 116, 117
Fairchild-Hiller 216-20
Fairey Aviation 170, 239
Fairey Reed 52
Fallen, Ted 239
Federal Express 191
Felsted, Bob 64, 65
Ferland, Roger 281
Fernandes, Joe 301
Ferry Command 44-6, 52
field service 52
Finklestein, Paul 297
Fitzwater, C.E. 217
Fleet Aircraft 35
Fleet Finch 38
Fletcher, Wendell 221
Fletcher FU-24 178, 221, 222
flight test 290-8
Florens, Nils 239
Fokker
  50 273
  F.14 28
  F-27 273
  Universal 17, 18, 27, 28, 34
Foley, Dennis 79
Fotheringham, John 104
Fort Rucker 144, 150, 151, 289
Fort William 35, 36
Fowler, R.H. 138, 146, 148, 194
Frakes, Fred 171, 172, 175, 178, 181
Frakes Mohawk 175, 298
Franklin, Benjamin W. 61
Funk, Bill 104

Gafhawk 284
Gagnon, Commissioner 64
Gagnon, Jacques 137
Gallacher, James 20, 25

Gander 50
Gannett, Jim 297, 298
Garratt, P.C. 68, 160
Garrett Air Research 128, 148, 155, 159, 160, 165, 168, 176, 177, 202, 220, 252, 257, 301, 304
Garrett/Allison 303, 304
Garrett/G.E. 303
Garton, Harry 137
gearbox (CL-44) 108, 109, 126
Gélineau, Gilles 127
General Electric 69, 93, 199
General Electric
  CT7 260, 265
  J85 116-9, 147
  T58 119, 234, 235
George, Ron 32
Gervais, Mike 236
Gibson, W.W. 12
Gillespie, Tom 157, 158
Gillies, Gerry 227
Gillies, Jack 68
Girn, Pritpal 309
Glasspoole, Fred 116, 121, 124
Godfrey, S/L A.E. 15, 17
Good, Guido 216
Goss, Arthur 116
government support 84, 94, 122, 123, 302
Graham, Jock 65, 76-78, 80, 230, 234, 243
Graham, Martin 23, 25, 28, 31, 33, 34
Granatelli, Andy 205
Gray, Harry 125, 185
Greff, George 290-2
Greffard, Michel 283
Grenier, André 243, 246
Grey, Gordon 239
Grossmith, Seth 239, 240, 243
Grumman
  Ag-Cat 178, 180-1
  Albatross 106, 107
  Goose 168, 170
  Gulfstream 102
  hydrofoil 199
  Mallard 77
  Panther 84
  Tracker 106, 283
  Turbo Goose 169-71
  Turbo Mallard 170-3
Guthrie, R.H. 85-6, 95, 99, 113, 115, 116, 121, 124, 131-3, 195, 205
Gwinn, Bill 84, 85, 88, 101, 136

Hale, Jerry 187
Halifax 276-8, 281
Halliburton Jet-Frac 209
Hamel, Pierre 279
Hamilton Standard 15, 28, 33, 45, 49, 52, 55-7, 63, 74, 98, 106, 109, 120, 264, 274, 304
Hamilton, S/L Bob 118
Hamilton, George 165
Hanchet, Dave 64, 65
Handley Page Hampden 35
Handy, Phil 297, 298
Harber, Betty 70

Harber, Ted 137
Harbin Aircraft Corp 228, 229
Harding, Dennis 126
Harding, Jim 127
Hardy, Gordon 109, 116, 117, 124, 127, 144, 156, 207, 223
Harned, Malcolm 251
Harris, Ted 137
Harrison, Tom 244
Hart, Dick 310
Hartford/East Hartford 11, 15, 32, 38, 44, 51, 83, 84, 87, 88, 95, 100, 104, 113, 116, 117, 122, 183, 184
"Hartford Team" 131-3
Hartigan, Bud 239
Hartman, Paul 308
Hartzell 182, 288
Hatton, V.J. 63
Hawke, Reg 135
Hawker Hurriance 35, 36
Hawker Siddeley 748 260
Haydon, Fred 57
Hayward, Tony 310
Heaslip, R.T. 78, 79
Hedrick, Frank 149, 150, 152, 158
Heenan, Webb 32
Helicopter and Systems Div. 234, 239, 244
Hill, Robert 158
Hill, Sam 58, 137
Hiller
   Ten99 133, 140-2
   UH-12 141, 143
Hiller, Stanley 141, 142
Hiscocks, R.D. 68, 73, 160
HMCS
   Algonquin 202
   Annapolis 202
   Assiniboine 238
   Athabaskan 202
   Bonaventure 240
   Bras d'Or 199, 200, 202
   Buckingham 80, 81
   Huron 202, 204, 205
   Iroquois 202, 205
   Labrador 80
   Magnificent 80
   Nipigon 202
   Shearwater 79
Hollinger Ungava Transport 98, 100
Hopson, Herb 32
Hornberger, Dwight 155
Horner, Jack 84, 101, 131, 133, 237
Houston, Rod 205
hovercraft 199-201
Howard, Dee 165, 168, 169, 294
Howard Super Ventura 102
Howe, C.D. 32, 33, 45, 54, 56, 61, 69, 83, 125, 129
Hubbard, Lloyd 256, 257, 290
Hudson Bay Mining and Smelting 69, 74, 77, 78
Hudson Strait Expedition 17, 18
Hughes OH-6 146
Hunt, John 127, 138, 140
Hunter, Bob 155
Hunter, Les 191

Hutt, Albert 33, 100
Hynes, Percy 127

IAI Arava 175, 229
icing trials 127, 155, 157
IMP 243, 244, 283
Industrial and Marine 195-214
Indy 500 205-8
inertial particle seperator 157
inspection 71
International Aero Engines V2500 294-8

Jacobs engines 34, 35
Jacques Cartier 87
James Aviation 221
Jobmaster 165, 169
John Bertram 24
Johnson, Philip G. 33
Johnson, Frank 236
Jones, Parnelli 206, 207
Jubinville, Shirley 309
Judge, Ben 127
Junkers 27

Kaman
   K-1125 145, 146
   Seasprite 234
Kaman, Charles H. 145
Kansas City 39
Keewatin Air 165
Kelly, Bart 230, 231
Kelly, Frank 47
Kelner Airways 190
Kemano 77
Kendrick, Dave 109
Kennedy, Ken 297
Kent, Peter 290
Kerwin-McCarthy, Kay 58
Ketelson, Dan 140, 290-2, 297
King Beaver 73
King, MacKenzie 31, 36
Kingsbury multiple drill 93
Klosowski, Stefan and Izabela 226
Koerner, Michael L. 211
Kongsberg Vapenfabrikk 197
Korea 200
Kroean War 75, 77, 88, 95, 97
Kowalczewski, Romy and Chantal 226
Krause, Willy 118
Krones, P.J. 119

Laberge, Louis 184
labour troubles 182-6
La Citière 210, 211
Lac la Pêche 17
LACV-30 200, 201
Lafontaine, Fernande 55
La Lumière, Lionel 236
Lamare, Jacques 58
Lampson, Bob 171
Landry, Claude 243
Langshur, Hugh 87, 109, 115, 116, 121, 132
Lank, Herb 211
Laporte, Gilles 184
LARC 199
Lawrance, Charles 15

Lauzier, Fernand 127
Lauton, Slim 226
Lavallé, Aimé 47
Lavigueur, Lucien 212
Learavia Lear Fan 282
Learjet 294
Lee, Bill 150, 152
Lees, Barry 127
Lefebvre, Jacques 127
Le Grave, Gerald L. 31, 45
Lennox, Ross 76-8, 239, 240, 243, 245, 246, 290, 294
Leslie, Dave 158, 187
LET 225
Lew, Jim 148
Lewis Airlines 161
Leyland 88
LHX 303, 304
Liberty engine 13-6, 100
Lilly, A.J. 63, 64
Lindbergh, Charles 11
Linnen, Gord 140
Littman, Daniel 301
Litton Systems Canada 205
Livesay, Frank 180
Lockheed
   10 32
   12 34
   14 33, 42, 43
   18 42, 43
   286 145, 230
   1011 193
   CL-475 144
   T-33 129
   XH-51A 144, 145, 289
   Hudson 45
   Jetstar 102, 119, 176
Local 510 183
LOH program 125, 133, 142, 146, 230
Longhurst, Bill 118
Longueuil 11, 27, 39, 52, 88, 106, 129, 131, 142, 281
Lord Beaverbrook 45
Losch, Bob 95, 104, 105, 118, 142, 146
Lowe, Donald C. 185, 228, 252, 273
Lowe, J.D. 79, 80
LRC 214
LTU 193
Lucas Rotax 115
Lycoming engines 200
Lynx engine 14
Lyon Air Service 181
Lyttle, C. 121

Mahon, M. 121
Malton 35, 68, 69, 104, 105, 119, 131
Mann, J. 121
manufacturing 83-94
Marlow, George 79, 80
Marmol, Ladi 218, 219
Marquis, Rocky 243, 246
Marr, Harold 23, 39, 47, 52, 54
Martinez, Juan 140
Masters, Sid 237, 239
Massé, Bert 58, 137

materials control laboratory 93
Matthias, Tony 301
MBB 299, 300
Ménard, Maurice 257
Mercury engine 47
Meyer, Paul 132
Meyer, Russ 191, 251
Meyer, W. 121
Mi-4 227, 228
Mid Canada Line 78, 98, 100, 106
Millar, Douglas 115, 116
Milard, Carl 168, 169, 253, 288
Miller, Donald S. 167, 177, 216, 223, 225, 227-9
Miller, F.R. 211
Miller, Mike 297
Millington, Dave 310
Mirabel 299
Mississauga 298, 299
MIT 86
Mitsubishi Diamond 254
Mochulsky, Harry 134, 135, 161, 217-9
Moffett, R.J. 55
Mont Joli 98
Montreal Locomotive Works 211
Montreal Olympics 184
Montreal Star 45
Moore, Bill 140
Morris, A.W. 121
Morris, Ron 257
Morrison Knudson Ltd. 75
"Mr. PT6" 156
MTU 302, 303
Muir, Jim 58
Munroe, Warren 283
Murton, Bert 47, 58
MacDonald, Betty 136, 281
MacDonald Bros. Aircraft 42
MacDonald, George 99
MacDougall, Frank 69
MacNeil, John 80, 81, 138, 140, 155, 239, 241, 243, 246, 290
McAlpine Expedition 27
McCarthy, Joseph F. 26
McClellan, Stephen A. 26
McClunie, Tom 221
McConachie, Grant 34
McConnell, J.W. 45
McCorky, Stu 237
McCormack, Russ 41, 52, 135, 142
McCurdy, J.A.D. 11, 123
McDonnell CF-101 119, 134, 135, 194
McGee, Harry 88, 121
McGill University 40, 56, 64, 86, 95, 115, 275
McGrath, Virginia 279
McGregor, C.M. 88, 94, 95, 191
McKaig, Leonard 114
McKee, J.D. 15-17
McKee Trophy 16, 100
McKenzie, A.M. 67
McKeown, Bill 58
McKinnon Goose 170
McLachlan, Dick 150, 158, 168, 222, 223
McLatchie, Bob 83, 88
McLean, George 236

McMann, Keith 170
McMaster, Ross H. 26, 101
McPhee, Jack 46, 47, 58, 130
McPhee, Isabelle 55
McPherson, Glen 224
McVicar, Don 99, 100

Nadeau, Philippe 94, 261
Napier Eland 91, 108, 109
NASA 145
Nash, Jim 135
National Air Transport 15
National Aviation Museum 13, 29, 34
National Research Council 69, 86-8, 115, 123-6, 128, 131, 133, 155, 239, 299
National Steel Car 35
NBAA 148, 152, 157, 164, 250
Neal, George 68, 74, 146
Nechako 199
NEDS 179
Nelson, Art 132
Nepal 217, 218
Nerriere, Max 68
Neveu, Jean 72
Newland, Allan 115, 116, 122, 124, 127, 129, 299
New Zealand 178, 221, 222
Nicholson, John 191
Nickel Belt Airways 67
Niles-Bement-Pond 24
Nimick, B. 121
Noorduyn Aircraft 30, 31, 36-9, 41, 66, 103
Nord 262, 189
Nordair 172
Norman NAC6 177
North American
 B-25 48
 OV-10 159, 160
 T-28 94
 Harvard 36, 37, 47, 52, 93, 94, 96, 97
 Saberliner 119, 176
Northern Consolidated Airlines 170, 172
Northwest Industries 66
Nova Corp. 209
NSIT 278

Ober, Eddie 137
Okanagan Air Service 75
Oligny, Léon 298
Ontario Provincial Air Service 28, 60, 67-9, 73, 148, 161
Ord, Lew 87
Orenda engines 68, 69, 73, 74, 104-6, 115, 116, 129-31
Orr, John 146
Ostiguy, Georges 225
Other Work Agreement 94, 95
Ottawa Car 35
Ouimet, Gilles 309
overhaul 26, 46, 47, 62, 74, 97-9, 308
Owen, Alf 213, 214

Pacific Airmotive 170

Page, Serge 217, 218, 225
Parent René 237
Paris Air Show 217
Parker, George 30, 31
Parkins, Wright 122, 131
Pascoe, Chris 205
Pashley, F.R. 121
Patendaude, Antonio 58
Patino, Marc 293
PBY Canso 54, 62
Peden, Paul 293
Pelletier, Joseph 58
Perks brothers 52, 58, 106
Pesco 46, 85
Peterson, Basil 158
Peterson, Gudmundur 115-7, 124, 232, 233
PEZETEL 225, 226
Phelan, Leo 137
Phillips, Fred 118
Phripp, Frank 125
Piaggio Avanti 281, 282
Piasecki
 16H 142, 144
 H-21 78, 79, 106, 142
 HUP 80
Pilatus
 PC-3 220
 PC-7 220, 221
 PC-6B 215-20, 285
 PC-9 190, 221, 285
Piper
 Cheyenne 176-8, 190
 Navajo 176, 177
Piper, W.T. 177
Plant 1 87, 93, 111, 113, 314
Plant 2 111-3, 239, 313, 314
Plant 5 239, 244, 291, 193
Plants 21 and 22 298-302, 307
Plant 41 276-8, 281
Poland 225, 226
Pollard, Art 207
Pond, Arthur 56, 137
Ponsford, George 60
Porto, Nick 132
Potez P841 215
Pottinger, Robert 302, 310
Pottinger, Mike 310
Power, Ed 127
Power Jets Inc. 69, 131
Pratt, Francis A. 15
Pratt & Whitney Aircraft Co. 11, 15, 21, 26
Pratt & Whitney Aircraft/Textron Lycoming 303
Pratt & Whitney Co. of Canada 24
Pratt & Whitney Tool Co. 15
Pratt & Whitney Aircraft
 eagle 88, 265
 FT4 202-5, 209-11
 FT12 203-5
 J42 83, 103, 104
 J48 83, 84, 104
 J57 119, 133-5, 150
 J58 104
 J75 104-6, 135
 JT3D 104, 135
 JT12/J60 115-9, 134, 145, 245
 R-985 42, 60, 67, 68, 95, 98, 103, 108, 168, 175, 182
 R-1340 Wasp 15-7, 19-32, 34, 36, 40, 47, 48, 52, 54, 73, 74, 79-81, 84-99, 103, 106, 113, 115, 180-2
 R-1535 43, 44, 47
 R-1830 43, 45-7, 54, 62, 63, 66, 95, 98-100, 175, 289
 R-2000 57, 95, 101, 102, 106, 109, 114, 173, 175
 R-2800 39, 54, 63, 99-102, 109, 114, 135, 175
 R-4360 114, 135, 173, 175
 R-4360 114, 135, 173, 175
 Hornet 15, 26, 28, 32, 42, 43, 54, 100
Pratt & Whitney Canada, Inc.
 DS-4J 125
 DS-5 121
 DS-10 120-2
 DS-30 249
 DS-32 249
 JT15D 183, 185, 249-59, 298, 299, 304, 306
 JT25D 257, 258
 PT6 (beginnings) 119-23, (configuration) 120, (financing) 122-3, (first run) 124, 127, 129, 131, (sales efforts) 125, (experimental) 125-9, (Hartford team) 131-3, (test flying) 138-51, 169, (PT6A-27/28) 163, ("Ag" engines) 177-82, (vs piston engine) 182, 192, (growth versions) 186-91, 281-90, (PT-6A-50) 189, (production changes) 191, 192, (Twin Pac) 186, 230-4
 PT7 261-3
 PW100 259-75, 290-2
 PW200 299, 300, 306
 PW205B 300
 PW300 300-3, 306
 PW400 281, 282
 PW901A 304, 305
 TDE-1 259, 260, 265
 "Y" Wasp 52
Pringle, T. 87
process sheets 88
propeller overhaul 52, 53, 174, 314
Prud'homme, Henri 20, 23, 25, 33
Purdue University 115
PZL-130T 308

Quality Plus 310, 312
Quarter Century Club 136, 137
Quebecair 109, 161
Quebec Federation of Labour 184, 185
Queen Charlotte Airlines 77
Quérel, Émile 137

Rankin, Jim 109, 116, 124
Ransome, Dawson 175
Rashad, Abdul 310
Raven, Bob 66, 236, 237, 277
Rawson, Barney 32, 217
RCAF 14, 17, 35-58, 62
RCN 79-83, 106

Recreation Club 136
red brick building 106, 311
Reid, Percy 12
Reid Rambler 28, 30
Reid, W.T. 28, 30
Reive, Bob 41, 47, 137
Renton, Doug 277
Rentschler, Frederick B. 14, 15, 18, 26, 136
Reynolds, Bill 104
Richardson, James 27, 28, 31
Richmond, R.D. 123, 131, 133, 135, 148, 152, 191, 231, 232, 236, 239
Riendeau, Thérèse 55
Riley, Ronald T. 37-41, 52, 55, 61, 65, 69, 83, 84, 86-8, 92, 94, 95, 97-9, 101, 104, 113-6, 125, 306
Rimfakse 150, 170, 196, 197
Rivest, Claude 290
Robbins, W.R. 101, 131
Robertson, George 205
Rockwell 181, 182
Rogers R.L. "Buck" 239, 240
Royal Military College 37
Rolls-Royce 14, 36, 57, 115, 116, 259
Rolls-Royce
 RB211 192, 193
 Conway 104
 Dart 104, 119, 260, 273
 Derwent 103
 Falcon 14
 Merlin 36, 43, 63, 69, 100, 101
 Nene 84
Romano, Doug 301
Rose, George 20, 23, 47, 52, 66, 137
Roskam, Del 250
Ross, James 25, 27, 28, 38, 39, 41, 45-7, 54, 60, 61, 97, 99, 101, 257, 281
Ross, John 58
Rutledge, Keith 244
Ryan, Claude 185, 280
Ryan, Dick 101
Ryan, Pat 205

SAAB 267, 271
Sa'ar, Dov 229
Sachs, Bob 226, 227
St. Laurent, PM 64
St. Hubert 27, 35, 45, 46, 291-4, 297
Saintsbury, John 154
Samoil, Larry 140
Sanders, George 86, 121
Santos, Frank 39, 47, 102, 137
São José dos Campos 222, 223
Saturday Night 43
Saunders, B.A. 121
Saunders, Mike 104, 110, 118, 150, 152, 219, 221
SCAN Marine 205
Scherrer, Gunther 132
Schinstock, Bill 283
Schweitzer, E.S. 37, 41, 43, 45, 52, 65, 68, 73, 100, 101, 103, 104, 135, 137, 158, 218, 275
Schweizer Ag aircraft 180, 182

APPENDICES 319

Scott, Lorne 297, 298
Seager, Charlie 65, 66, 239
Searle, John 240, 290, 292, 297, 298
Senez, Maurice 47
Senez, Normand 58
Senez, Raymond 58
Serv Aero 181
Sheer, Thomas 244
Short
   330 186, 187, 189, 286
   360 188, 189, 286
SIAI-Marchetti S.211 258, 259
Siers, Tommy 100
Sikorsky Aircraft
   ABC 248
   HO4S (S-55) 80, 81, 234, 238
   HSS-2 Sea King 80, 83, 191, 202, 234-46
   R-4 79
   R-6 64
   S-51 (H-5) 64-6, 75, 78-80
   S-52 75
   S-55 (H-19) 75-8
   S-58 (H-34) 79-82, 106, 230, 234
   S-58T 246-8
   S-62 83, 234
   S-64 245
   S-76 77, 248
Sikorsky, Igor 64, 65
Silver Dart 11
Silver Shark 196-8
Simatos restaurant 107
Simms, Gordon 172
Simmons, Shell 170
Skelding, Bill 20, 31
Smallritch, Walter 58
Smith, Elvie 115, 116, 119, 120, 122, 124, 130, 132, 136, 184, 191, 211, 250, 251, 273, 275, 277, 280
Smith, Les 47
Smith, Ted 94, 95
Snow, Leland 180
Solomides, Mike 297
Soloniewicz, Henry 226
Sosnkowski, Joe 239
Sortie, Jim 32
SPAR 243
Sparks, Bob 237
spare parts 95, 96, 113, 130, 135, 175
Specialized Aircraft 288
Spence, Howard 87, 88, 92, 95
Sperry Gyroscope 116
Spillane, Bob 132
Stamm, Rick 109, 126, 127, 132, 133, 135

Standard Aero Engines 98
standard procedures 94
Stearman 32, 38
Stedman, E.W. 17, 64, 69
Steel Co. of Canada 26, 88
Steinberg, Sam 56
Stephenson, Thor E. 99, 104, 123-5, 131-3, 141, 148, 152, 158, 185, 207, 222, 227, 244, 251, 306
Sterner Aero 179, 181
Stewart, Alex 247
Stewart, Thomas S. 25
Stimm, C. 130
Stoddard, Al 283
Stopps, Ralph 61, 62
Stoten, Mike 259
Stowell Screw Co. 52
STP 205, 207
Stringer, Alf 77, 78
Sullivan, Ken 73, 74, 104, 106-8, 119, 120, 146, 149, 156, 158, 167, 175, 211, 227, 228, 251, 257, 263, 266, 273
Sunbeam Arab 12
Supermarine Stranraer 35
Swearingen, Ed 166, 168, 169, 176

T800 303, 304
Tannock, Archie 310
Tatham, Claude 86, 121
Taylor, E.P. 56
Tennant, Dave 46
test cells 46, 72, 88, 91, 108, 111, 125, 126, 128, 133, 135, 301, 307
Thomas, Hugh 165, 170
Thomas, K.M. 121
Thompson, Peter 168
Thompson, R.F. 227
Thompson valves 88
Thorneycroft, Gerry 297
Thunderbird 197, 198
Titcomb, Allen 127
Todorovic, Martial 226
Tooley, J.F. 41
Torrell, Bruce 131-3, 250
Tracy, Peter 58
Trans-Canada Air Lines 16, 31-3, 40, 42, 43, 57, 63, 98, 104
Tring, Alan 290
Trott, Elsie 55
Trudeau, Roger 137
Trudel, Arthur 290
Trudel, Monique 290
TRUMP 204, 205
Tryon, Victor 87, 88, 99, 211
Turbo Research Ltd. 69, 86, 131
Turbo Firecat 283
Turbomeca Astazou/Bastan 120,
148, 189, 225, 229
Turbo Train 211-4
Turcotte, Donald 205
Turiff, Rob 135
Tusket 210
Tuttle, Steve 150
TwinPac 230-4

Uffen, Brian 310
utility projects 210, 211
"Utilizer" 207
Underwood brothers 12
United Aircraft of Canada 148
United Aircraft and Transportation 15, 26
United Air Lines 33
United Auto Workers 182-6
United States Army 69, 74, 109, 110, 119, 125, 144, 145, 148, 149, 199-201, 230
United States Navy 84, 93, 94
US Army Aviation Museum 151, 289
United Technologies Corp. 125, 148, 211, 258
University of Manitoba 115
University of Minnesota 131
University of Michigan 123
University of Saskatchewan 115
University of Toronto 73, 125

Vachon, Ferdie 31, 100
Vaillancourt, Jean-Paul and Dianne 226
Van Gelder, Leo 205
VASP 161
Vejins, Max 140
Versatile Vickers 205
Vertilneck, Bill 39, 40, 41, 47, 54, 58, 137, 152, 154-6, 180, 217
Vertilneck, Grace 217
Vertol 107 83
VIA Rail 214
Vickers Viscount 104, 189, 263, 264, 290, 292, 294, 296
Victory Aircraft 69
Victory Bonds 39, 57
Vincent, G. 121
Viner, Jimmy 64, 66, 67
Viper engine 14
Vought, Chance 15
Voyageur ACV 199, 200
Vrana, John 115, 116, 126

Waco 34
Wade, Steve 292-4
Walker, George 86, 88, 121
Wallnutt, T. 64, 65

Walmsley plant 21, 23, 25, 26, 46, 84, 87, 111
Waloka, John 172
Walser, Bert 226
War Assets Corp. 56, 59
Warren, Sid 239
Washburn, Joe 109, 121
Washingtonian 83
Watt, R.A. 239, 240
Watt, Robin 115
Webber, Jimmy 150
Webster, Capt. 297
Wein Alaska Airlines 216, 217
Weir, Stu 58
Welsford, H.G. 26, 101, 133
West Ed 163-5, 169
West, Oliver 40
Westinghouse
   J30 84
   J34 134
Weston, Eddie 58
Wheatley, William 17
White, Ed 205
Whitney, Amos 15
Whittle, Frank 69
Wichita, Kansas 108
Wick, Emile 217
Wicks, Omer 15
Wilcox, Jack 57
Wilkinson, Jimmy 55
Wilkinson, George 140
Wilson, John 310
wood chippers 207-9
Woodard, Ron 280
World-Wide Airways 99, 100
Wright engines 14, 15, 17, 18, 27, 28, 30, 31, 45, 54, 62
Wright
   R-1300 73, 79, 80, 181, 182
   R-1820 79, 94, 106, 107, 113, 181, 230, 247, 283
   R-3350 99, 100
Wrong, Colin 129-32, 227
WSK 225, 226
Wyman-Gordon 88
Wysocki, Bob, Helen and Tony 226
Y-12 228
Young, Bill 47
Young, James 20, 21, 24-7, 31, 37-40, 45, 54-9, 66, 68, 69, 83, 86, 92, 95, 97, 99, 101, 125, 133, 136, 306

Zbitnew, Larry 239
Zielinski, Ed 226
Zurakowski, Jan 105

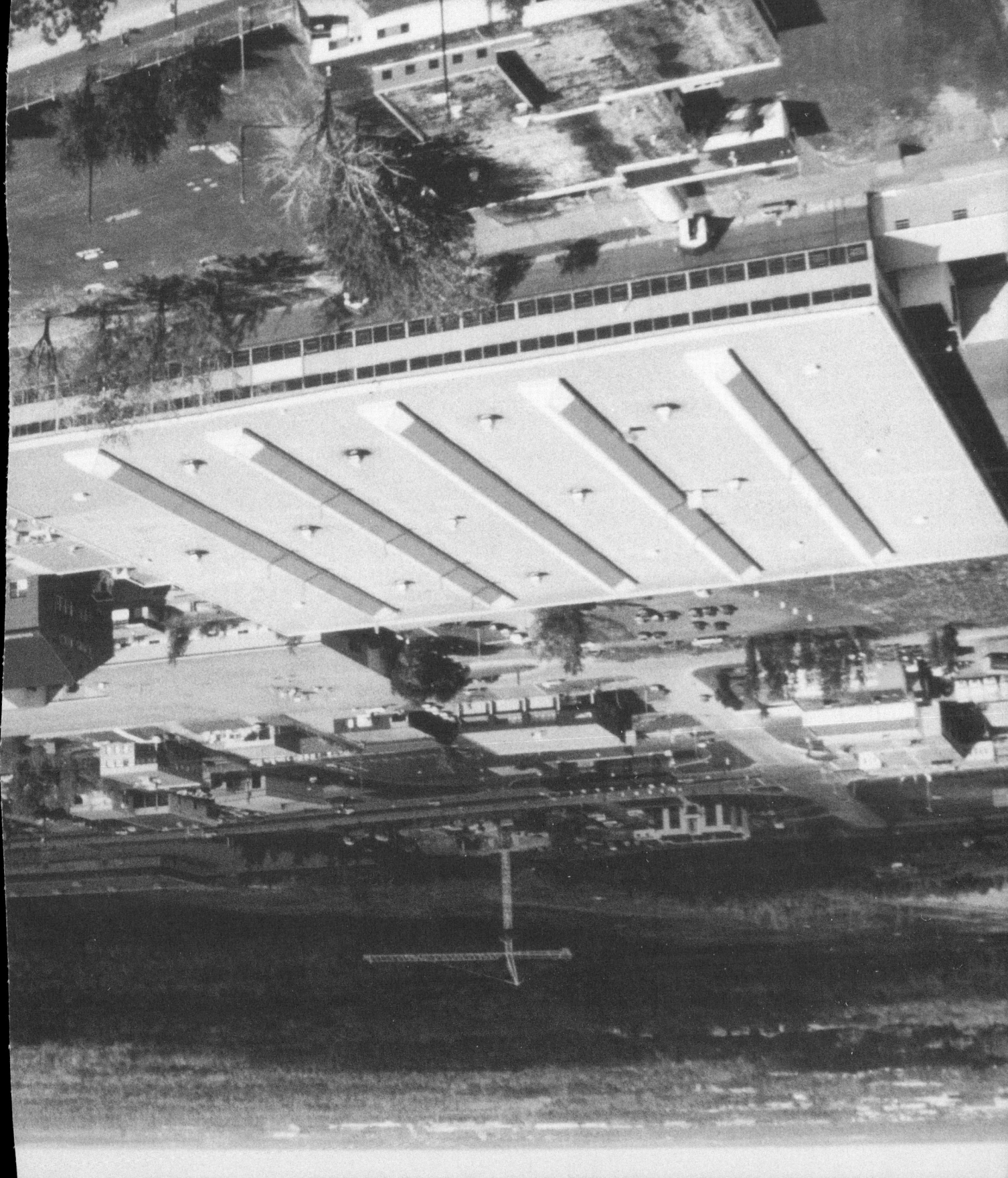